Exploring Life Phenomena with Statistical Mechanics of Molecular Liquids

Fumio Hirata
Toyota Physical and Chemical Research Institute (Toyota Riken)
Institute for Molecular Science
Nagakute, Aichi
Japan

CRC Press
Taylor & Francis Group
Boca Raton London New York

CRC Press is an imprint of the
Taylor & Francis Group, an **informa** business
A SCIENCE PUBLISHERS BOOK

Cover credit: Cover illustrations reproduced by kind courtesy of Prof. Norio Yoshida, Kyushu University, Japan.

CRC Press
Taylor & Francis Group
6000 Broken Sound Parkway NW, Suite 300
Boca Raton, FL 33487-2742

Version Date: 20190924

International Standard Book Number-13: 978-1-138-56388-9 (Hardback)

Library of Congress Cataloging-in-Publication Data

Names: Hirata, Fumio, author.
Title: Exploring life phenomena with statistical mechanics of molecular liquids / Fumio Hirata.
Description: Boca Raton : CRC Press, Taylor & Francis Group [2020] | Includes bibliographical references and index.
Identifiers: LCCN 2019037685 | ISBN 9781138563889 (hardcover : alk. paper)
Subjects: MESH: Molecular Conformation | Proteins--chemistry | Water--chemistry | Solutions--chemistry | Biochemical Phenomena | Models, Molecular | Models, Statistical
Classification: LCC QP551 | NLM QU 34 | DDC 572/.6--dc23
LC record available at https://lccn.loc.gov/2019037685

Visit the Taylor & Francis Web site at
http://www.taylorandfrancis.com

and the CRC Press Web site at
http://www.crcpress.com

Preface

About fifteen years ago, I edited a book titled "Molecular Theory of Solvation" (Kluwer 2003). The book was meant to provide the theoretical basis for the statistical mechanics of liquid, called the Reference Interaction Site Model (RISM), and its application to the fields of chemistry, physics, and molecular biology, where "solvation" phenomena play crucial roles. The main focus of the book was the solvent effect on chemical processes in solution. The book has been well accepted by many scientists and students, especially by those in the fields of the physical chemists or chemical physicists who are interested in analyzing the solvent effect in chemical processes in solution from thoroughly microscopic viewpoints. However, as the book has become popular among the scientists and students, especially in the field of life science, it has faced persistent requests for revision or rewriting. There are two reasons for the requests. Firstly, the book originally targeted those scientists who had some background in the statistical mechanics. Apparently, the requirement was too high for the scientists and students in the field of life science, pharmacology and medicine, since it is the worldwide standard of the university education-program in which a course of *statistical mechanics* is entirely disregarded. Secondly, the book does not reflect the progress made by the theory since the original book was published. The new progress includes the topics related to the *molecular recognition* and *self-organization*, the two most important aspects that characterize molecular processes in life phenomena. In 2005, right after the book was published, water molecules *recognized* by protein in its active site or cavity were probed with the 3D-RISM theory by Imai et al. The new finding triggered succeeding studies that are intimately related to the molecular recognition, such as enzymatic reactions, molecular channels, drug screening, and so on. The other aspect of life phenomena, or the self-organization, is concerned with the structural fluctuation of a molecule such as protein in solution. A molecular theory to be applied to such a process was formulated in 2012 by means of the generalized Langevin theory by Kim and Hirata. Although the computational application of the theory to actual processes in life phenomena has not been done yet, its conceptual robustness has been clarified.

The book was entirely rewritten to meet the two requests. In order to satisfy the first request, a chapter is devoted to describe the background related to the statistical mechanics, including the thermodynamics, the Boltzmann statistics, Gaussian distribution, and the functional differentiation. So, the students without having the knowledge of the field of the science will now be able to study the book by themselves. For replying to the second request, the three chapters in the last half of the book are devoted, which include the application of the RISM and 3D-RISM theories, to the topics related to the molecular recognition and structural fluctuation of protein. Those chapters will be of particular interest to those who are trying to discover a new drug based on the computer aided drug design (CADD), since the computational programs for the RISM and 3D-RISM calculations are now implemented in several software packages commercially available, such as AMBER, MOE, and MDF. The book may provide a background for those softwares.

In writing this book, I have benefited from the help of many scientists directly or indirectly. Prof. K. Arakawa and Prof. H.L. Friedman inspired me to study the statistical mechanics of liquids including water in my earlier career. A series of lectures delivered by Prof. Hiroike has given me the solid back ground on the liquid state theory. The articles written by Prof. D. Chandler taught me all about the statistical mechanics of molecular liquids. I thank all my collaborators for formulating and applying the theories related to the RISM and 3D-RISM theories as well as the non-equilibrium statistical mechanics of molecular liquids. I am indebted to Profs. M. Kataoka, N. Yoshida, Nishiyama, M. Sugita, M. Irisa, and Dr. Y. Ikuta for proof reading and/or providing some of the figures included in the book.

Nagakute
April, 2019

Fumio Hirata

Contents

General Introduction

In a living body, a variety of molecules are working in concerted manner to maintain their life, and to inherit the genetic information from generation to generation. In that respect, life embodies two aspects: one as molecules (or *matter*) and the other as *information*. Needless to say, protein, RNA, and DNA are among the main players of the *life theater*. Protein is a polymer generated from twenty kinds of amino-acids through the condensation reaction that plays a variety of roles to maintain our life: catalyzing chemical reactions (enzyme); permeating molecules such as ions through membrane (molecular channels); conveying a molecule from an organ to the other, such as hemoglobin, and so on. Such a variety of functions played by protein originated from the specific molecular structure, which in turn is attributed to a specific sequence of amino-acids consisting the macromolecule. DNA is also a polymer made from four different nucleotides, adenine, guanine, cytosine, thymine, which stores genetic information of each species of biosystems that is inherited from generation to generation. The genetic information encoded in DNA is decoded in protein in such a way that a nucleotide triplet corresponds to an amino-acid. DNA in a cell usually takes structure of helical double-strands, which are bound by hydrogen-bond between a pair of bases. The double strand wraps around protein called "histone" to make nucelosomes, which are further folded into the "chromosome." RNA is another biopolymer which is positioned between protein and DNA in the flow of the genetic information. RNA carries the sequence information for a protein, while DNA conveys the genetic information for all the proteins in a living body. The genetic information concerning a protein is transcribed from DNA to RNA synthesized by a protein called *RNA synthetase*. The information is used as is or with some modification for synthesizing a protein. The two types of RNA concerned in the two steps are referred to as transfer RNA (*t*-RNA) and messenger RNA (*m*-RNA). The well-regarded hypothesis called "Central Dogma" concerns the flow of *information* from DNA $\Longrightarrow$ RNA $\Longrightarrow$ protein [Crick 1970].

If one focuses on the other aspect of life as matter or molecules, one will find two elementary physicochemical processes which are universal in all the bioactivities: the *self-organization* and *molecular recognition* processes [Lehn 1990]. A typical example of the self-organization is the formation of cell membrane, which provides a stage for biomolecules to play their roles. A membrane is built spontaneously from many molecules of several kinds, including phospholipids and cholesterol. The other example is the protein folding. Although protein is a polymer made up from amino acids through polymerization, it forms a specific conformation depending on its amino-acid sequence under a proper thermodynamic condition, unlike artificial polymers such as polyethylene. The phenomenon is called "protein folding" [Anfinsen 1973]. It is well documented that the specific conformation is intimately related to its function. As in the two examples, the phenomena in which molecules make an assembly, or a molecule folds into a specific conformation spontaneously are referred to as "self-organization."

The other physico-chemical process universally seen in our body is "molecular recognition." A process in which an enzyme binds substrate molecules in its reaction pocket is a typical example of the molecular recognition. Another example of the molecular recognition is binding of an antibody to an antigen in an antigen-antibody reaction. A process in which an ion-channel accommodates an ion in its pore can also be regarded as molecular recognition. Namely, "molecular recognition" is an elementary process in life phenomena, through which a biomolecule should always pass, whenever the molecule performs its function [Yoshida et al. 2009].

The two elementary processes in living system do not occur spontaneously in a condition in which the molecules are isolated in *vacuum*. Let's consider first the self-organization, taking the membrane formation as an example. If such molecules exist in vacuum, assembling into a membrane would be quite difficult for two reasons. First, such a configuration in which molecules align closely to each other is energetically unfavorable, since the polar head-groups repel each other due to the electrostatic interaction. The configuration in which molecules are ordered as in membrane is unstable *entropically* as well. Then, why do the lipid molecules self-organize? It is because those molecules are dissolved in "aqueous solutions." In this case, water promotes the self-organization by inducing the hydrophobic interactions among the alkyl groups. Let's think about an enzymatic reaction as an example of the molecular recognition. An enzymatic reaction is characterized essentially with the two equilibrium constants: one that concerns binding of substrate molecules at a reaction pocket, or the molecular recognition. The other equilibrium constant concerns a chemical reaction in the pocket, which is associated with a change in the electronic structure. The theory of reaction rate by Michaelis and Menten takes the two chemical equilibriums into consideration [Michaelis and Menten 1913]. In general, the reaction pocket of an enzyme binds one or more water molecules and ions sometimes. Thereby, all or some of those water molecules and ions should be disposed from the pocket to bulk solution, in order for a substrate molecule to be accommodated in the active site. On the other hand, a substrate molecule itself is more likely to be hydrated by water molecules. Whether or not a substrate molecule is desolvated upon the molecular recognition is determined by the entire free-energy change concerning the binding process, including the entropy change associated with the desolvation process. It indicates that the molecular recognition process is also governed by water molecules.

It was emphasized so far that water molecules play essential roles in the two physicochemical processes governing life phenomena, "self-organization" and "molecular recognition." But, those are not only the roles water plays. In the enzymatic hydrolysis reaction, such as the hydrolysis of ATP, water molecules participate in the reaction as one of the reactants. Ions play crucial roles in many of the chemical reactions in our body, but they can exist only in aqueous or polar environment. A persistent belief that water does not play a crucial role for enzymatic reactions except for the "hydrolysis" is too naive. How many and in what configurations water molecules exist in a reaction pocket are essential factors to determine the reaction field of an enzyme. Any theory of chemical reactions disregarding the factors is senseless. The fact that approximately seventy percent of living body is occupied by water has a quite essential physicochemical significance.

The molecular recognition process is not only important for our understanding of life phenomena. It is also essential for practical purpose of developing a new technology for drug discovery. As is mentioned above, protein plays a variety of roles in order to maintain and inherit life. If one of the protein functions is lost or is too strong, a living body becomes sick, or dies in the worst case. What we call "drug" is a molecule (many of them are organic compounds) which inhibits or promotes activities of protein or DNA. The same statement applies to bacteria and virus. Protein and DNA play crucial roles in maintaining and inheriting their life, and the molecular recognition process works as an elementary process. If we can inhibit the molecular recognition process in those micro-organisms, which is vital for their existence and promotion, we may be able to kill the micro-organisms. Most of drug molecules used in treatment of disease binds to protein and DNA preferentially, and inhibits the function of the biomolecules. The preferential binding of a drug molecule to the active site of protein and DNA is nothing but the molecular recognition.

So far, we have emphasized the importance of water in biological processes. However, water itself has an own position in natural science [Franks 1972]. It has been a subject of intensive researches for the past century from the view point of physical chemistry [Eisenberg and Kauzman 1969]. It has been well documented that the outstanding features water exhibit in thermodynamic properties, such as negative thermal-expansibility below 4°C, have their origin in the unique liquid structure due to the hydrogen-bond network. The modern structural study of water has been initiated by Bernal and Fowler with the X-ray diffraction method, who first clarified existence of the ice-like tetrahedral coordination in liquid water [Bernal and Fowler 1933]. The negative thermal expansibility was attributed to the disruption of the bulky ice-like structure. The monumental paper has motivated succeeding studies, experimental as well as theoretical, for elucidating structure of water and its relation to thermodynamic and dynamic properties of the liquid. The combined X-ray and neutron diffraction study finally established a picture concerning water structure in terms of the atom-atom pair correlation functions, the essential feature of which remains valid [Thiessen and Narten 1982]. The pair correlation function between oxygen and hydrogen has a distinct peak at 1.8Å, which is assigned as the hydrogen-bond. The oxygen-oxygen (O-O) pair correlation function (PCF) has the second peak at the separation around $r = 1.63\sigma$, where σ is diameter of the molecule, which is characteristic of the ice-I-like tetrahedral coordination, and is regarded as the *finger print* of water structure. The first peak of O-O PCF has rather sharp peak, from which the coordination number has been determined as around 4.4 in the ambient condition. The number is very close to that in ice, four, rather than those typical to normal liquids, ten to eleven. Regarding an origin of the difference in the coordination number, 0.4, between ice and water, a variety of models and theories were proposed. The earlier theories of water assumed the existence of the tetrahedral ice-like coordination in one way or the other, and attributed the deviation in the coordination number from ice to some kind of imperfection in tetrahedral structure due to breaking or distortion of hydrogen-bonds. Roughly speaking, three models have been proposed to explain the deviation in the coordination number: the bent hydrogen-bond model, the interstitial model, and the mixture model represented, respectively, by Pople [Pople 1951], Samoilov [Samoilov 1965], and Nemethy and Scheraga [Nemethy and Scheraga 1962]. The bent-hydrogen bond model attributes the deviation to water molecules occupying the off-lattice space of ice due to distortion of the O-H $\cdots$ O angle from 180 degree. The interstitial model explains the deviation in terms of "interstitial molecules" which are removed from the ice lattice due to breaking of hydrogen-bonds. The mixture model sees water as a mixture of the two components, hydrogen-bonded clusters with ice-like structure and non-hydrogen bonded molecules, and interprets the deviation by increased fraction of non-hydrogen bonded components compared to the clusters. Each of those models could have explained the relation between the structural characteristics of water and its thermodynamic properties to some extent. However, those theories have never been able to answer a more challenging question: how is the characteristic structure of water formed from the intermolecular interactions? But, why it is important to clarify the structure of water from the intermolecular interaction? It is because such *ad hoc* models of water structure may not be applied to the microscopic processes occurring in biological systems described above, such as the self-organization and the molecular recognition. Microscopic description of the structure of water, and its application to biological processes, has become possible due to the development of two theoretical methodologies, the molecular simulation [Rahman and Stillinger 1971], and the statistical mechanics theory of molecular liquids: the RISM, XRISM, and 3D-RISM theories [Hirata 2003].

This book is devoted to elucidating life phenomena, which are *woven* by water and biomolecules, by means of the statistical mechanics molecular liquids. Chapter I is devoted to a general introduction of basic laws and concepts, including thermodynamics and statistical mechanics. The purpose of the chapter is to make the book self-contained, so that readers from the scientific fields other than chemistry and physics may be able to read the book by themselves. In Chapter II, the equilibrium theory of statistical mechanics of molecular liquids, or the RISM theory, and its application to water and aqueous solutions are outlined. Chapter III concerns the theories and concepts in the non-equilibrium statistical mechanics, and its application to aqueous solutions consisting of small molecules. Chapter IV is devoted to the theory of bimolecular solvation, or the 3D-RISM theory, and its application to the molecular recognition processes concerned with a variety of protein. In Chapter V, we develop a theory of the structural fluctuation of protein based on the generalized Langevin theory. The chapter does not include any application, thereby it should be considered as a future perspective. Chapter VI includes some attempts to apply the statistical mechanics to drug discovery.

The Literature

The book by Watson et al. [1987] provides comprehensive description of life phenomena taking place in living cells. The book can be used as a dictionary for each topics of living phenomena. For methodological aspects of the biophysical chemistry including experimental technique, the book written by Cantor and Schimmel [1971] is recommended. For physicochemical aspects of water, the book by Eisenberg and Kauzmann [1969] gives a good description concerning both the structure and thermodynamic properties of the material. The book edited by Franks [1972] provides comprehensive description of water including its relation to living phenomena. The book is a collection of articles concerning separate topics.

The topics related to the molecular simulation are almost entirely skipped from the book due to the following reasons. The methodology has essentially two elements. The first one is to sample a trajectory in the phase space. It involves considerable technologies concerning the computational science in order to accelerate the sampling, and to make the model as realistic as possible. Such computational technologies are completely out of scope of this book. For the molecular simulation, the books written by Harvey and McCammon [1987], Leach [2001], and Chipot and Pohorille [2007] are recommended. The second element is an analysis of the trajectory in order to calculate observables such as the thermodynamic properties and transport coefficients. It is the most essential part for the methodology. However, such analyses are carried out based on the statistical mechanics that is the main subject covered by this book. Therefore, the book serves as a theoretical basis for the molecular simulation, too.

CHAPTER I

Fundamentals: Basic Concepts Related to Statistical Mechanics

This chapter is devoted to review some basic concepts related to the statistical mechanics, which are crucial for the readers to understand contents in the succeeding chapters.

I-1 Thermodynamics

Thermodynamics is the most fundamental field of science on which all other fields of science, including the life science, is based. Here, we briefly review the field of science. The theory of thermodynamics was developed during the 19th century by many giants in science including Kelvin, Clasius, and Carnot, mainly motivated by improvement of the efficiency of the steam engine. A steam engine is a simple devise consisting of a cylinder and a piston, in which water is confined. When water inside the cylinder is heated up, it is evaporated to produce vapor in the cylinder. Since the vapor has large molar volume, it pushes the piston against the atmospheric pressure. The motion of the piston is mechanically transmitted to rotate the wheel of vehicle. Therefore, the principle of steam engine is to convert one kind of energy, or *heat*, to the other kind of energy called "work." So, the thermodynamics was originally developed for improving the efficiency of the engine to convert *heat* to *work*. There are essentially two laws in thermodynamics: the first law and the second law.

The First Law: The first law of thermodynamics is a statement of the principle of energy conservation. Suppose a system consisting of molecules included in a container, the volume of which is V, is in contact with an "environment." The system can exchange energy with the environment in two forms: *heat* and *work*. Let us write the exchange of energy as,

$$dE = d'Q + d'W \tag{I-1.1}$$

where E denotes the (internal) energy of the system, $d'Q$ and $d'W$ are the changes of energies in the forms of heat and work. Most typical form of the work is the mechanical work used in the steam engine, but it takes many different forms, such as an electro-magnetic work to drive electric motors.

$$d'W = -pdV \tag{I-1.2}$$

Since the heat can be converted into the work, and vice versa, $d'Q$ and $d'W$ are *inexact* derivatives. The "prime" put on the derivative indicates that the derivative is *inexact*. The sign of the energies is defined as *positive* if the system gains the energy.

***The Second Law*:** The second law of thermodynamics concerns direction of energy conversion between the two forms, *heat* and *work*. The *work* is converted into *heat* 100 percent. On the other hand, only a part of energy supplied in the form of *heat* can be converted into *work*. Of course, there is a part of energy lost due to *friction* between piston and cylinder in a real mechanical engine, but even in the case of ideal system in which the friction is ignored, the energy supplied in the form of *heat* cannot be converted into work a hundred percent. It is because the supplied energy is wasted partly by producing the other form of energy related to *entropy*.

In order to formulate the second law, we do a *thought* experiment in which the system under concern, for example vapor inside a cylinder of a steam engine, changes its state from A to B by expansion or compression. For such a process, we postulate the existence of an extensive function of state, $S(E,\mathbf{X})$, which is a monotonically increasing function of E. Suppose that the system is insulated completely from environment, namely an adiabatic process. For such a process, the postulate can be restated as

$$\Delta S_{adiabatic} \geq 0 \tag{I-1.3}$$

where $\Delta S_{adiabatic}$ denotes the change of $S(E,\mathbf{X})$ from state A to state B in an adiabatic process. The equality holds for reversible changes. The extensive variable is called "entropy." Equation (I-1.3) is an expression of the entropy maximum principle.

Now, let us consider differential of entropy.

$$\begin{aligned} dS &= \left(\frac{\partial S}{\partial E}\right)_{\mathbf{X}} dE + \left(\frac{\partial S}{\partial \mathbf{X}}\right)_{E} \cdot d\mathbf{X} \\ &= \left(\frac{\partial S}{\partial E}\right)_{\mathbf{X}} dE + \sum_{l} \left(\frac{\partial S}{\partial X_l}\right)_{E,m\neq l} dX_l \end{aligned} \tag{I-1.4}$$

where the subscript l in the second term specifies a *work*. So, the equation takes many different kind of works into consideration. For a reversible process, we have,

$$dE = d'Q_{rev} + \mathbf{f} \cdot d\mathbf{X} \tag{I-1.5}$$

Here, due to reversibility, the "force," **f**, is a property of the system. For example, in the case of steam engine, the external pressures due to atmosphere, p_{ext}, are the same as the pressure, p, of the system (water or vapor) inside the cylinder. From Eqs. (I-1.4) and (I-1.5), one gets,

$$dS = \left(\frac{\partial S}{\partial E}\right)_{\mathbf{X}} dQ'_{rev} + \left[\left(\frac{\partial S}{\partial \mathbf{X}}\right)_{E} + \left(\frac{\partial S}{\partial E}\right)_{\mathbf{X}} \mathbf{f}\right] \cdot d\mathbf{X} \tag{I-1.6}$$

If the process is *adiabatic* as well as *reversible*, both $d'Q_{rev}$ and dS are zero. Therefore, for an adiabatic as well as reversible process, the term in the square bracket should be zero, that leads to

$$\left(\frac{\partial S}{\partial \mathbf{X}}\right)_E = -\left(\frac{\partial S}{\partial E}\right)_{\mathbf{X}} \mathbf{f} \tag{I-1.7}$$

The equality holds for non-adiabatic as well as adiabatic processes, since the quantities involved in this equation are functions of state.

We postulate that S is the monotonically increasing function of E; that is, $(\partial S/\partial E)_X > 0$, or $(\partial E/\partial S)_X \geq 0$. We define the temperature by

$$T \equiv (\partial E/\partial S)_X \geq 0 \tag{I-1.8}$$

Since both E and S are extensive variables, the temperature is an *intensive* variable. That is, T is independent of the size of the system. Equations (I-1.7) and (I-1.8) give

$$\left(\frac{\partial S}{\partial \mathbf{X}}\right)_E = -\frac{\mathbf{f}}{T} \tag{I-1.9}$$

Since $dS = (\partial S/\partial E)_X dE - (\mathbf{f}/T) \cdot d\mathbf{X}$, we have

$$dS = \frac{1}{T}dE - \frac{\mathbf{f}}{T} \cdot d\mathbf{X} \qquad \text{or} \qquad dE = TdS + \mathbf{f}d\mathbf{X} \tag{I-1.10}$$

The equation implies that the energy of equilibrium state is characterized by S and $\boldsymbol{X}$. The equation is an expression of the second law of thermodynamics.

Variational Statement of Second Law: The entropy maximum principle can be formulated in terms of the variational principle concerning the response of the system to a perturbation. Let us imagine a system isolated from surroundings, and think of dividing the system into subsystems with heat conducting wall (Fig. I-1.1). Placing the conducting wall is a variation that imposes an internal constraint to the system. The second law tells us that the entropy change due to the perturbation is negative, that is,

$$S(E, \mathbf{X};\ \text{internal constraint}) - S(E, \mathbf{X}) < 0 \tag{I-1.11}$$

In other words, the equilibrium state is the state at which $S(E, \mathbf{X}$; internal constraint) has its global maximum. It is called the entropy maximum principle.

A principle that governs the equilibrium state of an isolated system can be also stated in terms of energy, that is, the energy minimum principle. It can be derived from the entropy maximum principle based on the thought experiment illustrated in Fig. (I-1.1). Let's think of a system confined in a container, the volume of which is $\mathbf{X}$, and the energy of which is E (picture on the left-hand side). The system is isolated from surroundings, so that there is no change in the energy and volume. Then, divide the system into the subsystems (1) and (2) by a heat-conducting wall (illustrated by a dashed line). Let the energy of the subsystems (1) and (2) be $E^{(1)}$ and $E^{(2)}$, and the volume be $\mathbf{X}^{(1)}$ and $\mathbf{X}^{(2)}$, respectively. Since the system is completely isolated from the surrounding, $E = E^{(1)} + E^{(2)}$ and $\mathbf{X} = \mathbf{X}^{(1)} + \mathbf{X}^{(2)}$. Suppose the system is equilibrated with the energies $E^{(1)}$, $E^{(2)}$ and the volumes $\mathbf{X}^{(1)}$, $\mathbf{X}^{(2)}$ in the respective subsystems.

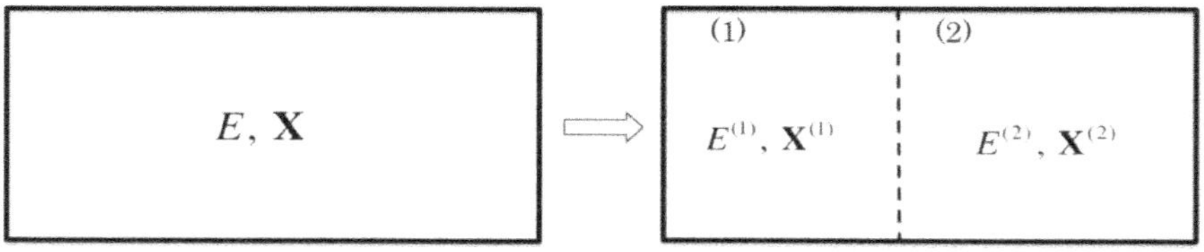

Fig. I-1.1: Illustration of the internal constraint.

Now, let us perturb the system by transferring a small amount of energy from the subsystem (1) to the subsystem (2). The perturbation or variation will bring the system into a fluctuated or non-equilibrium state. The entropy of such a process certainly satisfies the following inequality due to the entropy maximum principle stated above.

$$S(E^{(1)} - \Delta E, \mathbf{X}^{(1)}) + S(E^{(2)} + \Delta E, \mathbf{X}^{(2)}) < S(E^{(1)} + E^{(2)}, \mathbf{X}^{(1)} + \mathbf{X}^{(2)}), \qquad \text{(I-1.12)}$$

where $E^{(1)}$ and $E^{(2)}$ are the equilibrium partitioning of energy, and ΔE denotes an amount of energy transferred from subsystem 1 to subsystem 2. So, if you repartition the energy from their equilibrium state, then entropy of the partitioned system is necessarily decreased. If one tries to make the entropy before and after repartitioning to be unchanged, or

$$S(E^{(1)} - \Delta E, \mathbf{X}^{(1)}) + S(E^{(2)} + \Delta E, \mathbf{X}^{(2)}) = S(E^{(1)} + E^{(2)}, \mathbf{X}^{(1)} + \mathbf{X}^{(2)}), \qquad \text{(I-1.13)}$$

then, the inequality, $E < E^{(1)} + E^{(2)}$, should hold for the energy. In other words, one cannot apply internal constraints without increasing the total energy of the system. That is, $E(S, \mathbf{X})$ is a global minimum of $E(S, \mathbf{X}$; internal constraint), which is referred to as the energy minimum principle.

The energy minimum principle can be stated in terms of the variational principle. We may write the variation of energy from the equilibrium state, due to such a perturbation as the internal partitioning of the system, in terms of a Taylor series,

$$\begin{aligned}\Delta E &= E(S, \mathbf{X}; \delta Y) - E(S, \mathbf{X}; 0) \\ &= \left[\left(\frac{\partial E}{\partial Y}\right)_{S,\mathbf{X}}\right]_{Y=0} \delta Y + \left[\frac{1}{2}\left(\frac{\partial^2 E}{\partial Y^2}\right)_{S,\mathbf{X}}\right]_{Y=0} (\delta Y)^2 + \cdots \\ &= (\delta E)_{S,\mathbf{X}} + (\delta E^2)_{S,\mathbf{X}} + \cdots\end{aligned} \qquad \text{(I-1.14)}$$

where δY denote a perturbation applied to the system. The energy minimum principle is stated as,

$$(\delta E)_{S,\mathbf{X}} \geq 0 \qquad \text{(I-1.15)}$$

Similarly, $(\delta S)_{E,\mathbf{X}} \leq 0$ for entropy.

Those quantities determine the equilibrium state of the system. On the other hand, the second term in Eq. (I-1.14) is concerned with the fluctuation from the equilibrium state. It will be discussed in Section I-1.3 in this chapter.

***Thermodynamic Potentials*:** As we just saw, the energy and entropy play the role of determining the thermodynamic equilibrium state of a system. The role is similar to that played by the *potential* energy in a mechanical system, that is, the potential energy determines the most stable state of the system. In that respect, the energy and entropy are often called *thermodynamic potential*. One of the tasks of the thermodynamics is to find the most stable state of the system by controlling or adjusting the independent variables such as volume and temperature. For that purpose, the energy is not a favorite function because one of the independent variables of the quantity is entropy, which is not easy to be controlled in the ordinary experimental condition in a laboratory. It will be much easier to control other thermodynamic quantities such as pressure, volume, and concentration of solution. So, it will be desirable to transform the variables so that the independent variables may be controlled in a laboratory.

For the purpose, we write Eq. (I-1.10) in more concrete form that includes pressure and concentration, or mole number, in the general *work* term in the equation as,

$$\mathbf{f} \cdot d\mathbf{X} = -pdV + \sum_{i=1} \mu_i dN_i \tag{I-1.16}$$

where μ_i and N_i are the chemical potential of species i and its mole number. Then, Eq. (I-1.10) is written as

$$dE = TdS - pdV + \sum_{i=1} \mu_i dN_i \tag{I-1.17}$$

Now, we define the enthalpy (H), the Helmholz free energy (A), and the Gibbs free energy (G) by

$$H \equiv E + PV, \ A \equiv E - TS, \ G \equiv H - TS \tag{I-1.18}$$

Taking the derivative of those quantities, and taking Eq. (I-1.17) into consideration, it is easy to find the following relations,

$$dH = TdS + Vdp + \sum_{i=1} \mu_i dN_i \tag{I-1.19}$$

$$dA = -SdT - pdV + \sum_{i=1} \mu_i dN_i \tag{I-1.20}$$

$$dG = -SdT + VdP + \sum_{i=1} \mu_i dN_i \tag{I-1.21}$$

The relations for the free energy have (T, V, N_i) or (T, p, N_i) as independent variables, which can be readily controlled in experimental conditions. The variational principles can be readily derived to give,

$$(\delta H)_{S,p,N} \geq 0, \tag{I-1.22}$$

$$(\delta A)_{T,V,N} \geq 0, \tag{I-1.23}$$

$$(\delta G)_{T,p,N} \geq 0. \tag{I-1.24}$$

So, those quantities play the role of the thermodynamic potential to find an equilibrium condition. The transformation made here can be made in the more general framework of the Lagendre transformation.

From Eqs. (I-1.17) and (I-1.18) ~ (I-1.20), one finds another important relation concerning the ensemble invariance of the chemical potential, that is,

$$\left(\frac{\partial E}{\partial N_i}\right)_{S,V,N_{j\neq i}} = \left(\frac{\partial H}{\partial N_i}\right)_{S,p,N_{j\neq i}} = \left(\frac{\partial A}{\partial N_i}\right)_{T,V,N_{j\neq i}} = \left(\frac{\partial G}{\partial N_i}\right)_{T,p,N_{j\neq i}} \tag{I-1.25}$$

In experiments, the chemical potential is measured in a condition in which T and p are constant. On the other hand, it is calculated theoretically by means of the statistical mechanics in a condition in which T and V are constant. The relation implies that the chemical potentials obtained from the both methods are equivalent with each other in principle.

I-2 Statistical Mechanics

In the previous section, we briefly reviewed thermodynamics. Thermodynamics characterizes macroscopic properties of materials and living systems with few variables, such as temperature and pressure. However, it does not tell anything about microscopic states of molecules consisting the macroscopic system. Suppose one has a glass of water. There are ~ 10^{23} water molecules in the glass, which are characterized thermodynamically by a temperature T and a volume V (and/or pressure). The water molecules inside the glass are moving around changing their positions and velocities rapidly by colliding and interacting among others. Such microscopic states of water molecules can be described in principle either by classical mechanics or by quantum mechanics. Therefore, enormous numbers of different microscopic states correspond to a macroscopic state characterized by a small number of thermodynamic variables. It is the role of statistical mechanics to build a bridge between the two ways of describing properties of materials and living systems. It is the concept of *ensemble* that bridges the gap between the microscopic and macroscopic states.

(Microcanonical Ensemble)

Suppose we have a system consisting of N particles confined in a box, the volume and energy of which are fixed, respectively, as V and E. For such a system, Stephan Boltzmann has given a theorem (Boltzmann's theorem), that is

$$S = k_B \ln \Omega(N, V, E) \tag{I-2.1}$$

where S denotes entropy of the system, and $\Omega(N, V, E)$ is the number of microscopic states which are distinguishable. The great theorem has opened up the entire development of the statistical mechanics.

We assume that all the microscopic states are equally probable, so that the probability of finding the system in one of such states can be,

$$P_j = \frac{1}{\Omega(N,V,E)} \tag{I-2.2}$$

According to the thermodynamic definition of temperature (Eq. I-1.8),

$$\left(\frac{\partial S}{\partial E}\right)_{N,V} = \frac{1}{T} \qquad \text{(I-2.3)}$$

we can define temperature by taking the derivative of Eq. (I-2.1) with respect to energy as

$$\beta \equiv \frac{1}{k_B T} = \left(\frac{\partial \ln \Omega}{\partial E}\right)_{N,V} \qquad \text{(I-2.4)}$$

where k_B is a universal constant called Boltzmann's constant, and its value has been determined empirically to be

$$k_B = 1.380 \times 10^{-16} \text{ erg/deg}$$

(**Canonical Ensemble**)

Let us think a system consisting of N particles included in a container, volume of which is V. The system can exchange *heat* with an infinitely large heat bath, temperature of which is fixed to T. In the equilibrium state, the energy of the system is allowed to fluctuate due to the exchange of heat with the heat bath, but the sum $E = E_B + E_j$ is a constant. The fluctuating energy can be obtained as an eigen value of the Schrodinger equation, $H\Psi_j = E_j\Psi_j$. If the system is in one definite state j, the number of states accessible to the system plus the heat bath is $\Omega(E_B) = \Omega(E - E_j)$. If one applies the principle of equal weight, the probability of finding the system in the state j obeys

$$P_j \propto \Omega(E - E_j) = \exp[\ln \Omega(E - E_j)] \qquad \text{(I-2.5)}$$

Considering $E_j = E$, $\ln\Omega(E - E_j)$ can be expanded into the Taylor series to give,

$$\ln \Omega(E - E_j) = \ln \Omega(E) - E_j \left(\frac{d \ln \Omega}{dE}\right) + \cdots \qquad \text{(I-2.6)}$$

The second term in the right hand side is identified as an inverse of temperature, β, due to Eq. (I-2.4). Therefore, one finds,

$$\ln \Omega(E - E_j) \approx \ln \Omega(E) - \beta E_j \qquad \text{(I-2.7)}$$

Combining Eq. (I-2.7) with (I-2.5)), the probability of finding the system in the j-th state is written as,

$$P_j \propto \exp\left(-\beta E_j\right) \qquad \text{(I-2.8)}$$

By considering the normalization $\sum_j P_j = 1$, one gets the expression for the normalized probability as,

$$P_j = Q^{-1} \exp(-\beta E_j) \qquad \text{(I-2.9)}$$

where Q is the normalization constant defined by

$$Q(\beta, N, V) = \sum_j \exp(-\beta E_j) \tag{I-2.10}$$

The function Q is called "Canonical partition function," and is related to the Helmholtz free energy as follows. By taking the derivative Eq. (I-2.10) with respect to β, one finds

$$\frac{\partial \ln Q}{\partial \beta} = \frac{\sum_j (-E_j)\exp(-\beta E_j)}{\sum_j \exp(-\beta E_j)} = -\langle E \rangle \tag{I-2.11}$$

On the other hand, there is a thermodynamic relation that relates the Helmholtz free energy F with (internal) energy E, called the Gibbs-Helmholtz equation,

$$E = \frac{\partial(\beta F)}{\partial \beta} \tag{I-2.12}$$

Comparing those two equations, one finds,

$$A = -kT \ln Q \tag{I-2.13}$$

(Generalized Ensembles)

Here, we consider a system that exchanges energy with environment, or a reservoir, not only in the form of heat, but also in the form of a *generalized* work. The generalized work includes the mechanical work due to pressure, the work by the osmotic pressure or chemical potential, and so on. The entropy of the entire system including the reservoir can be defined due to Boltzmann's theorem.

$$S = k_B \ln \Omega(E, X) \tag{I-2.14}$$

where E and X denote the energy of the entire system including the system under consideration and the extensive variables conjugated to the general intensive variables: for example, the volume and the number of molecules. The derivative of Eq. (I-2.14) reads,

$$k_B^{-1} dS = \beta dE + \xi dX \tag{I-2.15}$$

where

$$\left(\frac{\partial \ln \Omega}{\partial E}\right)_X = \beta, \tag{I-2.16}$$

$$\left(\frac{\partial \ln \Omega}{\partial X}\right)_\beta = \xi \tag{I-2.17}$$

The extensive variables of the system are fluctuating by exchanging with the reservoir. Let us denote one of the fluctuating state with subscript "j". The probability of finding the system in the j state is proportional to $\Omega(E - E_j; X - X_j)$, namely,

$$P_j \propto \exp\left[\ln \Omega(E - E_j, X - X_j)\right] \tag{I-2.18}$$

Now let us expand the $\ln\Omega(E - E_j, X - X_j)$ into a Taylor series,

$$\ln\Omega(E - E_j; X - X_j) = \ln\Omega(E; X) - E_j\left(\frac{\partial \ln\Omega}{\partial E}\right) - X_j\left(\frac{\partial \ln\Omega}{\partial X}\right) + \cdots \qquad \text{(I-2.19)}$$

Considering Eqs. (I-2.19) with (I-2.16) and (I-2.17), one finds

$$\ln\Omega \approx \ln\Omega(E, X) - \beta E_j - \xi X_j \qquad \text{(I-2.20)}$$

and

$$P_j \propto \exp(-\beta E_j - \xi X_j) \qquad \text{(I-2.21)}$$

The normalized probability is

$$P_j = \exp(-\beta E_j - \xi X_j)/\Xi \qquad \text{(I-2.22)}$$

where the normalization constant Ξ is a partition function of the generalized ensemble.

$$\Xi = \sum_j \exp(-\beta E_j - \xi X_j) \qquad \text{(I-2.23)}$$

(Gibbs formula of entropy)

Let us define a function φ by

$$\varphi \equiv -k_B \sum P_j \ln P_j \qquad \text{(I-2.24)}$$

The function turns out to be entropy as shown in the following.

First, we take the derivative of Eq. (I-2.23)

$$d\ln\Xi = -\langle E\rangle d\beta - \langle X\rangle d\xi \qquad \text{(I-2.25)}$$

where

$$\langle E\rangle = \sum_j P_j E_j = \left[\frac{\partial \ln\Xi}{\partial(-\beta)}\right]_{\xi,Y} \qquad \langle X\rangle = \sum_j P_j X_j = \left[\frac{\partial \ln\Xi}{\partial(-\xi)}\right]_{\beta,Y}$$

Substituting the expression of (Eq. (I-2.22)) into Eq. (I-2.24), one finds

$$\begin{aligned}\varphi &= -k_B \sum_j P_j\left[-\ln\Xi - \beta E_j - \xi X_j\right] \\ &= k_B\left\{\ln\Xi + \beta\langle E\rangle + \xi\langle X\rangle\right\}\end{aligned} \qquad \text{(I-2.26)}$$

Taking the derivative, one gets,

$$d\varphi = \beta k_B d\langle E\rangle + \xi k_B d\langle X\rangle$$

Comparing the equation with corresponding thermodynamic expression for the case of mechanical work (Eq. I-1.10),

$$dS = \frac{1}{T}dE - \frac{p}{T}dV \tag{I-2.27}$$

φ should have the meaning of entropy.

$$S = -k_B \sum_j P_j \ln P_j \tag{I-2.28}$$

The expression is called the Gibbs entropy formula.

(Grand Canonical Ensemble)

The formula obtained in the previous section can be used to derive so called *grand canonical partition function.* In this case, the number of molecules in the system is allowed to fluctuate by exchanging the molecules with the reservoir. The fluctuation is controlled by the difference in the chemical potential of the system. In this case, the expression for the probability distribution of the j state reads,

$$P_j = \Xi^{-1} \exp\left(-\beta E_j + \beta\mu N\right) \tag{I-2.29}$$

where μ is the chemical potential of the system, and N is the number of molecules in the j-state of the system. The normalization constant

$$\Xi = \sum_{N\geq 0} \sum_j \exp\left(-\beta E_j + \beta\mu N\right) \tag{I-2.30}$$

is referred to as the "grand canonical partition function." The equation can be also written in terms of the canonical partition function, defined by Eq. (I-2.10), as

$$\Xi = \sum_{N\geq 0} \exp(\beta\mu N) Q_N \tag{I-2.31}$$

where $Q_N \equiv Q(N, V, T)$ defined by Eq. (I-2.10).

Using the Gibbs entropy formula, one finds

$$\begin{aligned} S &\equiv -k_B \sum_j P_j \ln P_j \\ &= -k_B \sum_j P_j \left[-\ln \Xi - \beta E_j + \beta\mu N_j\right] \\ &= -k_B \left[-\ln \Xi - \beta\langle E\rangle + \beta\mu\langle N\rangle\right] \end{aligned} \tag{I-2.32}$$

Comparing the result with corresponding thermodynamic expression, one finds,

$$\ln \Xi = \beta p V \tag{I-2.33}$$

where pV is the thermodynamic potential to measure the stability of the system in which the energy and the number of particles are allowed to fluctuate.

(Classical partition functions)

So far, we have formulated the statistical mechanics in the general framework of the quantum mechanics in which the energy state is discrete. However, most events we are interested in here are those taking place in the ambient condition, or at room temperature, where the quantum effect is negligible, except for chemical reactions in which change of the electronic structure takes place. Therefore, we work with the classical Hamiltonian (energy) such as

$$E(\mathbf{r}_1,\mathbf{r}_2,\cdots,\mathbf{r}_N;\mathbf{p}_1,\mathbf{p}_2,\cdots,\mathbf{p}_N)=\sum_i \frac{1}{2m}\mathbf{p}_i^2+U(\mathbf{r}_1,\mathbf{r}_2,\cdots,\mathbf{r}_N) \qquad \text{(I-2.34)}$$

where the first term is the kinetic energy, and m and $\mathbf{p}_i$ are the mass and momentum of a molecule in the system, respectively. $U(\mathbf{r}_1, \mathbf{r}_2,\cdots, \mathbf{r}_N)$ is the interaction potential energy among molecules. The transformation of the quantum expression of the partition function to classical one can be made by changing the *sum* over quantum state j to the *integration* over the phase space spanned by momentum and positional coordinates.

$$\sum_j \rightarrow \int\cdots\int d\mathbf{r}_1 d\mathbf{r}_2\cdots d\mathbf{r}_N d\mathbf{p}_1 d\mathbf{p}_2\cdots d\mathbf{p}_N$$

There are two points to be noted for the transformation. One is the Heisenberg uncertainty principle, $\Delta x\Delta p_x \geq h$. Namely, the differential volume-element in the phase space cannot be less than h^{3N}. The other point concerns the indistinguishability of particles of same species. In the integration above, there are many cases in which the position and momentum of two identical particles are just exchanged. Those cases make redundant contributions to the state sum. So, in order to remove the redundancy, we should divide the integral by a factor $N!$. Keep the notes in mind, one get the classical expression for the canonical partition function.

$$Q_N=\frac{1}{h^{3N}N!}\int\cdots\int \exp\left(-\frac{E(\mathbf{r}_1,\mathbf{r}_2,\cdots,\mathbf{r}_N;\mathbf{p}_1,\mathbf{p}_2,\cdots,\mathbf{p}_N)}{k_BT}\right)d\mathbf{r}_1 d\mathbf{r}_2\cdots d\mathbf{r}_N d\mathbf{p}_1 d\mathbf{p}_2\cdots d\mathbf{p}_N \qquad \text{(I-2.35)}$$

Unlike a quantum system, there is no correlation between the positional and momentum coordinates in a classical system. Therefore, we can split the integral in Eq. (I-2.35) into those over positions and those over momentums. The integral over the positions is a many-body integral concerning the Boltzmann factor, which is referred to as "configuration integral," defined by

$$Z_N=\int_V\int_V\cdots\int_V \exp[-\beta U(\mathbf{r}_1,\mathbf{r}_2,\cdots\mathbf{r}_N)]d\mathbf{r}_1 d\mathbf{r}_2\cdots d\mathbf{r}_N \qquad \text{(I-2.36)}$$

The integral over the momentum coordinates can be performed readily in terms of Gaussain integrals to produce a factor, $(2\pi mkT)^{3N/2}$. The canonical partition function can be written in a compact form as

$$Q_N=\frac{1}{N!\Lambda^{3N}}Z_N \qquad \text{(I-2.37)}$$

where Λ defined by the following equation is called "de Blogie thermal wave length."

$$\Lambda=\left(\frac{h^2}{2\pi mkT}\right)^{1/2} \qquad \text{(I-2.38)}$$

It will be instructive to show the relation between the statistical mechanics and the thermodynamics in the simple case of ideal gas. The ideal gas is characterized by a system in which there is no interaction among molecules, or $U(\mathbf{r}_1, \mathbf{r}_2, \cdots, \mathbf{r}_N)$ in Eq. (I-2.36) is zero. Therefore, $Z_N = V^N$, and

$$Q_N = \frac{1}{N!\Lambda^{3N}} V^N \tag{I-2.39}$$

The canonical partition function is related to the Helmholtz free energy as $A = -kT \log Q_N$ (Eq. I-2.13). Putting the expression into the thermodynamic relation, or $p = -(\partial F/\partial V)_T$, one finds the equation of state of ideal gas.

$$pV = NkT \tag{I-2.40}$$

(Grand canonical ensemble and fluctuation theorem)

Using the definition of the canonical partition function, the grand canonical partition function (Eq. I-2.31) can be rewritten as

$$\Xi = \sum_{N=0} \frac{z^N}{N!} Z_N$$

where z is referred to as "activity," and is defined by $z = \exp[\mu/k_BT]/\Lambda^3$. The expression is useful to derive the fluctuation formula for density as follows.

Let us take the derivative of $\ln \Xi$ with respect to the $\ln z$, which gives the average number of molecules as follows.

$$\frac{\partial \ln \Xi}{\partial \ln z} = \frac{z}{\Xi}\frac{\partial \Xi}{\partial z} = \frac{1}{\Xi}\sum_{N=0} N \frac{z^N}{N!} Z_N = \langle N \rangle \tag{I-2.41}$$

Then, we take the derivative of $\langle N \rangle$ with respect to $\ln z$, that leads the density fluctuation as follows.

$$\begin{aligned} \frac{\partial \langle N \rangle}{\partial \ln z} &= z\frac{\partial}{\partial z}\left[\frac{1}{\Xi}\sum_{N=0} N \frac{z^N}{N!} Z_N\right] \\ &= -\frac{z}{\Xi^2}\frac{\partial \Xi}{\partial z}\sum_{N=0} N \frac{z^N}{N!} Z_N + \frac{z}{\Xi}\sum_{N=0} N^2 \frac{z^{N-1}}{N!} Z_N \\ &= -\left(\frac{z}{\Xi}\frac{\partial \Xi}{\partial z}\right)\frac{1}{\Xi}\sum_{N=0} N \frac{z^N}{N!} Z_N + \langle N^2 \rangle \\ &= -\langle N \rangle\langle N \rangle + \langle N^2 \rangle \end{aligned} \tag{I-2.42}$$

Dividing both sides of the equation by V, and defining the density ρ by $\rho = \langle N \rangle / V$, and taking $d\log z = d\mu$ into consideration, one gets

$$kT\left(\frac{\partial \rho}{\partial \mu}\right)_T = \frac{\langle N^2 \rangle - \langle N \rangle^2}{V} \tag{I-2.43}$$

Using the Gibbs-Duhem relation, or, $d\mu = \rho^{-1} dp - s dT$, the left hand side of Eq. (I-2.43) can be rewritten thermodynamically as

$$\left(\frac{\partial \rho}{\partial \mu}\right)_T = \left(\frac{\partial \rho}{\partial p}\right)_T \left(\frac{\partial p}{\partial \mu}\right)_T = -\frac{\langle N \rangle}{V^2}\left(\frac{\partial V}{\partial p}\right)_T \left(\frac{\partial p}{\partial \mu}\right)_T = \frac{\langle N \rangle}{V}\kappa_T \left(\frac{\partial p}{\partial \mu}\right)_T = \rho \kappa_T \rho = \rho^2 \kappa_T \quad \text{(I-2.44)}$$

where $\kappa_T = -V^{-1}(\partial V/\partial T)_T$ is the isothermal compressibility. Equation (I-2.43) and Eq. (I-2.44) give the famous expression that relates the density fluctuation to the isothermal compressibility.

$$kT\rho^2 \kappa_T = \frac{\langle N^2 \rangle - \langle N \rangle^2}{V} \quad \text{(I-2.45)}$$

I-3 Gaussian Distribution

In the preceding section, the method to build a bridge between the microscopic quantities, or the partition function, and thermodynamic properties such as the Helmholtz free energy was outlined. The thermodynamic properties concerned in this case is equilibrium quantities, meaning that the properties are those averaged over spatial and temporal fluctuations of microscopic states. Such description is very important to understand the stability of matter, such as structure of protein in water. However, the description concerning equilibrium properties is not complete to understand the entire properties of matter and living phenomena. A *fluctuation* around an equilibrium state plays a crucial role, especially in a living body. A typical example of such events is the *molecular recognition*.

It has been well documented that randomly fluctuating variables takes the Gaussian distribution, or the normal distribution, around its average value, when deviations (fluctuation) of the variables from the average are not extraordinary large. The rigorously proved natural law has been known as the "central limiting theorem" (Chandrasekhar 1943, Kubo 1991).

$$w(\Delta x) = \sqrt{\frac{C}{2\pi}} \exp\left[-\frac{1}{2}C\Delta x^2\right] \quad \text{(I-3.1)}$$

The half-width of the distribution is a variance that measures the magnitude of a fluctuation.

$$C = \frac{1}{\langle \Delta x^2 \rangle} \quad \text{(I-3.2)}$$

where $\Delta x (= x - \langle x \rangle)$ is the fluctuation of a random variable x from its average $\langle x \rangle$, and

$$\langle \Delta x^2 \rangle = \int_{-\infty}^{\infty} \Delta x^2 w(\Delta x) dx \quad \text{(I-3.3)}$$

So, $(1/C)^{1/2}$ is the standard deviation or the mean square fluctuation.

The equation can be readily generalized to the case where several fluctuating variables are concerned.

$$w(\Delta x_1, \Delta x_2, \ldots, \Delta x_n) = \sqrt{\frac{C}{(2\pi)^n}} \exp\left[-\frac{1}{2}\sum_i \sum_j C_{ij} \Delta x_i \Delta x_j\right] \quad \text{(I-3.4)}$$

$$\Delta x_i = x_i - \langle x_i \rangle$$

where Δx_i is the fluctuation of the *i*-th random variable, C_{ij} is related to the variance-covariance matrix $\langle \Delta x_i \Delta x_j \rangle$ of the fluctuation by

$$C_{ij} = \frac{1}{\langle \Delta x_i \Delta x_j \rangle} \quad \text{(I-3.5)}$$

$$\langle \Delta x_i \Delta x_j \rangle = \int_{-\infty}^{\infty} \int_{-\infty}^{\infty} \Delta x_i \Delta x_j w(\Delta x_1, \Delta x_2, \ldots, \Delta x_n) dx_i dx_j$$

and C is the determinant of the matrix $\{C_{ij}\}$.

Some examples of Gaussian distribution in physics:

(Thermodynamic fluctuation)

Here, we consider a thermodynamic system of constant volume (V) and temperature (T). The equilibrium state of the system is defined by the point where the Helmholtz free energy becomes minimum due to the variational principle (Section I-1), or

$$\left(\frac{\partial A}{\partial T} \right)_V = 0, \qquad \left(\frac{\partial A}{\partial V} \right)_T = 0 \quad \text{(I-3.6)}$$

Suppose the thermodynamic variables are fluctuating around the equilibrium state. The deviation of the free energy can be expanded around the equilibrium state up to the second order as

$$\Delta A = \frac{1}{2} \left(\frac{\partial^2 A}{\partial V^2} \right)_T \Delta V^2 + \frac{1}{2} \left(\frac{\partial^2 A}{\partial V \partial T} \right) \Delta V \Delta T + \frac{1}{2} \left(\frac{\partial^2 A}{\partial T^2} \right)_{V,} \Delta T^2 \quad \text{(I-3.7)}$$

where $\Delta A = A - \langle F \rangle$, $\Delta T = T - \langle T \rangle$, $\Delta V = V - \langle V \rangle$ are the fluctuation of the Helmholtz free energy, volume, and temperature, respectively. The first order term in the expansion disappears due to Eq. (I-3.6). The equation can be transformed into the following equation by the standard manupiration in thermodynamics,

$$\Delta A = \frac{1}{2V \chi_T} \Delta V^2 - \frac{1}{2} \beta \Delta V \Delta T - \frac{1}{2} \frac{C_V}{T} \Delta T^2 \quad \text{(I-3.8)}$$

where χ_T, β, and Cv are defined, respectively, by

$$\kappa_T = -\frac{1}{V} \left(\frac{\partial V}{\partial p} \right)_T, \quad \beta = \left(\frac{\partial p}{\partial T} \right)_V, \quad C_V = T \left(\frac{\partial S}{\partial T} \right)_V \quad \text{(I-3.9)}$$

The probability distribution of the free-energy fluctuation can be written as

$$w(\Delta A) \propto \exp\left[-\frac{\Delta A}{k_B T} \right] \quad \text{(I-3.10)}$$

or

$$w(\Delta A) \propto \exp\left[-\frac{1}{2Vk_BT\chi_T}\Delta V^2 + \frac{\beta}{2k_BT}\Delta V\Delta T + \frac{1}{2}\frac{C_V}{k_BT^2}\Delta T^2\right] \tag{I-3.11}$$

Applying the theorems concerning the Gaussian distribution, described in the previous section, one can get the relation between the thermodynamic variables and the variance-covariance matrix of the fluctuation as,

$$\langle \Delta V^2 \rangle = Vk_BT\chi_T, \quad \langle \Delta V \Delta T \rangle = \frac{k_BT}{\beta}, \quad \langle \Delta T^2 \rangle = \frac{k_BT^2}{C_v} \tag{I-3.12}$$

The relations originated by Landau scored a great landmark in the history of thermodynamics, and any phenomenological fluctuation should follow the general theorem (Landau 1964).

Similar relations for other statistical ensembles such as constant pressure and temperature can be derived as follows, by assuming the Gaussian fluctuation for the Gibbs free energy. If one expand the Gibbs free energy around its equilibrium state in a Taylor series of its independent variables, temperature and pressure,

$$\begin{aligned} G(T,P) &= G_{eq}(T,p) + \left(\frac{\partial G}{\partial T}\right)_p \Delta T + \left(\frac{\partial G}{\partial p}\right)_T \Delta p \\ &+ \frac{1}{2}\left(\frac{\partial^2 G}{\partial p^2}\right)_T \Delta p^2 + \frac{1}{2}\left(\frac{\partial^2 G}{\partial p \partial T}\right)\Delta p \Delta T + \frac{1}{2}\left(\frac{\partial^2 G}{\partial T^2}\right)_{V,} \Delta T^2 + \cdots \end{aligned} \tag{I-3.13}$$

Since $(\partial G/\partial T)_p = (\partial G/\partial p)_T = 0$ at an equilibrium, (I-3.13) becomes up to the second order of ΔT and Δp

$$\Delta G \equiv G(T,p) - G_{eq}(T,p) = \frac{1}{2}\left(\frac{\partial^2 G}{\partial p^2}\right)_T \Delta p^2 + \frac{1}{2}\left(\frac{\partial^2 G}{\partial p \partial T}\right)\Delta p \Delta T + \frac{1}{2}\left(\frac{\partial^2 G}{\partial T^2}\right)_{V,} \Delta T^2 \tag{I-3.14}$$

where ΔG represents the fluctuation of the Gibbs free energy. Equation (I-3.14) implies that the probability distribution of the fluctuation of the Gibbs free energy is a Gaussian function as,

$$w(\Delta G) \propto \exp\left[-\frac{1}{2\langle \Delta p^2 \rangle}\Delta p^2 - \frac{1}{2\langle \Delta p \Delta T \rangle}\Delta p \Delta T - \frac{1}{2\langle \Delta T^2 \rangle}\Delta T^2\right] \tag{I-3.15}$$

The second derivative in Eq. (I-3.14) can be related to the compressibility, thermal expansion, and the heat capacity at constant pressure as

$$\left(\frac{\partial^2 G}{\partial P \partial T}\right) = \left(\frac{\partial V}{\partial T}\right) = \alpha_T V \tag{I-3.16}$$

From Eqs. (I-3.14) and (I-3.16), one gets

$$w(\Delta G) \propto \exp\left[\frac{1}{2}\frac{V\kappa_T}{k_BT}\Delta p^2 - \frac{1}{2}\frac{\alpha_T V}{k_BT}\Delta p \Delta T - \frac{1}{2}\frac{C_p}{k_BT}\Delta T^2\right] \tag{I-3.17}$$

Comparing (I-3.15) with (I-3.17), one finds the following formula that relate the fluctuation of pressure and temperature with the compressibility, thermal expansibility, and the heat capacity.

$$\kappa_T = -\frac{k_B T}{V\langle \Delta P^2 \rangle}, \quad \alpha_T = \frac{k_B T}{\langle \Delta P \Delta T \rangle}, \quad C_P = \frac{k_B T}{\langle \Delta T^2 \rangle}. \tag{I-3.18}$$

(Random walk)

Imagine motion of a macromolecule, such as protein, in solutions of a living cell. The particle may not be sitting still at a certain position, but in contiguous erratic motion, bombarded by surrounding molecules including water. Such a motion was observed first by R. Brown in 1827 for a pollen particle in solution, and is called "Brownian motion." The motion is not deterministic, but probabilistic or *stochastic*, meaning that the position after some errupsed time, given a definite initial position, may not be predicted for sure, but can be predicted only in some probability, just like weather forecast. One and the simplest model to describe the Brownian motion is the *random walk* model. The two dimensional case of the model is illustrated in Fig. I-3.1. Let's think of a particle that is sitting at a position in a square lattice, the unit length of which is l. The particle moves from a lattice position to the next along x- or y-direction, in forward or backward, with the same probability, 1/4.

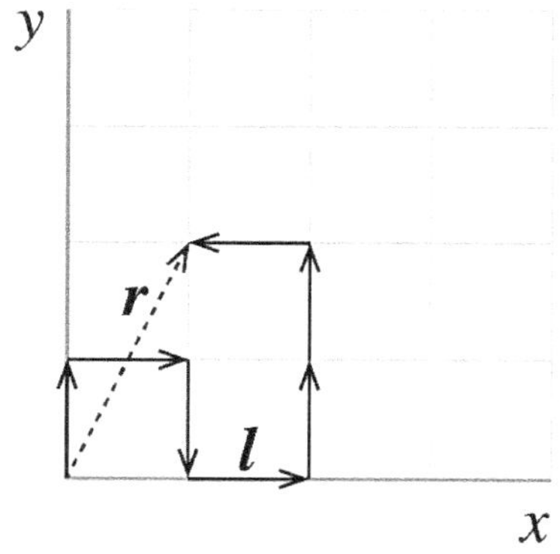

Fig. I-3.1: Illustration of the random walk model.

Here, we treat the random walk in one-dimensional space for simplicity. The model can be treated by primitive algebra, but captures an essential nature of the motion. Think of a particle in motion, *walking* along the x-axis starting from an initial position, say $x = 0$. Suppose the particle can walk just a step, l, at a time in either backward or forward direction. Now, ask *what is the probability of finding the particle at a position, x, on the axis*, after N steps consisting of N_+-forward and N_--backward steps. So, obviously

$$N = N_+ + N_- \tag{I-3.19}$$

The position on the axis after N steps can be defined by the difference between the forward and backward steps, or $x = ml$, where l denotes the distance of a single step, and

$$m = N_+ - N_- \tag{I-3.20}$$

From Eqs. (I-3.18) and (I-3.19),

$$N_+ = \frac{N+m}{2} \qquad N_- = \frac{N-m}{2} \tag{I-3.21}$$

The number of possible ways of walks, which brings the particle to the position at $x = ml$ is

$$\frac{N!}{[(N+m)/2]![(N-m)/2]!}$$

If one assumes that the forward and backward steps occurs in the same probability, one gets

$$w(m,N) = \frac{N!}{[(N+m)/2]![(N-m)/2]!}\left(\frac{1}{2}\right)^N \tag{I-3.22}$$

for the probability of finding the particle at x after the N steps.

The factorials appearing in the above expressions can be approximated by Stirling's approximation. Stirling's approximation says for a large number n,

$$\log n! \approx n \log n - n + \log(2\pi n)^{1/2} \tag{I-3.23}$$

We also expand the logarithm into a Taylor series, and approximate it by

$$\log\left(1 \pm \frac{m}{N}\right) = \pm\frac{m}{N} - \frac{m^2}{2N^2} + O(m^2/N^2) \tag{I-3.24}$$

Using the two approximations, Eq. (I-3.21) is written as

$$\log w(m,N);-\frac{1}{2}\log N + \log 2 - \frac{1}{2}\log 2\pi - \frac{m^2}{2N} \tag{I-3.25}$$

or

$$w(m,N) = \left(\frac{2}{\pi N}\right)^{1/2} \exp\left(\frac{-m^2}{2N}\right) \tag{I-3.26}$$

The result is a Gaussian distribution, but it should be noted that the distribution is derived by making the assumption $N >> 1$.

We have so far used discretized variable m for defining the position. However, we in fact need to find the probability density at a position between x and $x + \Delta x$, that is a continuous variable. The transformation from the discretized to continuous variables can be made by

$$w(x, N)\Delta x = w(m, N)(\Delta x/2l) \tag{I-3.27}$$

Then, the probability density at x can be written as

$$w(x,N) = \frac{1}{(2\pi Nl^2)^{1/2}} \exp\left(-\frac{x^2}{2Nl^2}\right) \tag{I-3.28}$$

and the probability for finding the particle between x and $x + \Delta x$ is

$$w(x,t)\Delta x = \frac{1}{2(\pi Dt)^{1/2}} \exp\left(-\frac{x^2}{4Dt}\right)\Delta x \tag{I-3.29}$$

where D defined as

$$D = \frac{1}{2} nl^2 \tag{I-3.30}$$

has the meaning of the diffusion constant.

(Statistical polymer chain)

Physicochemical characterization of a polymer has been an important field of science and technology. The most common maneuver to characterize a polymer chain is the distribution of the end-to-end distance. The end-to-end distance (ETED) of a polymer is defined by the distance between two atoms or segments at the two termini of a polymer chain. The distance is not definite but fluctuating, because the structure of a polymer is not rigid. Therefore, the quantity should be described by a language of statistics, the *probability distribution* of the distance.

The most primitive description that captures the essential physics of the ETED distribution can be made using so-called *freely-jointed* model (Fig. I-3.2). The model describes a polymer in terms of a chain consisting of many segments, in which the *bond lengths* between two segments are fixed, but the bending and torsional angles among the segments are freely varied. The physics of such a model of polymer can be mapped onto the one-dimensional random walk model of the Brownian motion described above.

Let the fixed bond-length between two segments be l. The projection of l onto the x-axis is $l\cos\theta$, where θ is the polar angle of the vector $\mathbf{l}$ around the x-axis. Then, the mean square length $\langle l_x^2 \rangle$ and its root mean square can be expressed by the orientational average of l_x^2 over the polar angle as,

$$\langle l_x^2 \rangle = \frac{\int_0^\pi l^2 \cos^2\theta (2\pi l^2 \sin\theta) d\theta}{\int_0^\pi 2\pi l^2 \sin\theta d\theta} = l^2/3$$

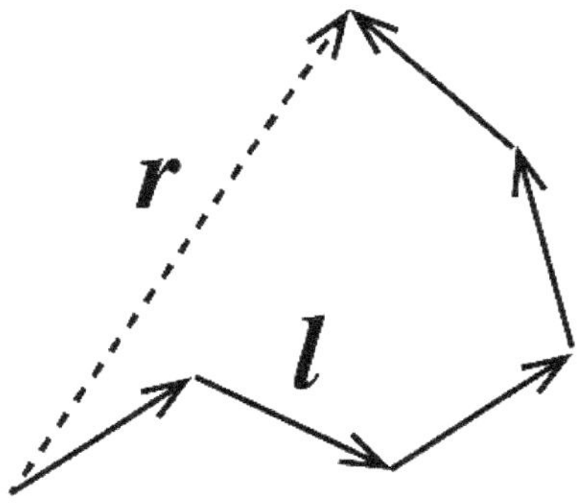

Fig. I-3.2: Illustration of the statistical chain.

and

$$\sqrt{\langle l_x^2 \rangle} = l/\sqrt{3} \tag{I-3.31}$$

Since the orientation of **l** around the x-axis is completely random, the projection of **l** onto the x-axis can be either positive or negative. The positive and negative projection can be mathematically equivalent to the steps in the positive and negative directions of the random walks in the Brownian motion. Let the numbers of positive and negative projections be N_+ and N_-. Then, the end-to-end distance after N_+ positive steps and N_- negative steps can be given by

$$x = (N_+ - N_-)l/\sqrt{3} \tag{I-3.32}$$

Using the same definitions for m and N with Eqs. (I-3.18) and (I-3.19), the problem can be mapped on to the random walk model, and one gets the following equation for the probability distribution for $m = N_+ - N_-$.

$$w(m,N) = \left(\frac{2}{\pi N}\right)^{1/2} \exp\left(\frac{-m^2}{2N}\right) \tag{I-3.33}$$

Using Eq. (I-3.32), m is replaced by x, and one gets the following expression for the ETED distribution.

$$\begin{aligned} w(x,N) &= \frac{\sqrt{3}}{2l} w(m,N) \\ &= \sqrt{\frac{3}{2\pi n l^2}} \exp\left(-\frac{3x^2}{2nl^2}\right) \end{aligned} \tag{I-3.34}$$

The expression can be readily generalized to the three dimensional model of freely jointed model as

$$w(x,y,z;N)dxdydz = \left(\frac{\beta}{\pi^{1/2}}\right)^3 \exp\left(-\beta r^2\right)dxdydz \tag{I-3.35}$$

$$\beta = \left(\frac{3}{2Nl^2}\right)^{1/2} \tag{I-3.36}$$

(Gaussian distribution and Central limiting theorem)

As been explained in the two examples, many phenomena in the nature exhibit the normal distribution, or the Gaussian distribution. Here, let us summarize the properties of the Gaussian distribution.

It is convenient to introduce a function called the characteristic function of a probability distribution function $w(x)$ by

$$\Phi(\xi) = \langle e^{i\xi x} \rangle \tag{I-3.37}$$

where $\Phi(\xi)$ is defined by a Fourier transform of $w(x)$ as

$$\Phi(\xi) = \int_{-\infty}^{\infty} e^{i\xi x} w(x)dx \tag{I-3.38}$$

For independent events x and y,

$$\left\langle e^{i\xi(x+y)}\right\rangle = \left\langle e^{i\xi x}\right\rangle \left\langle e^{i\xi y}\right\rangle \tag{I-3.39}$$

Suppose $\Phi(\xi)$ can be expanded in a Taylor series,

$$\Phi(\xi) = \sum_{n=0}^{\infty} \frac{(i\xi)^n}{n!} \left\langle x^n \right\rangle \tag{I-3.40}$$

or

$$\begin{aligned}\Phi(\xi) &= \left\langle e^{i\xi x}\right\rangle \\ &= \left\langle 1 + i\xi x + \frac{1}{2}(i\xi x)^2 + \cdots \right\rangle \\ &= 1 + i\xi\left\langle x\right\rangle + \frac{1}{2}(i\xi)^2\left\langle x^2\right\rangle + \frac{1}{3!}(i\xi)^3\left\langle x^3\right\rangle + \cdots\end{aligned} \tag{I-3.41}$$

where $\langle x^n \rangle$ is the nth moment of the distribution function. Now, we define another characteristic function $\Psi(\xi)$ by

$$\Psi(\xi) \equiv \ln \Phi(\xi) \tag{I-3.42}$$

Let us expand $\Psi(\xi)$ in a series of $i\xi$ as

$$\Psi(\xi) = C_1(i\xi) + C_2(i\xi)^2 + \cdots \tag{I-3.43}$$

$\Phi(\xi)$ in Eq. (I-3.42) can be expanded in terms of $\Psi(\xi)$ as,

$$\begin{aligned}\Phi(\xi) &= e^{\psi(\xi)} \\ &= 1 + \Psi(\xi) + \frac{1}{2}\Psi(\xi)^2 + \cdots \\ &= 1 + (C_1(i\xi) + C_2(i\xi)^2 + \cdots) + \frac{1}{2}(C_1(i\xi) + C_2(i\xi)^2 + \cdots)^2 + \frac{1}{3!}(C_1(i\xi) + C_2(i\xi)^2 + \cdots)^3 + \cdots \\ &= 1 + C_1(i\xi) + (C_2 + \frac{1}{2}C_1^2)(i\xi)^2 + (C_3 + \frac{1}{2}C_1C_2 + \frac{1}{3!}C_1^3)(i\xi)^3 + \cdots\end{aligned} \tag{I-3.44}$$

Comparing the coefficients of corresponding terms in Eqs. (I-3.41) and (I-3.44), one finds for C_j

$$\begin{aligned}C_1 &= \left\langle x\right\rangle \equiv \left\langle x\right\rangle_c \\ C_2 &= \left\langle x^2\right\rangle - \left\langle x\right\rangle^2 \equiv \left\langle x^2\right\rangle_c \\ C_3 &= \left\langle x^3\right\rangle - 3\left\langle x\right\rangle^2\left\langle x\right\rangle + 2\left\langle x\right\rangle^3 \equiv \left\langle x^3\right\rangle_c \\ &\cdots\cdots\cdots\end{aligned} \tag{I-3.45}$$

where $\langle x^n \rangle_c$ is called "nth cumulant."

Central limiting theorem

Let us consider a stochastic process X_n as a superposition of n randomly varying variables, or

$$X_n = \sum_{j=1}^{n} \Delta X_j \tag{I-3.46}$$

where ΔX_j is a randomly varying displacement, and for each ΔX_j,

$$\left\langle \Delta X_j \right\rangle = 0 \tag{I-3.47}$$

Therefore, the average of the total displacement also vanishes,

$$\left\langle X_n \right\rangle = 0 \tag{I-3.48}$$

We define a characteristic function for the process by

$$\Phi(\xi) = \left\langle e^{i\xi X_n} \right\rangle = \prod_{j=1}^{n} \left\langle e^{i\xi \Delta X_j} \right\rangle \tag{I-3.49}$$

Let's the cumulant functions of $\Delta X_j (j = 1, 2, \cdots, n)$ and X_n be $\psi_j(\xi)$ and $\Psi_n(\xi)$, respectively. Then, we get from Eq. (I-3.46),

$$\Psi_n(\xi) = \sum_{j=1}^{n} \psi_j(\xi) \tag{I-3.50}$$

Suppose the process has the cumulants of all orders, then $\psi_j(\xi)$ can be expanded in the series,

$$\psi_j(\xi) = i\xi \left\langle \Delta X_j \right\rangle_c + \frac{1}{2}(i\xi)^2 \left\langle \Delta X_j^2 \right\rangle_c + \frac{1}{3!}(i\xi)^3 \left\langle \Delta X_j^3 \right\rangle_c + \cdots \tag{I-3.51}$$

The fluctuation is taking place around an average value, therefore,

$$\left\langle \Delta X \right\rangle_c = \left\langle \Delta X \right\rangle = 0 \tag{I-3.52}$$

With the definition of the dispersion (Eq. I-3.45), one may rewrite Eq. (I-3.50) as

$$\Psi_n(\xi) = -\frac{\xi^2}{2} s_n^2 + \frac{1}{3!}(i\xi)^3 \sum \left\langle \Delta X_j^3 \right\rangle_c + \cdots \tag{I-3.53}$$

Let us consider the characteristic function for Y_n instead of X_n,

$$\left\langle e^{i\eta Y_n} \right\rangle = \left\langle e^{i\eta X_n/\varepsilon_n} \right\rangle \tag{I-3.54}$$

Replacing ξ in Eq. (I-3.53) by η/s_n, one gets,

$$\left\langle e^{i\eta Y_n} \right\rangle = \exp\left(-\frac{1}{2}\eta^2 + \frac{(i\eta)^3}{3!} \sum_{j=1}^{n} \frac{\left\langle \Delta X_j^3 \right\rangle_c}{s_n^3} + \cdots \right) \tag{I-3.55}$$

Suppose the mth moment of ΔX_j is all finite and similar in magnitude. s_n^2 increases in the order of n with increasing n. Therefore, the term of the mth cumulant becomes,

$$\frac{O(n)}{O(n^{m/2})} \to 0$$

That is, each term in $m \geq 3$ approaches zero with increasing n. Therefore,

$$\left\langle e^{i\eta Y_n} \right\rangle \to e^{-\eta^2/2} \tag{I-3.56}$$

I-4 Functional Differentiation

I-4.1 Theorems in Functional Differentiation

In the derivation of equations in the liquid state theory, we use a mathematical tool called "Functional differentiation." Since some readers in the field of bioscience may not be familiar with the mathematics, a brief explanation of the mathematics is provided here. Functional differentiation is a differentiation of a *functional*. Then, what is "functional"? A functional is a *function of functions*. Let's think of a function $f(x)$ that is a function of a variable x. If one expands the function in a series,

$$f(x_0 + dx) = f(x_0) + \left(\frac{df}{dx}\right)_{x=x_0} dx + \frac{1}{2}\left(\frac{d^2 f}{dx^2}\right)_{x=x_0} dx^2 + \cdots \tag{I-4.1}$$

when the deviation dx is very small, the value of the function at $x = x_0 + dx$ can be approximated by the first two terms of the series. The coefficient of the second term, or linear term, is called differentiation or a derivative of function $f(x)$.

Let's define a functional of a function $f(x)$ by $\Gamma \equiv \Gamma[f(x)]$. As an illustrative example of the functional, let Γ be an integral of a function $f(x)$ over the range of x from $x = a$ to $x = b$, or

$$\Gamma[f] = \int_a^b f(x)dx.$$

If the function $f(x)$ is changed a little bit between $x = a$ and $x = b$, the result of integration will be changed (Fig. I-4.1).

The change of functional Γ due to the change of the function $f(x)$ can be also expanded in a series as

$$\Gamma[f + \delta f] = \Gamma[f] + \int_a^b \left[\frac{\delta \Gamma}{\delta f}\right] \delta f(x)dx + \frac{1}{2}\int_a^b \left[\frac{\delta^2 \Gamma}{\delta f^2}\right] \delta f(x)^2 dx + \cdots \tag{I-4.2}$$

For a small change of $f(x)$, $\Gamma[f + \delta f]$ can be approximated by the first two terms of the series, just like a series expansion of a function. The second term in the right hand side is called the "first variation," and the coefficient $\delta\Gamma/\delta f$ is called the *functional differentiation* or *functional derivative*. The uniqueness of the definition can be proved based on the following assumption.

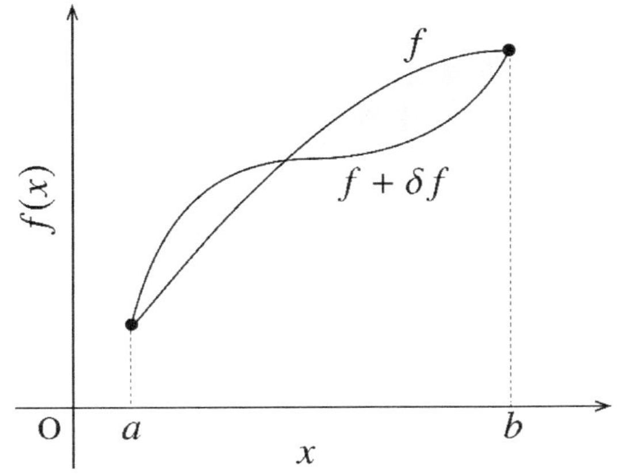

Fig. I-4.1: Illustrative example of a functional.

(**Assumption 1**) "If an arbitrary set of functions, $f_1(r)$, $f_2(r)$, ... , $f_n(r)$, satisfies the following relation,

$$\int \cdots \int F(r_1, r_2, \cdots, r_n) \delta f_1(r_1) \delta f_2(r_2) \cdots \delta f_n(r_n) dr_1 dr_2 \cdots dr_n$$

then, we assume that $F(r_1, r_2, \ldots, r_n) \equiv 0$."

(***Proof of the uniqueness of the functional differentiation***)

Suppose there are two definitions for the functional differentiation. Then,

$$\delta\Gamma \equiv \Gamma[f + \delta f] - \Gamma[f] = \int \left[\frac{\delta\Gamma}{\delta f(r)}\right]_1 \delta f(r) dr = \int \left[\frac{\delta\Gamma}{\delta f(r)}\right]_2 \delta f(r) dr \tag{I-4.3}$$

or

$$\int \left\{ \left[\frac{\delta\Gamma}{\delta f(r)}\right]_1 - \left[\frac{\delta\Gamma}{\delta f(r)}\right]_2 \right\} \delta f(r) dr = 0 \tag{I-4.4}$$

where subscript 1 and 2 indicate there are two ways of defining the functional differentiation. From Assumption 1, it is proved that

$$\left[\frac{\delta\Gamma}{\delta f(r)}\right]_1 - \left[\frac{\delta\Gamma}{\delta f(r)}\right]_2 \equiv 0$$

Theorems of functional differentiation: There are useful theorems with respect to the functional differentiation, which are similar to those concerning the differentiation of functions. The theorems are listed in Table I-4.1 along with corresponding theorems in the differentiation of functions, side by side. Most of the theorems are readily proved. For example, the last theorem,

$$\frac{\delta(1/\Gamma)}{\delta f(r)} = -\frac{1}{\Gamma^2}\frac{\delta(\Gamma)}{\delta f(r)} \tag{I-4.5}$$

in the table can be proved as follows.

$$\begin{aligned}
\frac{1}{\Gamma[f+\delta f]} - \frac{1}{\Gamma[f]} &\equiv \int \frac{\delta(1/\Gamma)}{\delta f(r)}\delta f(r)dr \\
&= \frac{1}{\Gamma+\delta\Gamma+O(\delta f)^2} - \frac{1}{\Gamma} \\
&= \frac{-\left\{\delta\Gamma+O(\delta f)^2\right\}}{\Gamma^2+\Gamma\delta\Gamma+O(\delta f)^2\Gamma} \\
&= -\frac{\delta\Gamma}{\Gamma^2} \\
&= -\frac{1}{\Gamma^2}\int\left[\frac{\delta\Gamma}{\delta f}\right]\delta f(r)dr \\
&= \int\left(-\frac{1}{\Gamma^2}\frac{\delta\Gamma}{\delta f}\right)\delta f(r)dr
\end{aligned}$$

Table I-4.1: Theorem of functional differentiation.

Functional derivatives	Derivative of functions
$\frac{\delta(c\Gamma)}{\delta f(r)} = c\frac{\delta(\Gamma)}{\delta f(r)}$	$\frac{d(cf(r))}{dr} = c\frac{df(r)}{dr}$
$\frac{\delta(\Gamma_1+\Gamma_2)}{\delta f(r)} = \frac{\delta(\Gamma_1)}{\delta f(r)} + \frac{\delta(\Gamma_2)}{\delta f(r)}$	$\frac{d(f(r)+g(r))}{dr} = \frac{df(r)}{dr} + \frac{dg(r)}{dr}$
$\frac{\delta(\Gamma_1\Gamma_2)}{\delta f(r)} = \frac{\delta(\Gamma_1)}{\delta f(r)}\Gamma_2 + \frac{\delta(\Gamma_2)}{\delta f(r)}\Gamma_1$	$\frac{d(f(r)g(r))}{dr} = \frac{df(r)}{dr}g(r) + \frac{dg(r)}{dr}f(r)$
$\frac{\delta(1/\Gamma)}{\delta f(r)} = -\frac{1}{\Gamma^2}\frac{\delta(\Gamma)}{\delta f(r)}$	$\frac{d(1/f(r))}{dr} = -\frac{1}{f(r)^2}\frac{df(r)}{dr}$

Functional differentiation of a functional: Let Γ be a functional of $f(r)$, and $f(r)$ be a functional of $g(r)$, that is,

$$\Gamma = \Gamma[f], \quad f(r) = f[g:r]$$

Then, the following relation called "chain rule" can be proved.

$$\frac{\delta\Gamma}{\delta g(r)} = \int \frac{\delta\Gamma}{\delta f(r')}\frac{\delta f(r')}{\delta g(r)}dr' \tag{I-4.6}$$

(Proof)

Since Γ is a functional of $f(r)$, and $f(r)$ is a functional of $g(r)$

$$\delta\Gamma = \int \frac{\delta\Gamma}{\delta f(r')}\delta f(r')dr' \tag{I-4.7}$$

$$\delta f(r') = \int \frac{\delta f(r')}{\delta g(r)}\delta g(r)dr \tag{I-4.8}$$

Therefore,

$$\delta\Gamma = \int\left\{\int \frac{\delta\Gamma}{\delta f(r')}\cdot\frac{\delta f(r')}{\delta g(r)}dr'\right\}\delta g(r)dr \tag{I-4.9}$$

On the other hand, Γ is also a functional $g(r)$, or $\Gamma = \Gamma\,[g(r)]$, thereby,

$$\delta\Gamma = \int \frac{\delta\Gamma}{\delta g(r)}\delta g(r)dr \tag{I-4.10}$$

From (I-4.9) and (I-4.10),

$$\frac{\delta\Gamma}{\delta g(r)} = \int \frac{\delta\Gamma}{\delta f(r')}\cdot\frac{\delta f(r')}{\delta g(r)}dr' \quad \text{(chain rule)} \tag{I-4.11}$$

As a corollary of the theorem, the following relation can be derived, that is extremely useful in the derivation of the Ornstein-Zernike type equations. Suppose $f(r)$ is a functional of $g(r)$, or $f(r) = f[g; r]$, while $g(r)$ is a functional of $h(r)$, or $g(r) = g[h; r]$. Then,

$$\frac{\delta f(r)}{\delta h(r')} = \int \frac{\delta f(r)}{\delta g(r'')}\cdot\frac{\delta g(r'')}{\delta h(r')}dr'' \tag{I-4.12}$$

If one thinks $f(r)$ is a functional of $f(r)$ itself, then, by definition,

$$\delta f(r) = \int \frac{\delta f(r)}{\delta f(r')}\delta f(r')dr' \tag{I-4.13}$$

Comparing the equation with the theorem concerning the Dirac delta-function, or

$$\int \delta(r-r')\delta f(r')dr' = \delta f(r) \tag{I-4.14}$$

one finds a very important relation concerning the functional differentiation, that is,

$$\frac{\delta f(r)}{\delta f(r')} = \delta(r - r') \tag{I-4.15}$$

If $h = f$ in Eq. (I-4.12), the relation becomes

$$\int \frac{\delta f(r)}{\delta g(r'')} \cdot \frac{\delta g(r'')}{\delta f(r')} dr'' = \delta(r - r') \tag{I-4.16}$$

The relation corresponds to the relation holding for the differentiation of ordinary functions,

$$\frac{dy}{dx}\frac{dx}{dy} = 1 \tag{I-4.17}$$

where y is a function of x, and x is a function of y.

I-4.2 Some Applications of Functional Differentiation in Physics

Functional differentiation has been employed to derive some fundamental equations in physics based on the *variational principle*.

Principle of least action and Lagrange equation of motion in mechanics

Let us define a function called *Larangian* by

$$L = T - U \tag{I-4.18}$$

where U and T are functions, respectively, of generalized spatial coordinates $(q_1, q_2, \cdots, q_s)$ and of their corresponding velocities $(\dot{q}_1, \dot{q}_2, \cdots, \dot{q}_s)$. Now we define a functional S called "*action integral*" by

$$S \equiv \int_{t_1}^{t_2} L(\mathbf{q}, \dot{\mathbf{q}}; t) dt \tag{I-4.19}$$

In the equation, t_1 and t_2 are the initial and final times of the motion at which the trajectory coincides, or $\mathbf{q}(t_1) = \mathbf{q}(t_2)$ and $\dot{\mathbf{q}}(t_1) = \dot{\mathbf{q}}(t_2)$. The principle of least action states that a classical system takes only the trajectory $(\mathbf{q}, \dot{\mathbf{q}})$, or path, that makes the action integral S minimum or least. (A quantum system does not necessarily follow the least action path. Generalization of the action integral to take account for all possible paths is the essential physics of the *path integral* formulation of quantum mechanics, originated by R. Feynman) [Feynman and Hibbs 1965]. The statement is mathematically translated into the variational principle as

$$\delta S = \delta \int_{t_1}^{t_2} L(\mathbf{q}, \dot{\mathbf{q}}; t) dt = 0 \tag{I-4.20}$$

By applying the theorem of the functional differentiation, one has

$$\int_{t_1}^{t_2} \left[\frac{\partial L}{\partial q_i} \delta q_i + \frac{\partial L}{\partial \dot{q}_i} \delta \dot{q}_i \right] dt = 0 \tag{I-4.21}$$

Integrating the second term of the equation by parts, one gets

$$\delta S = \left[\frac{\partial L}{\partial \dot{q}_i} \delta q_i \right]_{t_1}^{t_2} + \int_{t_1}^{t_2} \left(\frac{\partial L}{\partial q_i} - \frac{d}{dt} \frac{\partial L}{\partial \dot{q}_i} \right) \delta q dt = 0 \tag{I-4.22}$$

The first term in the right hand side of the equation vanishes because $\delta q_i(t_1) = \delta q_i(t_2) = 0$. Therefore, we find the following equation called the Lagrange equation of motion.

$$\frac{d}{dt}\left(\frac{\partial L}{\partial \dot{q}_i} \right) - \frac{\partial L}{\partial q_i} = 0 \quad (i = 1, 2, \cdots, s) \tag{I-4.23}$$

If the system is described by the Cartesian-coordinate system, the Lagrangian is defined by

$$L = \sum_i \frac{1}{2} m \dot{q}_i^2 - U(q_1, q_2 \cdots, q_N) \tag{I-4.24}$$

It leads to the familiar equation of motion by Newton, or

$$m \ddot{q}_i = -\frac{\partial}{\partial q_i} U(q_1, q_2, \cdots, q_s) \tag{I-4.25}$$

Some readers may have a question concerning the physical meaning of the principle of least action. R. Feynman has answered the question, referring to his high school teacher, in his famous textbook of physics [Feynman 1963]. According to the explanation, a classical system draws a trajectory that makes the kinetic energy and potential energy equal in their average. The statement can be readily realized by actually integrating (I-4.24) over the range between t_1 and t_2 as,

$$\frac{1}{t_2 - t_1} \int_{t_1}^{t_2} T[\dot{q}(t)] dt = \frac{1}{t_2 - t_1} \int_{t_1}^{t_2} U[q(t)] dt \tag{I-4.26}$$

The left hand side is an average of the kinetic energy during the time t_1 to t_2, while the right hand side is that of the potential energy in the same duration.

Hamilton's equation of motion

The formulation based on the Lagrangian allows to describe the mechanical sate of a system in terms of generalized coordinates and velocities. However, there is another formulation of mechanics that employs generalized coordinates and momentum, instead of velocities. The formulation plays important roles in describing the time evolution of a system consisting of many particles that are interacting with each other. The transformation of the description are made by defining a new function called Hamiltonian by

$$H(p, q, t) \equiv \sum_i p_i \dot{q}_i - L \tag{I-4.27}$$

Let us first take the total derivative of the Lagrangian as

$$dL = \sum_i \frac{\partial L}{\partial q_i} dq_i + \sum_i \frac{\partial L}{\partial \dot{q}_i} d\dot{q}_i \tag{I-4.28}$$

The derivatives $\partial L/\partial \dot{q}_i$ and $\partial L/\partial q_i$ in Eq. (I-4.28) are generalized momenta and force in the light of the Lagrange equation of motion, and one may write as

$$\frac{\partial L}{\partial q_i} = \dot{p}_i \tag{I-4.29}$$

$$\frac{\partial L}{\partial \dot{q}_i} = p_i \tag{I-4.30}$$

Therefore, Eq. (I-4.28) may be written as

$$dL = \sum_i \dot{p}_i dq_i + \sum_i p_i d\dot{q}_i \tag{I-4.31}$$

Writing the second term as $\sum_i \dot{q}_i dp_i = d\sum_i p_i\dot{q}_i - \sum_I p_i d\dot{q}_i$, one gets

$$d(\sum p_i\dot{q}_i - L) = -\sum \dot{p}_i dq_i + \sum \dot{q}_i dp_i \tag{I-4.32}$$

With the definition of Hamiltonian, Eq. (I-4.27), we find

$$dH = -\sum \dot{p}_i dq_i + \sum \dot{q}_i dp_i \tag{I-4.33}$$

From the equation, we identify the following equation of motion,

$$\dot{q}_i = \frac{\partial H}{\partial p_i} \tag{I-4.34}$$

$$\dot{p}_i = -\frac{\partial H}{\partial q_i} \tag{I-4.35}$$

In the simple case of the Cartesian coordinate system, it is readily seen that the above definition of the Hamiltonian coincides with the familiar definition, that is, $H = T + U$, since $\sum p_i\dot{q}_i = 2T$.

The transformation expressed by Eq. (I-4.27) is a special case of the mathematical procedure referred to as the *Langendre transformation.*

Hamilton's equation of motion plays a crucial role for formulating the Liouville equation that governs the time evolution of many body systems, including water and protein. Chapter III and Chapter V of this book are based on the description provided here. Of course, the molecular dynamics simulation is based entirely on the equations derived here.

Formulation of Schrodinger equation based on the variational principle

The functional differentiation can be also applied to derive the Schrodinger equation based on the variational principle. Schrodinger equation is written as

$$\mathcal{H}\varphi = \varepsilon\varphi \tag{I-4.36}$$

where $\mathcal{H}$, φ, and ε are the Hamiltonian of a quantum system, its wave function, and energy.

The equation can be derived based on the variational principle as follows. Let us define the energy of a system by an average of the Hamiltonian over all the quantum states as,

$$I = \int \varphi^* \mathcal{H} \varphi dq \tag{I-4.37}$$

The problem is to find the wave functions, φ, that minimizes I. The wave function may be expanded by a complete set of basis functions in the functional space as

$$\varphi = c_1 u_1 + c_2 u_2 + c_3 u_3 + \cdots \tag{I-4.38}$$

The wave function should satisfy the orthonormal condition,

$$(\varphi, \varphi) \equiv \int \varphi^* \varphi dq = 1 \tag{I-4.39}$$

Equation (I-4.37) can be written as

$$I = \sum_i \sum_j c_i^* c_j H_{ij} \tag{I-4.40}$$

where $H_{ij} = \int \varphi_i^* \mathcal{H} \phi_j dq$.

Now, let us define a functional by $I - \varepsilon J$ with the Lagrange undetermined multiplier ε, where J is a function to impose the orthonormal condition.

$$J \equiv (\varphi, \varphi) - 1 = \sum_i c_i^* c_i - 1 \tag{I-4.41}$$

The variational principle may be stated as

$$\frac{\delta}{\delta c_i^*}(I - \varepsilon J) = 0, \quad \text{and} \quad \frac{\delta}{\delta c_i}(I - \varepsilon J) = 0 \tag{I-4.42}$$

Putting Eqs. (I-4.40) and (I-4.41) into Eq. (I-4.42), one finds

$$\sum_j (H_{ij} - \varepsilon \delta_{ij}) c_j = 0, \quad \text{and} \quad \sum_i c_i^* (H_{ij} - \varepsilon \delta_{ij}) = 0 \tag{I-4.43}$$

Those equations are the matrix representation of the Schrodinger equation with the wave functions expanded in a complete set of the basis functions.

Later in this book (Chapter II), we use this variational principle to derive the RISM-SCF equation that enables us to evaluate the electronic structure of a molecule solvated in solutions.

The Literature

Sections I-1 and I-2 are based on the book by Chandler (1987), although many topics related to the stability of phases are discarded. The book by Chandler provides a modern and the most advanced introduction to the thermodynamics and statistical mechanics. Section I-3 is written referring to the books by Landau and Lifshitz (1957), Kubo et al. (1972), Cantor and Schmimmel (1980). The mathematical basics of the functional differentiation in Section I-4 is based on the lecture by Hiroike delivered in Hokkaido University in 1972. The application of the functional differentiation to the derivation of the least action law can be found in many text books. Here, the author consulted the books by Landau and Lifshitz (1957), and Fyneman et al. (1963). The derivation of Schrodinger equation by means of the variation principle is found in many standard text books of quantum chemistries, but here the book by Koide (1969) is referred to.

CHAPTER II

Statistical Mechanics of Liquid and Solutions

As was emphasized in the introduction of the book, water is essential for living bodies on the earth. It is because all the biomolecules, including protein and DNA, in our body can maintain their structure, and perform their functions only in aqueous environment. It was also emphasized that microscopic structure of water plays crucial role in the two aspects of biomolecular activity, the self-organization and the molecular recognition. Then, what is the structure of liquids including water, and how it can be characterized? This is the main theme of this chapter.

II-1 Structure of Liquid and Ornstein-Zernike Equation

II-1.1 What is the Structure of Liquid?

It is not difficult to define structure of a molecule. For example, the structure of a water molecule may be characterized by a triangle made of an oxygen atom and two hydrogen atoms, giving the three lengths among O-O, O-H, H-H atoms, or the two O-H-H angles and the H-H length, or the two O-H length and the H-O-H and angle (Fig. II-1.1a). The situation is more or less similar for the case of a solid crystal. Crystal structure can be characterized by giving lattice constant of a unit lattice (Fig. II-1.1b).

However, the liquid structure may not be characterized by the language of geometry, such as bond length and lattice constant, because molecules in liquid are in contiguous thermal

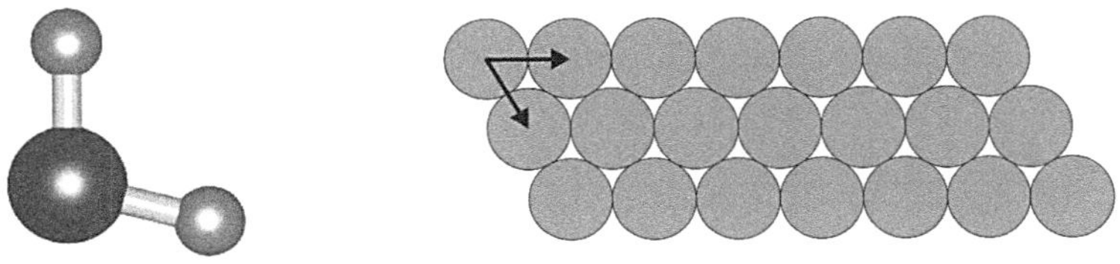

(a) (b)

Fig. II-1.1: Illustration of the structure of a molecule and crystal.

motion, and geometrical configuration among molecules are not well defined. Then, how can one characterize the structure of liquids? The answer to the question is "in terms of a statistical language."

Let us set up a thought experiment in which many water molecules are put in a box or a container, where the number of water molecules is N, and the volume of the box is V. All water molecules are changing their position and orientation contiguously by diffusive or thermal motion. Now, let's try to find an oxygen atom of a water molecule at position $\mathbf{r}$ in the box. You may be able to find an oxygen atom at the position by chance, or in some probability. The probability should be constant if there is no preference for a position in the box, and it should be proportional to the (local) average density of water molecules in the box, or N/V. There is no structural information involved. However, what if we ask what is the probability of finding two molecules at two positions $\mathbf{r}$ and $\mathbf{r}'$, or $\rho(\mathbf{r},\mathbf{r}')$? It is not a constant anymore, but a function of distance $|\mathbf{r} - \mathbf{r}'|$ between the two positions. For example, if the distance is too short, or $|\mathbf{r} - \mathbf{r}'| \leq \sigma$ (σ: diameter of a molecule), $\rho(\mathbf{r},\mathbf{r}')$ should be zero, because two atoms cannot exist simultaneously in such short distance due to core repulsion between the two atoms. On the other hand, if the distance is very large, or $|\mathrm{r} - \mathrm{r}'| >> 1$, $\rho(\mathbf{r},\mathbf{r}')$ should be a product of the *average* density at the two positions, or $\rho(\mathbf{r},\mathbf{r}') = \rho(\mathbf{r})\rho(\mathbf{r}')$, because densities at the two positions become independent of each other. What happens between the two extreme cases? It will show more complicated dependence on the distance $|\mathbf{r} - \mathbf{r}'|$ between the two atoms because the density at one position will have an influence upon that of the other position, and vice versa. We refer to the influence as "correlation." Taking the correlation into consideration, a thermal average of a product of densities at two positions can be written as,

$$\rho(\mathbf{r},\mathbf{r}') = \rho(\mathbf{r})\rho(\mathbf{r}')g(\mathbf{r},\mathbf{r}') \tag{II-1.1}$$

where $g(\mathbf{r},\mathbf{r}')$ represents the correlation. When a liquid system is not subject to any external field, such as an electrostatic field, the average density is a constant, or $\rho(\mathbf{r}) = \rho(\equiv N/V)$, and the correlation function $g(\mathbf{r},\mathbf{r}')$ becomes a function of distance between the two positions, or $g(\mathbf{r},\mathbf{r}') = g(|\mathbf{r} - \mathbf{r}'|)$. The function $g(\mathbf{r},\mathbf{r}')$ is referred to as the pair correlation function, and $g(|\mathbf{r} - \mathbf{r}'|)$ as the *radial distribution function* (RDF) for obvious reason. It is this function that conveys the structural characteristics of liquids, since it depends sensitively upon interactions between molecules in liquids. More rigorous definition of the (local) average density $\rho(\mathbf{r})$ and the correlation function $g(\mathbf{r},\mathbf{r}')$ is given in terms of statistical mechanics in the succeeding sections.

In Fig. II-1.2, illustrated are the radial distribution functions between the two oxygen atoms in water, which are compared with that of liquid "neon", as an example. Neon is mimicked by an oxygen atom without a partial charge. Both curves are the results from the calculation based on the theory explained in the present chapter, which is known to qualitatively reproduce the experimental results. The example clearly demonstrates that the correlation functions or the radial distributions express the characteristics of liquid structure. Depicted by the solid line is RDF of the O-O atom pairs in water plotted against the radial distance, while the dotted line is that of a pair of Ne atoms in liquid neon. The RDF of neon has the first and second peaks around $r \sim \sigma$ and $r \sim 2\sigma$ (σ : diameter) that correspond roughly to the first and second coordination shells around a neon molecule (atom), which is illustrated in Fig. II-1.3(a). On the other hand, the RDF of O-O pair in water has the first and second peak around $r \sim \sigma$ and $r \sim 1.6\sigma$, roughly speaking (Fig. II-1.3b). The first peak is considerably sharper compared to that corresponding

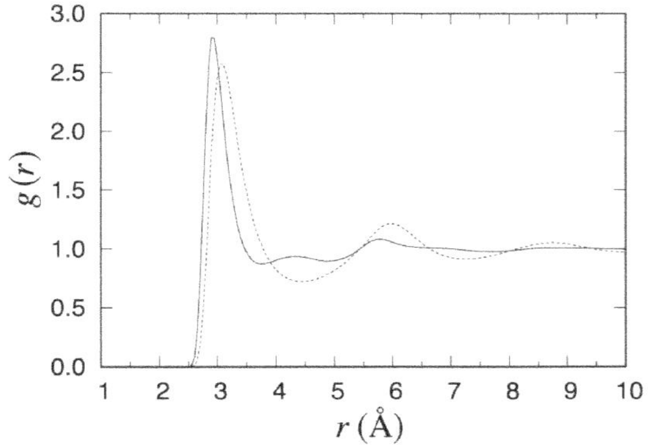

Fig. II-1.2: The pair correlation function between two oxygen-atoms in water (solid line) in comparison with that of neon (dotted line). Both plots are theoretical results calculated based on the liquid state theory explained in this chapter.

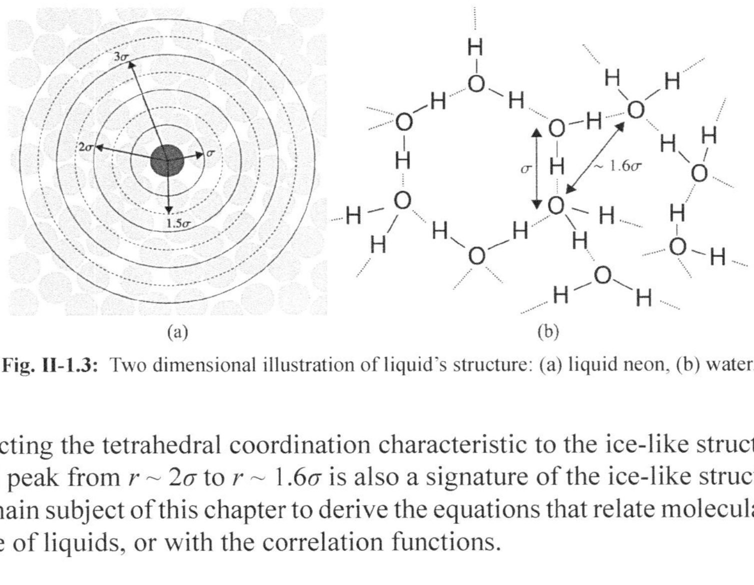

Fig. II-1.3: Two dimensional illustration of liquid's structure: (a) liquid neon, (b) water.

to neon, reflecting the tetrahedral coordination characteristic to the ice-like structure. The shift of the second peak from $r \sim 2\sigma$ to $r \sim 1.6\sigma$ is also a signature of the ice-like structure.

It is the main subject of this chapter to derive the equations that relate molecular interactions with structure of liquids, or with the correlation functions.

II-1.2 Density Field and the Density Distribution Functions

In the preceding section, we have talked about density of liquid at a *position or a point*. But, what does "the density at a *position*" mean? The position is a point with "no volume." How can it be described mathematically? A quick answer to the question is the "Dirac delta-function." But, how it can be the Dirac delta-function? Let's find out how.

In Fig. II-1.4, illustrated by a cartoon is a molecule fixed at position x_i in one dimensional space along the x-axis. Let us denote the density of a molecule at position x by $\nu(x)$, and take an integral of $\nu(x)$ over a range L defined along the x-axis. Then, the integral will be unity if the range L includes the position x_i, otherwise it is zero, that is,

$$\begin{aligned} \int_L \nu(x)dx &= 1, \quad x_i \in L \\ &= 0 \qquad x_i \notin L \end{aligned} \tag{II-1.2}$$

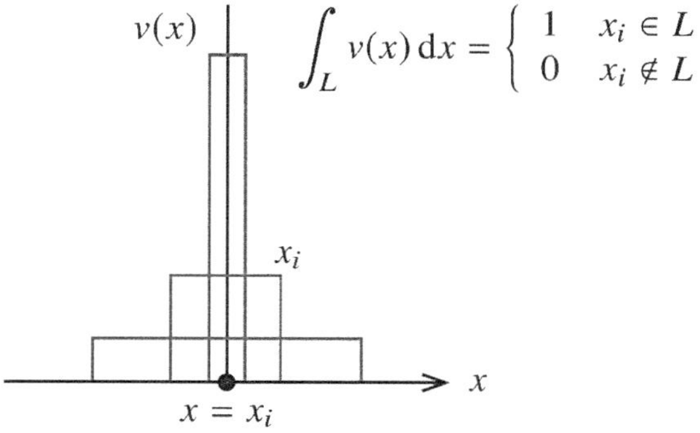

Fig. II-1.4: Illustration of $v(x)$.

Now, let the range L be smaller and smaller sustaining the above condition as illustrated in Fig. II-1.4, and bring it to a limit of $L = 0$. A function which satisfies such a condition is nothing but the Dirac delta-function, or

$$v^{(1)}(x) = \sum_{i=1}^{N} \delta(x - x_i) \tag{II-1.3}$$

The definition can be readily generalized to the case where N molecules are fixed in three dimensional space,

$$v^{(1)}(\mathbf{r}) = \sum_{i=1}^{N} \delta(\mathbf{r} - \mathbf{r}_i) \tag{II-1.4}$$

where $v^{(1)}(\mathbf{r})$ denotes the density of molecules at $\mathbf{r}$, and $\mathbf{r}_i$ is the position of molecules fixed at a three dimensional space.

In a similar manner, let's think about the density of molecules at two positions at the same time. The density of a pair of atoms will be defined by the following equation

$$v^{(2)}(\mathbf{r},\mathbf{r}') = \sum_{i=1}\sum_{j\neq i} \delta(\mathbf{r} - \mathbf{r}_i)\delta(\mathbf{r}' - \mathbf{r}_j) \tag{II-1.5}$$

where $v^{(2)}(\mathbf{r},\mathbf{r}')$ denotes the density of molecules at two positions $\mathbf{r}$ and $\mathbf{r}'$, and $\mathbf{r}_i$ and $\mathbf{r}_j$ are the positions of two different molecules fixed in the space.

The *density* defined so far, which is referred to as "density field," applies to such a system in which all the molecules are fixed at positions in space. However, in real liquids, molecules are moving around a space due to the thermal motion. Therefore, the density field should be averaged over the ensemble produced by the thermal motion (see Chapter I for the meaning of ensemble). Here, we employ the grand canonical ensemble to take the thermal average of the density field. As is reviewed in Chapter I, the grand canonical ensemble average of a variable $A(\mathbf{r}_1, \mathbf{r}_2, \cdots, \mathbf{r}_N)$ depending on the configuration or the position of molecules can be taken by the following equation

$$\langle A \rangle = \frac{1}{\Xi}\sum_{N=0}\frac{1}{N!}\int_V \cdots \int_V A(\mathbf{r}_1,\mathbf{r}_2,\cdots,\mathbf{r}_N)\prod_{i=1}^{N} z(\mathbf{r}_i)\exp\left[-U(\mathbf{r}_1,\mathbf{r}_2,\cdots,\mathbf{r}_N)/k_BT\right]d\mathbf{r}_1 d\mathbf{r}_2\cdots d\mathbf{r}_N \tag{II-1.6}$$

In the equation, Ξ is the grand partition function defined by,

$$\Xi = \sum_{N=0} \frac{1}{N!}\int_V \cdots \int_V \prod_{i=1}^{N} z(\mathbf{r}_i)\exp\left[-U(\mathbf{r}_1,\mathbf{r}_2,\cdots,\mathbf{r}_N)/k_BT\right]d\mathbf{r}_1 d\mathbf{r}_2\cdots d\mathbf{r}_N \tag{II-1.7}$$

where $U(\mathbf{r}_1,\mathbf{r}_2,\ldots,\mathbf{r}_N)$ denotes the interaction potential energy among the molecules, and $z(\mathbf{r})$ is the generalized *activity* defined by the following equation

$$z(\mathbf{r}_i) = z\exp[-\beta\Psi(\mathbf{r}_i)] \tag{II-1.8}$$

In the equation, z is the activity defined by $z \equiv \exp(\beta\mu N)/\Lambda^3$, and $\Psi(\mathbf{r}_i)$ the potential energy due to an external field acting on the particle placed at $\mathbf{r}_i$.

Now, let's take the thermal average of the density fields defined by Eq. (II-1.4) and Eq. (II-1.5) using Eq. (II-1.6) or the grand canonical ensemble average. In the case of Eq. (II-1.4),

$$\left\langle \nu^{(1)}(\mathbf{r})\right\rangle = \frac{1}{\Xi}\sum_{N=0}\frac{1}{N!}\int_V \cdots \int_V \nu^{(1)}(\mathbf{r})\prod_{i=1}^{N} z(\mathbf{r}_i)\exp\left[-U(\mathbf{r}_1,\mathbf{r}_2,\cdots,\mathbf{r}_N)/k_BT\right]d\mathbf{r}_1 d\mathbf{r}_2\cdots d\mathbf{r}_N \tag{II-1.9}$$

Putting Eq. (II-1.4) into Eq. (II-1.9),

$$\left\langle \nu^{(1)}(\mathbf{r})\right\rangle = \frac{1}{\Xi}\sum_{N=0}\frac{1}{N!}\int_V \cdots \int_V \sum_{i=1}^{N}\delta(\mathbf{r}-\mathbf{r}_i)\prod_{i=1}^{N} z(\mathbf{r}_i)\exp\left[-U(\mathbf{r}_1,\mathbf{r}_2,\cdots,\mathbf{r}_N)/k_BT\right]d\mathbf{r}_1 d\mathbf{r}_2\cdots d\mathbf{r}_N$$

After some manipulation using the property of the Dirac delta-function (see Chapter I), the equation reduces to the definition of the *density distribution function* $\rho^{(1)}(\mathbf{r})$ in the standard text book of statistical mechanics, that is,

$$\begin{aligned}\left\langle \nu^{(1)}(\mathbf{r})\right\rangle &= \frac{1}{\Xi}\sum_{N=1}\frac{1}{(N-1)!}\int_V \cdots \int_V \prod_{i=1}^{N} z(\mathbf{r}_i)\exp\left[-U(\mathbf{r},\mathbf{r}_2,\cdots,\mathbf{r}_N)/k_BT\right]d\mathbf{r}_2\cdots d\mathbf{r}_N \\ &\equiv \rho^{(1)}(\mathbf{r})\end{aligned} \tag{II-1.10}$$

The equation gives a clear physical meaning to the density distribution function. It is an average of the density field over configurations or an ensemble produced by the thermal motion. The quantity is proportional to the probability for finding a molecule at position $\mathbf{r}$.

In a similar manner, the expression for the density pair distribution function $\rho^{(2)}(\mathbf{r},\mathbf{r}')$ can be derived as a thermal average of the pair density field $\nu^{(2)}(\mathbf{r},\mathbf{r}')$ defined in Eq. (II-1.5), with respect to the grand canonical ensemble.

$$\begin{aligned}\left\langle \nu^{(2)}(\mathbf{r},\mathbf{r}')\right\rangle &= \frac{1}{\Xi}\sum_{N=0}\frac{1}{N!}\int_V \cdots \int_V \nu^{(2)}(\mathbf{r},\mathbf{r}')\prod_{i=1}^{N} z(\mathbf{r}_i)\exp\left[-U(\mathbf{r}_1,\mathbf{r}_2,\cdots,\mathbf{r}_N)/k_BT\right]d\mathbf{r}_1 d\mathbf{r}_2\cdots d\mathbf{r}_N \\ &= \frac{1}{\Xi}\sum_{N=0}\frac{1}{N!}\int_V \cdots \int_V \sum_{i=1}\sum_{j\neq i}\delta(\mathbf{r}-\mathbf{r}_i)\delta(\mathbf{r}'-\mathbf{r}_j)\prod_{i=1}^{N} z(\mathbf{r}_i)\exp\left[-U(\mathbf{r}_1,\mathbf{r}_2,\cdots,\mathbf{r}_N)/k_BT\right]d\mathbf{r}_1 d\mathbf{r}_2\cdots d\mathbf{r}_N \\ &= \frac{1}{\Xi}\sum_{N=2}\frac{1}{(N-2)!}\int_V \cdots \int_V \prod_{i=1}^{N} z(\mathbf{r}_i)\exp\left[-U(\mathbf{r},\mathbf{r}',\cdots,\mathbf{r}_N)/k_BT\right]d\mathbf{r}_3\cdots d\mathbf{r}_N \\ &\equiv \rho^{(2)}(\mathbf{r},\mathbf{r}')\end{aligned} \tag{II-1.11}$$

II-1.3 Density Distribution Functions in Terms of the Functional Derivatives

The density distribution functions derived in the previous section can be defined in terms of the functional derivatives (Theorems concerning the functional derivative are summarized in Chapter I). Let us take the functional derivative of the grand partition function Ξ[z] defined by Eq. (II-1.7) with respect to z(**r**),

$$\frac{\delta\Xi[z]}{\delta z(\mathbf{r})}=\sum_{N=0}\frac{1}{N!}\int_V\cdots\int_V\sum_{i=1}\frac{\delta z(\mathbf{r}_i)}{\delta z(\mathbf{r})}\prod_{i=2}^{N}z(\mathbf{r}_i)\exp\left[-U(\mathbf{r}_1,\mathbf{r}_2,\cdots,\mathbf{r}_N)/k_BT\right]d\mathbf{r}_1d\mathbf{r}_2\cdots d\mathbf{r}_N \quad \text{(II-1.12)}$$

According to the theorem 1 in the appendix concerning the functional derivative, the derivative in the integrand is

$$\frac{\delta z(\mathbf{r}_i)}{\delta z(\mathbf{r})}=\delta(\mathbf{r}-\mathbf{r}_i) \quad \text{(II-1.13)}$$

Therefore, Eq. (II-1.12) becomes,

$$\frac{\delta\Xi[z]}{\delta z(\mathbf{r})}=\sum_{N=0}\frac{1}{N!}\int_V\cdots\int_V\sum_{i=1}^{N}\delta(\mathbf{r}-\mathbf{r}_i)\prod_{i=2}^{N}z(\mathbf{r}_i)\exp\left[-U(\mathbf{r}_1,\mathbf{r}_2,\cdots,\mathbf{r}_N)/k_BT\right]d\mathbf{r}_1d\mathbf{r}_2\cdots d\mathbf{r}_N$$

The integration over the variables ($\mathbf{r}_1$, $\mathbf{r}_2$, · · · , $\mathbf{r}_i$, · · · $\mathbf{r}_N$) concerning the Dirac delta-function can be readily carried out due to the theorem, and gives

$$\frac{\delta\Xi[z]}{\delta z(\mathbf{r})}=\sum_{N=1}\frac{1}{(N-1)!}\int_V\cdots\int_V\prod_{i=2}^{N}z(\mathbf{r}_i)\exp\left[-U(\mathbf{r},\mathbf{r}_2,\cdots,\mathbf{r}_N)/k_BT\right]d\mathbf{r}_2\cdots d\mathbf{r}_N$$

Multiplying both sides of the equation by $z(\mathbf{r})/\Xi[z]$, one gets,

$$\frac{z(\mathbf{r})}{\Xi[z]}\frac{\delta\Xi[z]}{\delta z(\mathbf{r})}=\frac{1}{\Xi[z]}\sum_{N=1}\frac{1}{(N-1)!}\int_V\cdots\int_V\prod_{i=1}^{N}z(\mathbf{r}_i)\exp\left[-U(\mathbf{r},\mathbf{r}_2,\cdots,\mathbf{r}_N)/k_BT\right]d\mathbf{r}_2\cdots d\mathbf{r}_N$$

The right-hand side of the equation is nothing but the expression of the (single particle) density distribution function $\rho^{(1)}(\mathbf{r})$ defined by Eq. (II-1.10), thereby,

$$\rho^{(1)}(\mathbf{r})=\frac{z(\mathbf{r})}{\Xi}\frac{\delta\Xi[z]}{\delta z(\mathbf{r})} \quad \text{(II-1.14)}$$

In a similar way, the expression for the density pair distribution function $\rho^{(2)}(\mathbf{r},\mathbf{r}')$ in terms of the functional derivative can be also found as follows

$$\begin{aligned}\frac{\delta^2\Xi[z]}{\delta z(\mathbf{r})\delta z(\mathbf{r}')}&=\sum_{N=2}\frac{N-1}{(N-1)!}\int_V\cdots\int_V\delta(\mathbf{r}'-\mathbf{r}_2)\prod_{i=3}^{N}z(\mathbf{r}_i)\exp\left[-U(\mathbf{r},\mathbf{r}_2,\cdots,\mathbf{r}_N)/k_BT\right]d\mathbf{r}_2\cdots d\mathbf{r}_N\\&=\sum_{N=2}\frac{1}{(N-2)!}\int_V\cdots\int_V\prod_{i=3}^{N}z(\mathbf{r}_i)\exp\left[-U(\mathbf{r},\mathbf{r}',\cdots,\mathbf{r}_N)/k_BT\right]d\mathbf{r}_3\cdots d\mathbf{r}_N\end{aligned} \quad \text{(II-1.15)}$$

$$\begin{aligned}\frac{z(\mathbf{r})z(\mathbf{r}')}{\Xi}\frac{\delta^2\Xi[z]}{\delta z(\mathbf{r})\delta z(\mathbf{r}')}&=\frac{1}{\Xi[z]}\sum_{N=2}\frac{1}{(N-2)!}\int_V\cdots\int_V\prod_{i=3}^{N}z(\mathbf{r}_i)\exp\left[-U(\mathbf{r},\mathbf{r}',\cdots,\mathbf{r}_N)/k_BT\right]d\mathbf{r}_3\cdots d\mathbf{r}_N\\&=\rho^{(2)}(\mathbf{r},\mathbf{r}')\end{aligned}$$

II-1.4 Ornstein-Zernike Equation

Now, we are ready to *derive* the Ornstein-Zernike equation, an integral equation governing the density pair correlation functions in liquids. Let us take a functional derivative of $\rho^{(1)}(\mathbf{r})$ with respect to $\ln z(\mathbf{r})$. With the definitions of the density distribution functions (II-1.14) and (II-1.15), the derivative gives,

$$\begin{aligned}
\frac{\delta\rho^{(1)}(\mathbf{r})}{\delta \ln z(\mathbf{r}')} &= z(\mathbf{r}')\frac{\delta}{\delta z(\mathbf{r}')}\left(\frac{z(\mathbf{r})}{\Xi}\frac{\delta\Xi}{\delta z(\mathbf{r})}\right) \\
&= z(\mathbf{r}')\left[\frac{\delta z(\mathbf{r})}{\delta z(\mathbf{r}')}\left(\frac{1}{\Xi}\frac{\delta\Xi}{\delta z(\mathbf{r}')}\right) - z(\mathbf{r})\frac{1}{\Xi^2}\frac{\delta\Xi}{\delta z(\mathbf{r}')}\frac{\delta\Xi}{\delta z(\mathbf{r})} + \frac{z(\mathbf{r})}{\Xi}\frac{\delta^2\Xi}{\delta z(\mathbf{r})\delta z(\mathbf{r}')}\right] \\
&= \delta(\mathbf{r}\text{-}\mathbf{r}')\frac{1}{\Xi}\frac{\delta\Xi}{\delta \ln z(\mathbf{r}')} - \left(\frac{1}{\Xi}\frac{\delta\Xi}{\delta \ln z(\mathbf{r})}\right)\left(\frac{1}{\Xi}\frac{\delta\Xi}{\delta \ln z(\mathbf{r}')}\right) + \frac{\delta^2 \ln\Xi}{\delta \ln z(\mathbf{r})\delta \ln z(\mathbf{r}')} \\
&= \delta(\mathbf{r}\text{-}\mathbf{r}')\rho^{(1)}(\mathbf{r}) - \rho^{(1)}(\mathbf{r})\rho^{(1)}(\mathbf{r}') + \rho^{(2)}(\mathbf{r},\mathbf{r}')
\end{aligned} \qquad \text{(II-1.16)}$$

We define an inverse of the above relation with a function called "direct correlation function" as follows,

$$\frac{\delta \ln z(\mathbf{r})}{\delta\rho^{(1)}(\mathbf{r}')} = \frac{\delta(\mathbf{r}-\mathbf{r}')}{\rho^{(1)}(\mathbf{r})} - c(\mathbf{r},\mathbf{r}') \qquad \text{(II-1.17)}$$

Substituting Eqs. (II-1.16) and (II-1.17) into the chain rule of the functional derivative (see Chapter I),

$$\int \frac{\delta\rho^{(1)}(\mathbf{r})}{\delta \ln z(\mathbf{r}'')}\frac{\delta \ln z(\mathbf{r}'')}{\delta\rho^{(1)}(\mathbf{r}')}d\mathbf{r}'' = \delta(\mathbf{r}-\mathbf{r}')$$

one finds the Ornstein-Zernike equation after some manipulation with the theorems concerning the functional derivative (see Chapter I)

$$h(\mathbf{r},\mathbf{r}') = c(\mathbf{r},\mathbf{r}') + \int_V c(\mathbf{r},\mathbf{r}'')\rho(\mathbf{r}'')h(\mathbf{r}'',\mathbf{r}')d\mathbf{r}'' \qquad \text{(II-1.18)}$$

It will be worthwhile to consider the physical meaning of the direct correlation function defined by Eq. (II-1.17). From the definition of $z(\mathbf{r})$ given in Eq. (II-1.8), $\ln z(\mathbf{r})$ is essentially the interaction energy of a molecule placed at position $\mathbf{r}$ with all other molecules, while $\rho^{(1)}(\mathbf{r}')$ is the density at position $\mathbf{r}'$. A perturbation of $\rho^{(1)}(\mathbf{r}')$ at position $\mathbf{r}'$ will induce a change (or response) in $\ln z(\mathbf{r})$ at position $\mathbf{r}$. Thus, $\delta \ln z(\mathbf{r})$ is the response of the interaction energy to the perturbation in the liquid density, $\delta\rho^{(1)}(\mathbf{r}')$, at position $\mathbf{r}'$, and $\delta \ln z(\mathbf{r})/\delta\rho^{(1)}(\mathbf{r}')$ signifies a response function, main part of which is $c(\mathbf{r},\mathbf{r}')$. The first term in Eq. (II-1.17) signifies just the self-correlation.

II-1.5 Closure Relations

In the previous section, we have *derived* the Ornstein-Zernike equation. By solving the equation, we may be able to obtain the density pair correlation functions, which of course represents structure of liquids as is explained in Section I-1. However, the equation cannot be solved by itself, since it includes two unknown variables, $h(\mathbf{r},\mathbf{r}')$ and $c(\mathbf{r},\mathbf{r}')$. It is necessary to

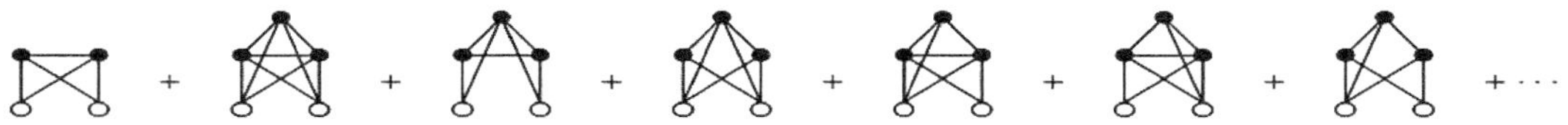

Fig. II-1.5: Bridge diagrams.

find another equation to relate those two variables from an alternative route. Such equations are referred to as a "closure" in the liquid state theory. In the historical view points, there are two ways to derive the closure relation. One is based on the diagrammatic expansion of the correlation functions. According to the theory, the pair correlation functions are expanded in an infinite series of diagrams in the order of density, similar to the cluster expansion of J. Mayer, consisting of so called Mayer f-function or $f(\mathbf{r}, \mathbf{r}') = \exp[-u(\mathbf{r}, \mathbf{r}')/k_BT]-1$ and the density as a function of position. The diagrams are classified into several categories according to topological connectivity. Most of the closure relations proposed so far is derived by ignoring some set of diagrams. For example, a closure called the *Hyper-Netted-Chain* (HNC) equation is derived by ignoring a set of diagrams called "bridge diagrams." Few of the bridge diagrams in the lowest order is illustrated in Fig. II-1.5. The feature of the diagrams is that those diagrams cannot be reduced to the convolution integrals by any means. The feature prohibits further renormalization of the diagrams into a tractable form. So, the only way to treat the diagrams is to carry out the many-body integrals in a straightforward manner. However, such integrations are impossible to be performed for the higher order terms of the density expansion. For example, the lowest-order bridge-diagram in the density expansion of a simple liquid involves a six-fold integral, which can be only performed numerically using, for example, the Monte-Carlo procedure.

So, it is a conventional approximation in the liquid state theory to ignore the diagrams at all. The approximation leads to the HNC closure.

$$h(\mathbf{r},\mathbf{r}') = \exp\left[-\beta u(\mathbf{r},\mathbf{r}') + h(\mathbf{r},\mathbf{r}') - c(\mathbf{r},\mathbf{r}')\right] - 1 \quad \text{(HNC)} \qquad \text{(II-1.19)}$$

The HNC closure can be derived from alternative way based on a functional Taylor expansion of a generating functional, and from a theorem called "Percus trick." The Percus trick asserts the following identity

$$\rho^{(2)}(\mathbf{r},\mathbf{r}') = \rho^{(1)}(\mathbf{r}')\rho^{(1)}\left(\mathbf{r}\middle|\Psi(\mathbf{r}')\right) \qquad \text{(II-1.20)}$$

where $\rho^{(1)}(\mathbf{r})$ and $\rho^{(2)}(\mathbf{r},\mathbf{r}')$ are the density distribution functions defined in Eq. (II-1.14) and Eq. (II-1.15), respectively. $\rho^{(1)}(\mathbf{r}|\Psi(\mathbf{r}'))$ is a density distribution function at position $\mathbf{r}$, subject to a condition that a molecule is fixed at position $\mathbf{r}'$, and $\Psi(\mathbf{r}')$ is the interaction of a molecule fixed at position $\mathbf{r}'$ with all other molecules. Therefore, the Percus trick asserts that the density pair distribution function at positions $\mathbf{r}$ and $\mathbf{r}'$ is equivalent to the single density distribution function at position $\mathbf{r}'$ multiplied by the single density distribution function at position $\mathbf{r}$, subject to the condition that a molecule is fixed at the position $\mathbf{r}$.

In order to prove the theorem, let us consider a system consisting of $N+1$ molecules in which one molecule is fixed at position $\mathbf{r}$, while positions of the other N molecules are not fixed. The interaction energy of the entire set may be written as

$$U_{N+1}\left(\mathbf{r}_1,\mathbf{r}_2,\cdots,\mathbf{r}_N,\mathbf{r}\right) = U\left(\mathbf{r}_1,\mathbf{r}_2,\cdots,\mathbf{r}_N\right) + \sum_{i=1}^{N}\Psi(\mathbf{r}_i;\mathbf{r}) \qquad \text{(II-1.21)}$$

In the equation, $U(\mathbf{r}_1,\mathbf{r}_2,..,\mathbf{r}_N)$ denotes the interaction energy among the unfixed molecules, and $\Psi(\mathbf{r}_i;\mathbf{r})$ the interaction energy between the i-th unfixed molecule and the fixed molecule. With the interaction energy, the grand partition function of the system can be defined as a

$$\Xi[\Psi]=\sum_{N=0}\frac{z^N}{N!}\int_V\cdots\int_V\exp\left[-\beta\left\{U_N(\mathbf{r}_1,\mathbf{r}_2,\cdots,\mathbf{r}_N)+\sum_{i=1}^{N}\Psi(\mathbf{r}_i;\mathbf{r})\right\}\right]d\mathbf{r}_1d\mathbf{r}_2\cdots d\mathbf{r}_N \quad \text{(II-1.22)}$$

For the grand partition function, the following theorem can be proved

$$\Xi[\Psi]=\frac{\Xi[0]}{z}\rho^{(1)}(\mathbf{r}) \qquad \text{(Theorem 1)} \qquad \text{(II-1.23)}$$

where $\Xi[0]$ is defined by

$$\Xi[0]=\sum_{N=0}\frac{z^N}{N!}\int_V\cdots\int_V\exp\left[-\beta\left\{U_N(\mathbf{r}_1,\mathbf{r}_2,\cdots,\mathbf{r}_N)\right\}\right]d\mathbf{r}_1d\mathbf{r}_2\cdots d\mathbf{r}_N \qquad \text{(II-1.24)}$$

Equation (II-1.23) multiplied by $z/\Xi[z]$ and divided by the same quantity gives

$$\Xi[\Psi]=\frac{\Xi[0]}{z}\frac{1}{\Xi[0]}\sum_{N=0}\frac{z^{N+1}}{N!}\int_V\cdots\int_V\exp\left[-\beta\left\{U_N(\mathbf{r}_1,\mathbf{r}_2,\cdots,\mathbf{r}_N)+\sum_{i=1}^{N}\Psi(\mathbf{r}_i;\mathbf{r})\right\}\right]d\mathbf{r}_1d\mathbf{r}_2\cdots d\mathbf{r}_N$$

Changing the numbering index from N to $M-1$, the equation becomes

$$\Xi[\Psi]=\frac{\Xi[0]}{z}\frac{1}{\Xi[0]}\sum_{M=1}\frac{z^{M}}{(M-1)!}\int_V\cdots\int_V\exp\left[-\beta\left\{U_{M-1}(\mathbf{r}_1,\mathbf{r}_2,\cdots,\mathbf{r}_{M-1})+\sum_{i=1}^{M-1}\Psi(\mathbf{r}_i;\mathbf{r})\right\}\right]d\mathbf{r}_1d\mathbf{r}_2\cdots d\mathbf{r}_{M-1}$$

By taking the definition of $\rho^{(1)}(\mathbf{r})$ into consideration, the theorem 1 (Eq. II-1.23) is proved.

With theorem 1 just proved, the Percus trick can be proved as follows. The single density distribution function with a molecule fixed at position $\mathbf{r}$ is defined by

$$\rho^{(1)}(\mathbf{r}|\Psi)=\frac{1}{\Xi[\Psi]}\sum_{N=1}\frac{z^{N}}{(N-1)!}\int_V\cdots\int_V\exp\left[-\beta\left\{U_N(\mathbf{r},\mathbf{r}_2,\cdots,\mathbf{r}_N)+\sum_{i=1}^{N-1}\Psi(\mathbf{r}_i;\mathbf{r}')\right\}\right]d\mathbf{r}_2\cdots d\mathbf{r}_N$$

Changing the numbering index N to $M-1$ leads to

$$\rho^{(1)}(\mathbf{r}|\Psi)=\frac{1}{\Xi[\Psi]}\sum_{M\geq 2}\frac{z^{M-1}}{(M-2)!}\int_V\cdots\int_V\exp\left[-\beta\left\{U_N(\mathbf{r},\mathbf{r}_2,\cdots,\mathbf{r}_{M-1})+\sum_{i=1}^{M-2}\Psi(\mathbf{r}_i;\mathbf{r}')\right\}\right]d\mathbf{r}_2\cdots d\mathbf{r}_{M-1}$$

Taking the theorem 1 for $\Xi[z]$ into consideration, the Percus trick can be proved as follows

$$\begin{aligned}\rho^{(1)}(\mathbf{r}|\Psi)&=\frac{z}{\Xi[0]\rho^{(1)}(\mathbf{r}')}\sum_{M\geq 2}\frac{z^{M-1}}{(M-2)!}\int_V\cdots\int_V\exp\left[-\beta\left\{U_N(\mathbf{r},\mathbf{r}_2,\cdots,\mathbf{r}_{M-1})+\sum_{i=1}^{M-1}\Psi(\mathbf{r}_i;\mathbf{r}')\right\}\right]d\mathbf{r}_2\cdots d\mathbf{r}_{M-1}\\&=\frac{1}{\rho^{(1)}(\mathbf{r}')}\frac{1}{\Xi[0]}\sum_{M\geq 2}\frac{z^{M}}{(M-2)!}\int_V\cdots\int_V\exp\left[-\beta\left\{U_N(\mathbf{r},\mathbf{r}',\mathbf{r}_2,\cdots,\mathbf{r}_{M-1})\right\}\right]d\mathbf{r}_2\cdots d\mathbf{r}_{M-1}\\&=\frac{1}{\rho^{(1)}(\mathbf{r}')}\rho^{(2)}(\mathbf{r},\mathbf{r}')\end{aligned}$$

HNC closure from functional Taylor expansion

Let us define the following functional as a generating functional.

$$\Gamma[\Psi] = \ln\left[\rho^{(1)}(\mathbf{r}|\Psi)e^{\beta\Psi(\mathbf{r},\mathbf{r}')}\right] \tag{II-1.25}$$

In order to figure out the physical meaning of the functional, take a derivative of $\ln\rho^{(1)}(\mathbf{r}|\Psi)$ with respect to the position $\mathbf{r}$.

$$\frac{\partial}{\partial \mathbf{r}}\ln\rho^{(1)}(\mathbf{r}|\Psi) = \frac{\sum_{N=1}\frac{z^N}{(N-1)!}\int_V\cdots\int_V\left(-\beta\frac{\partial U_N}{\partial \mathbf{r}}\right)\exp\left[-\beta\left\{U_N(\mathbf{r},\mathbf{r}_2,\cdots,\mathbf{r}_N)+\sum_{i=1}^{N-1}\Psi(\mathbf{r}_i;\mathbf{r}')\right\}\right]d\mathbf{r}_2\cdots d\mathbf{r}_N}{\sum_{N=1}\frac{z^N}{(N-1)!}\int_V\cdots\int_V\exp\left[-\beta\left\{U_N(\mathbf{r},\mathbf{r}_2,\cdots,\mathbf{r}_N)+\sum_{i=1}^{N-1}\Psi(\mathbf{r}_i;\mathbf{r}')\right\}\right]d\mathbf{r}_2\cdots d\mathbf{r}_N}$$

The right-hand-side of the equation signifies essentially the mean force acting on a molecule at $\mathbf{r}$ from the molecule fixed at $\mathbf{r}'$, averaged over the configuration of all other molecules, $\mathbf{r}_2$ to $\mathbf{r}_N$. Therefore, $-(1/\beta)\ln\rho^{(1)}(\mathbf{r}|\Psi)$ is the *potential of mean force* between the two molecules. If the direct interaction between the molecules at $\mathbf{r}$ and $\mathbf{r}'$ is subtracted from the potential of mean force, it gives $\Gamma[\Psi]$.

Now, let's take a functional Taylor expansion of $\Gamma[\Psi]$ around $\Psi = 0$, that is, the case there is no molecule fixed at the position $\mathbf{r}'$.

$$\Gamma[\Psi] = \Gamma[0] + \int\left\{\frac{\delta\Gamma}{\delta\rho^{(1)}(\mathbf{r}''|\Psi)}\right\}_{\Psi=0}\left\{\rho^{(1)}(\mathbf{r}''|\Psi) - \rho^{(1)}(\mathbf{r}''|\Psi=0)\right\}d\mathbf{r}'' \tag{II-1.26}$$

$$+ \text{ (second order in } \Delta\rho)$$

The functional derivative in the integrand can be identified as the direct correlation function defined by Eq. (I-1.17) as follows

$$\frac{\delta\Gamma}{\delta\rho^{(1)}(\mathbf{r}''|\Psi)} = \frac{\delta}{\delta\rho^{(1)}(\mathbf{r}''|\Psi)}\left\{\ln\rho^{(1)}(\mathbf{r}|\Psi) + \ln\frac{z}{z(\mathbf{r})}\right\}$$

$$= \frac{1}{\rho^{(1)}(\mathbf{r}|\Psi)}\delta(\mathbf{r}-\mathbf{r}'') - \frac{\delta\ln z(\mathbf{r})}{\delta\rho^{(1)}(\mathbf{r}''|\Psi)}$$

The right-hand-side of the equation agrees with the definition of the direct correlation function.

$$\left\{\frac{\delta\Gamma}{\delta\rho^{(1)}(\mathbf{r}''|\Psi)}\right\}_{\Psi=0} = c(\mathbf{r},\mathbf{r}'') \tag{II-1.27}$$

From Eqs. (II-1.25), (II-1.26), and (II-1.27), one finds,

$$\ln\left[\rho^{(1)}(\mathbf{r}|\Psi)e^{\beta\Psi(\mathbf{r},\mathbf{r}')}\right] = \ln\left[\rho^{(1)}(\mathbf{r}|\Psi=0)e^{\beta\Psi(\mathbf{r},\mathbf{r}')}\right] + \int c(\mathbf{r},\mathbf{r}'')\left\{\rho^{(1)}(\mathbf{r}''|\Psi) - \rho^{(1)}(\mathbf{r}''|\Psi=0)\right\}d\mathbf{r}'' \tag{II-1.28}$$

Here, $\rho^{(1)}(\mathbf{r}|\Psi)$ is related to the pair correlation function $g(\mathbf{r},\mathbf{r}')$ due to the Percus trick as

$$\rho^{(1)}\left(\mathbf{r}\middle|\Psi\right) = \rho(\mathbf{r})g(\mathbf{r},\mathbf{r}') \tag{II-1.29}$$

$$\rho^{(1)}\left(\mathbf{r}\middle|\Psi = 0\right) = \rho(\mathbf{r}) \tag{II-1.30}$$

Then, Eq. (II-1.28) becomes

$$\ln g(\mathbf{r},\mathbf{r}') + u(\mathbf{r},\mathbf{r}')/kT = \int c(\mathbf{r},\mathbf{r}'')\rho(\mathbf{r}'')[g(\mathbf{r}'',\mathbf{r}') - 1]d\mathbf{r}' \tag{II-1.31}$$

Taking the definition of $h(\mathbf{r},\mathbf{r}')$ and the Ornstein-Zernike Eq. (II-1.18) into consideration, HNC closure (II-1.19) can be obtained.

So far, we talked about the HNC closure, which is of course an approximation. There have been considerable efforts to improve the approximation. Among those efforts, an *ad hoc* modification of the HNC closure, named "Kovalenko-Hirata (KH)" closure, may deserve attention because it is popularly employed in the community of the computational life science. The approximation reads

$$g(r)\begin{cases} \exp[d(r)] & \text{for} \quad d(r) \le 0 \\ 1 + d(r) & \text{for} \quad d(r) > 0 \end{cases}, \tag{II-1.32}$$

where $d(r) = -\beta u(\mathrm{r}) + h(r) - c(r)$.

II-1.6 Iterative Solution of Ornstein-Zernike Equation and Renormalization

It is a non-trivial problem to solve the Ornstein-Zernike (OZ) equation that is essentially a non-linear integral equation. Here, we explain a mathematical technique called "Renormalization" in order to solve the OZ equation. Let us substitute the entire right-hand-side of (1.1) into $h(\mathbf{r}'', \mathbf{r}')$, iteratively, that will produce an infinite series illustrated by the following diagram. The sum of the diagrams is called "chain sum."

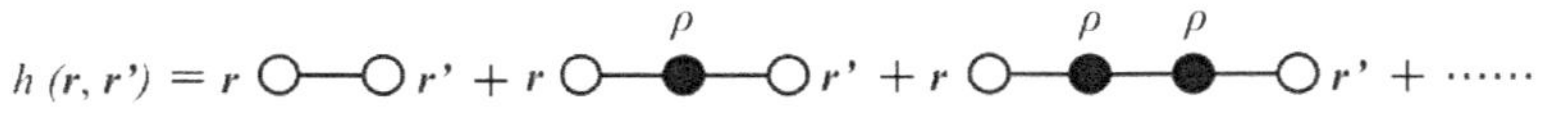

$$= \mathrm{C}[c\,|\,\rho] \tag{II-1.33}$$

In the equation, a black circle denotes an integral over the entire space as follows.

$$= \rho\int c(\mathbf{r},\mathbf{r}'')c(\mathbf{r}'',\mathbf{r}')d\mathbf{r}'' \tag{I-1.34}$$

$C[c|\rho]$ is a book-keeping device to represent the chain sum. Each term in the infinite series has a form of the *convolution integral*, the Fourier transform of which becomes a simple product of the Fourier transform of functions consisting the integrand, and the infinite series becomes a geometric series. The sum of geometric series can be readily calculated as

$$\begin{aligned} \tilde{h}(\mathbf{k}) &= \tilde{c}(\mathbf{k}) + \rho\tilde{c}(\mathbf{k})^2 + \rho^2\tilde{c}(\mathbf{k})^3 + \cdots \\ &= \frac{\tilde{c}(\mathbf{k})}{1 - \rho\tilde{c}(\mathbf{k})} \end{aligned} \tag{II-1.35}$$

where $\tilde{h}(\mathbf{k})$, $\tilde{c}(\mathbf{k})$ are, respectively, the Fourier transform of $h(\mathbf{r})$ and $c(\mathbf{r})$, defined as

$$\tilde{c}(\mathbf{k}) = \int c(r)\exp[i\mathbf{k}\cdot\mathbf{r}]d\mathbf{r} \quad \text{and} \quad \tilde{h}(\mathbf{k}) = \int h(\mathbf{r})\exp[i\mathbf{k}\cdot\mathbf{r}]d\mathbf{r}$$

In Fig. II-1.6, a standard method to solve the integral equation by means of an iterative procedure is shown.

Although Fig. II-1.6 shows a standard algorithm to solve an integral equation appearing in the liquid state theory, more important is that it represents a typical example for a general concept in physics of many-body problem, referred to as "Renormalization." In this algorithm, the functional form of $c(\mathbf{r}, \mathbf{r'})$, or

$$c(\mathbf{r},\mathbf{r}') = \exp[-\beta u(\mathbf{r},\mathbf{r}') + t(\mathbf{r},\mathbf{r}')] - 1 + t(\mathbf{r},\mathbf{r}')$$

stays unchanged as the iteration progresses. However, the more complicated many-body interactions are *renormalized* into $c(\mathbf{r}, \mathbf{r'})$ as the iteration proceeds. Since each black circle corresponds to an integral in the coordinate space, the sum of the geometric series involves multiple integrals, multiplicity of which extends to infinity. It means that this procedure corresponds to sampling of the phase space that is infinitely large. Nevertheless, actual integration we should perform in this algorithm is just a single integral associated with the Fourier transform.

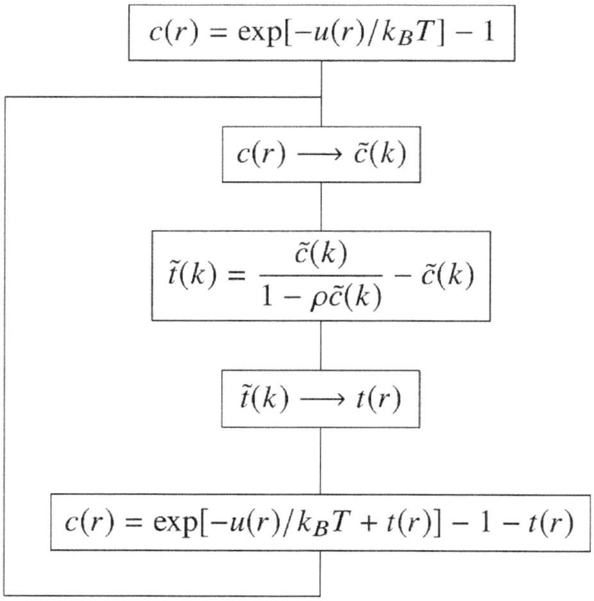

Fig. II-1.6: Iterative solution of the O-Z equation: the iteration is initiated by calculating initial values of $c(r)$ from the intermolecular interaction $u(r)$. The function $c(r)$ is Fourier-transformed into $\tilde{c}(k)$. The sum of the infinite series is calculated to get $\tilde{t}(k)$. $\tilde{t}(k)$ is inverse transformed to get $c(r)$ in the real space. The process is iterated until $c(r)$ converges.

Renormalization of Coulomb interactions: The theory described above can be applied to a liquid system consisting of spherical molecules to calculate the pair correlation function. However, there is a tumbling block if one tries to apply the theory in a naive way to an electrolyte solution. The problem is the divergence of Coulomb interaction. The equation described above involves the integration of interactions over the space extending to infinity. The integration of the Coulomb interaction over the infinite space diverges as

$$\phi(r) \sim \frac{1}{r} \qquad \int_0^{\infty} \phi(r) 4\pi r^2 dr \to \infty$$

J. Mayer, who is famous with the method called "cluster expansion" of non-ideal gas, solved the problem [3]. He found that the sum of the infinite series (chain sum) below converges, in spite that the integral of the Coulomb interaction by itself diverges.

$C[\phi|\rho] =$
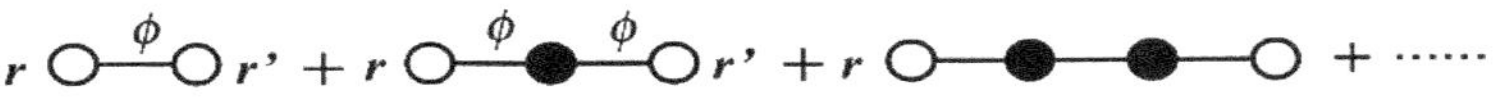
(II.1-36)

$$= \phi(r)\exp[-\kappa r] \equiv q(r)$$

In the above equation, κ denotes inverse of the Debye screening length that is proportional to the square root of the ionic strength. Using the function, Mayer renormalized the cluster expansion (or virial expansion) of electrolyte solutions. (The Debye-Huckel limiting law of electrolyte solutions corresponds to the first order term of the cluster expansion.) The Ornstein-Zernike equation can also be rewritten in terms of the "renormalized Coulomb interaction" $q(r)$ as

$$h = C\left[c - \phi \middle| \bar{q}\right] + q \qquad \text{(II-1.37)}$$

where $\bar{q}(r) = \rho\delta(r) + \rho^2 q(r)$.

II-2 Reference Interaction-Site Model (RISM) Theory

The theory described so far, which is concerned with a system consisting of spherical molecules, cannot be applied to such liquid systems as water and alcohol that are of chemical interests. It is because the interaction between molecules in such liquids, referred to as molecular liquids, depends not only on their distance but also on their mutual orientation such as

$$u(1, 2) = u(\mathbf{R}_1, \boldsymbol{\Omega}_1; \mathbf{R}_2, \boldsymbol{\Omega}_2) \qquad \text{(II-2.1)}$$

where "1" on the left-hand-side represents the position $\mathbf{R}_1$ and the orientation $\boldsymbol{\Omega}_1$ of the first molecule, and "2" does the position $\mathbf{R}_2$ and the orientation $\boldsymbol{\Omega}_2$ of the second molecule. It is not so sophisticated to generalize the Ornstein-Zernike equation to a system in which the interaction potential may be represented by Eq. (II-2.1). In fact, it can be made just conceptually by changing the arguments of correlation functions in Eq. (II-1.18) from those denoting the positions ($\mathbf{r}$, $\mathbf{r}'$) to those representing both positions and orientations ("1" and "2").

$$h(1,2) = c(1,2) + \left(\frac{\rho}{\Omega}\right)\int c(1,3)h(3,2)d(3) \qquad \text{(II-2.2)}$$

where Ω denotes the solid angle, and (ρ/Ω) is the normalization constant. The equation is called the Molecular Ornstein-Zernike (MOZ) equation. It is not so easy, however, to solve the equation either analytically or numerically. It is because the integration involved in Eq. (II-2.2) is no way to be reduced to a one dimensional integral as in the case of the usual Ornstein-Zernike equation for atomic liquids. There is a way to solve the equation by expanding the correlation into the spherical harmonics, similar to solving the Schrodinger equation for a hydrogen atom. The method is called the "spherical invariance." However, an application of the method is largely limited to such a system in which the expansion converges quickly.

Another way to look at the molecular liquid is the interaction site model (ISM), proposed by D. Chandler and H.C. Andersen [Chandler and Andersen 1972]. The most important feature of the ISM is to express the interaction between a pair of molecules by the sum of interactions among atoms (or interaction sites) in the molecules as

$$u(1,2) = \sum_{\alpha=1}^{n}\sum_{\gamma=1}^{n} u\left(\left|\mathbf{r}_1^{\alpha} - \mathbf{r}_2^{\gamma}\right|\right) \tag{II-2.3}$$

where $u\left(\left|\mathbf{r}_1^{\alpha} - \mathbf{r}_2^{\gamma}\right|\right)$ on the right hand side denotes the interaction between atom α in the molecule 1 and atom γ in the molecule 2. That is, the left hand side represents the interaction between the two molecules "1" and "2" by a function of the center-to-center distance and the orientation, while the right hand side by the sum of atom-atom interactions. The model is illustrated by a cartoon in Fig. (II-2.1). The model is essentially the same with that employed in the conventional molecular simulation.

Chandler and Andersen reformulated the MOZ equation in terms of the interaction-site model. The theory is called the "Reference Interaction Site Model (RISM)" theory. Here, the idea to derive the equation is briefly sketched. The RISM equation is derived from the MOZ Eq. (II-2.2). Let us substitute the entire right-hand-side of the equation into $h(1, 2)$, iteratively, that will produce an infinite series of the chain-sum, similar to the O-Z equation (Eq. II-1.31) for spherical molecules. Using the chain sum notation, one gets

$$\begin{aligned} h(1,2) &= C\left[c\middle|(\rho/\Omega)\right] \\ &= \sum_{n=0}^{\infty} C^{(n)}\left[c\middle|(\rho/\Omega)\right] \end{aligned} \tag{II-2.4}$$

where we denoted the n-th term in the chain sum as $C^{(n)}[c|(\rho/\Omega)]$. First few terms of the series are,

$$C^{(0)}\left[c\middle|(\rho/\Omega)\right] = c(1,2) \tag{II-2.5}$$

$$C^{(1)}\left[c\middle|(\rho/\Omega)\right] = \frac{\rho}{\Omega}\int c(1,3)\; c(3,2)\; \mathrm{d}(3) \tag{II-2.6}$$

$$C^{(2)}\left[c\middle|(\rho/\Omega)\right] = \left(\frac{\rho}{\Omega}\right)^2 \int c(1,3)\; c(3,4)c(4,2)\; \mathrm{d}(3)\mathrm{d}(4) \tag{II-2.7}$$

....

Now, we try to take the average of Eq. (II-2.4) over the orientation. The average can be represented by the following equation,

$$f_{\alpha\gamma}(r) = \frac{1}{\Omega^2}\int f(1,2)\delta\left(\mathbf{R}_1 + \mathbf{l}_1^{\alpha}\right)\delta\left(\mathbf{R}_2 + \mathbf{l}_2^{\gamma} - \mathbf{r}\right)d(1)d(2), \tag{II-2.8}$$

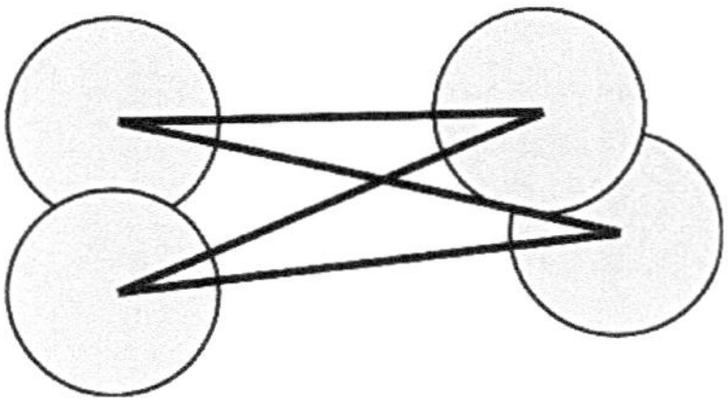

Fig. II-2.1: Illustration of the interaction site model: spheres, atoms, lines, interactions.

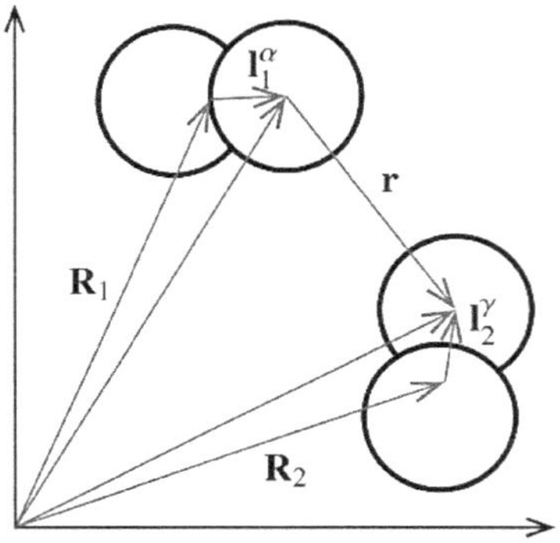

Fig. II-2.2: Coordinates for the interaction-site model of liquids.

where $\mathbf{R}_1$ and $\mathbf{R}_2$ denote the positions of the center of the molecules, and $\mathbf{l}_1^\alpha$ and $\mathbf{l}_2^\gamma$ are the vectors pointing the site α and γ of the molecules 1 and 2 from the center of mass, respectively, and $\delta(x)$ signifies the Dirac delta-function (Fig. II-2.2).

The transformation corresponds to an average of the function $f(1,2)$ over the angular space by fixing the position $(\mathbf{R}_1 + \mathbf{l}_1^\alpha)$ of atom a of the molecule 1 at the origin, while the position $(\mathbf{R}_2 + \mathbf{l}_2^\gamma)$ of the molecule 2 at the position $\mathbf{r}$ (Fig. II-2.2). We referred to the averaging as "Chandler-Andersen" transformation. Unfortunately, we cannot apply the transformation directly to Eq. (II-2.4). We must make a dramatic approximation to the direct correlation function, $c(1, 2)$. The approximation is expressed by the following equation [Chandler and Andersen 1972]

$$c(1,2) = \sum_{\alpha,\gamma} c_{\alpha\gamma}\left(\left|\mathbf{r}_1^\alpha - \mathbf{r}_2^\gamma\right|\right) \tag{II-2.9}$$

where $\mathbf{r}_1^\alpha$ and $\mathbf{r}_2^\gamma$ denote the positions of the α-th site of molecule 1 and the γ-th site of molecule 2, respectively, and $c_{\alpha\gamma}(r)$ denotes the direct correlation function between the two sites. That is, the direct correlation functions between two molecules are approximated by a sum of the direct correlation function between the interaction sites or atoms. Anyway, the approximation allows one to make the transformation analytically, as follows.

$\rho^2 h(\mathbf{r},\mathbf{r}') =$ 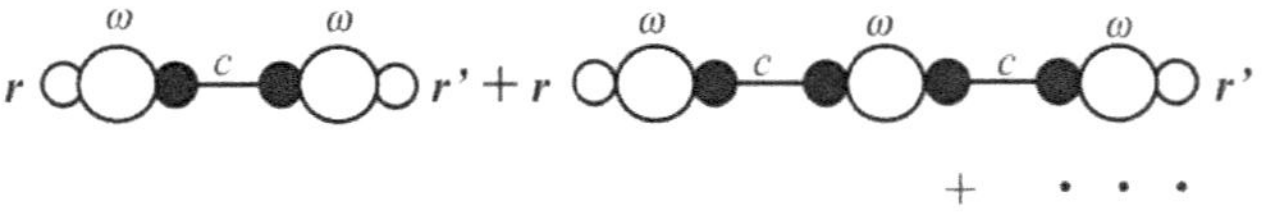 (II-2.10) where c denotes the direct correlation function (matrix) $c(\mathbf{r}, \mathbf{r}')$ between atoms (interaction sites), the black circles represent the integrals with respect to the coordinates $\mathbf{r}$ and $\mathbf{r}'$. ω is the intra-molecular correlation function (matrix) defined by

$$\omega_{\alpha\gamma}(r) = \rho\delta_{\alpha\gamma}\delta(r) + \rho(1-\delta_{\alpha\gamma})s_{\alpha\gamma}(r) \qquad \text{(II-2.11)}$$

$$s_{\alpha\beta}(r) = \frac{1}{4\pi L_{\alpha\gamma}^2}\delta\left(r - L_{\alpha\gamma}\right) \qquad \text{(II-2.12)}$$

In the expressions, $L_{\alpha\gamma}(\equiv|\mathbf{l}_i^\alpha - \mathbf{l}_i^\gamma|)$ denotes the distance between atoms α and γ, and $\delta(r - L_{\alpha\gamma})$ represents the Dirac delta-function. This corresponds to fixing the distance between atoms α and γ in a molecule to $L_{\alpha\gamma}$. If one provides all the distance $L_{\alpha\gamma}$ between atoms in a molecule, one can fix the structure or geometry of the molecule: for example, in the case of a water molecule, L_{OH}, L_{OH}, L_{HH}. The derivation of the equation is provided in Appendix B at the end of this Chapter.

Shown in Fig. (II-2.3) is the iterative algorithm to solve the RISM equation. It is again based on the renormalization scheme.

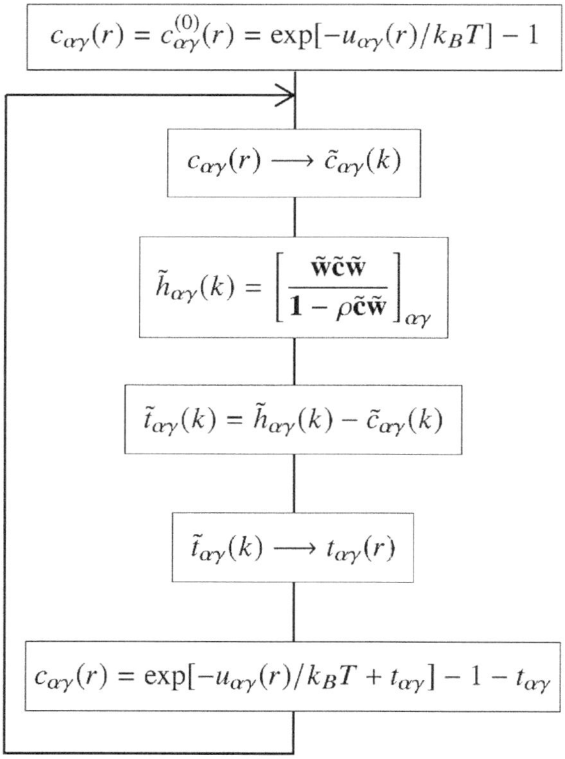

Fig. II-2.3: Algorithm to solve the RISM equation.

II-3 Extended RISM Theory and its Application to Structure of Molecular Liquids

***Extended RISM (XRISM) theory*:** A theory to handle liquids consisting of non-spherical molecules was completed by the RISM theory. However, it was not yet enough to treat such liquids as water and alcohol in which molecules are interacting with hydrogen-bonds. It is because the hydrogen-bond is essentially an electrostatic interaction, and the problem associated with divergence of the Coulomb integral stays unsolved (see II-1.6). It is the "extended RISM theory" that solved the divergence problem to be applied to almost any type of liquids and solutions including water and alcohol [Hirata and Rossky 1981, Hirata et al. 1982].

The extended RISM (XRISM) equation was derived from the RISM theory by renormalizing the Coulomb interaction in a way similar to the case of the O-Z equation for electrolyte solutions. Let us express the site-site interaction by a sum of the short-range interaction and the Coulomb interaction as,

$$u_{\alpha\gamma}(r) = u^{*}_{\alpha\gamma}(r) + \frac{Z_\alpha Z_\gamma}{r}, \tag{II-3.1}$$

where $u_{\alpha\gamma}(r)$ denotes the short range interaction, and Z_α and Z_γ denote the partial charge carried by the atoms α and γ. The RISM equation can be written using the chain-sum convention in the matrix form as,

$$\boldsymbol{\rho}\mathbf{h}\boldsymbol{\rho} = C\left[\mathbf{c}\middle|\boldsymbol{\omega}\right] \tag{II-3.2}$$

where the elements of the matrices $\mathbf{h}$, $\mathbf{c}$, and $\boldsymbol{\omega}$ are $h_{\alpha\gamma}(r)$, $c_{\alpha\gamma}(r)$, and $\omega_{\alpha\gamma}(r)$. Now, we split the direct correlation function into two contributions as follows,

$$c_{\alpha\gamma}(r) = c^{s}_{\alpha\gamma}(r) + \phi_{\alpha\gamma}(r), \tag{II-3.3}$$

where $c^{s}_{\alpha\gamma}(r)$ and $\phi_{\alpha\gamma}(r)$ are defined as

$$c^{s}_{\alpha\gamma}(r) = c_{\alpha\gamma}(r) - \phi_{\alpha\gamma}(r) \tag{II-3.4}$$

$$\phi_{\alpha\gamma}(r) = -\beta(Z_\alpha Z_\gamma / r) \tag{II-3.5}$$

Then, applying the chain-sum algebra described in Appendix A in this chapter, one finds (in matrix form),

$$\begin{aligned} \mathbf{t} \equiv C\left[\mathbf{c}\middle|\boldsymbol{\omega}\right] - \mathbf{c} &= C'\left[\mathbf{c}^{s} + \phi\middle|\boldsymbol{\omega}\right] \\ &= C'\left[\phi\middle|\boldsymbol{\omega}\right] + C'\left[\mathbf{c}^{s}\middle|\boldsymbol{\omega} + \rho C\left[\phi\middle|\boldsymbol{\omega}\right]\rho\right] \end{aligned} \tag{II-3.6}$$

Let us define new renormalized functions $\mathbf{Q}$, $\boldsymbol{\tau}$, and $\boldsymbol{\Omega}$ by

$$\mathbf{Q} = C\left[\phi\middle|\boldsymbol{\omega}\right] = C'\left[\phi\middle|\boldsymbol{\omega}\right] + \phi, \tag{II-3.7}$$

$$\boldsymbol{\tau} = \mathbf{t} - C'\left[\phi\middle|\boldsymbol{\omega}\right] = \mathbf{t} - \mathbf{Q} + \phi \tag{II-3.8}$$

$$\boldsymbol{\Omega} = \boldsymbol{\omega} + \rho C\left[\phi\middle|\boldsymbol{\omega}\right] = \mathbf{t} - \mathbf{Q} + \phi\,. \tag{II-3.9}$$

Then, Eq. (II-3.6) reduces to

$$\tau = C'\left[\mathbf{c}_s \middle| \boldsymbol{\Omega}\right] \tag{II-3.10}$$

or

$$\rho\mathbf{h}\rho = \mathbf{c}_s + \mathbf{Q} + \tau = C\left[\mathbf{c}_s \middle| \boldsymbol{\Omega}\right] + \mathbf{Q} \tag{II-3.11}$$

In the equations, $Q_{\alpha\gamma}(r)$ (II-3.7) is a function resulting from applying the renormalization of Coulomb interaction to the RISM theory. It signifies a *renormalized Coulomb interaction between partial charges bound by intramolecular constraints*. It can be readily proved that $Q_{\alpha\gamma}(r)$ decays asymptotically as r^{-6} at large r (Hirata et al. 1982). Therefore, the problem of divergence associated with the integral of Coulomb interaction in infinite space is solved.

***XRISM theory for liquid mixtures*:** So far, we have derived the XRISM equation for molecular liquids that involve a single chemical species. However, in ordinary chemistry, we face a liquid system that contains more than one component, usually called "solution." Especially, the solution in a living cell contains many different chemical species, water, of course, a variety of ions, protein, DNA, and so on. Major and minor components of solution are conventionally referred to as "solvent" and "solute." The vanishing limit of solute concentration, or the infinite dilution, is of particular interest because it purely reflects the nature of solute-solvent interactions. The word "solvation" is most commonly used for describing properties concerning solute-solvent interactions at the infinite dilution limit. Here, we generalize the RISM equation to solutions [Hirata et al. 1983].

The RISM equation for a liquid mixture can be written in a matrix form as

$$\rho\mathbf{h}\rho = \omega * \mathbf{c} * \omega + \omega * \mathbf{c} * \rho\mathbf{h}\rho, \tag{II-3.12}$$

where "*" denotes a convolution integral as well as a matrix product, and $\mathbf{h}$, $\mathbf{c}$, ω are, respectively, the matrices of site-site total, direct, and intra-molecular correlation functions. The elements of those matrices are $h_{\alpha_M\gamma_{M'}}$, $c_{\alpha_M\gamma_{M'}}$, $\omega_{\alpha_M\gamma_{M'}}$, where the suffix α_M labels the α-th atom in a molecular species M. ρ is a diagonal matrix consisting of density of each molecular species, or $\rho_{\alpha_M\gamma_{M'}} = \delta_{\alpha\gamma}\delta_{MM'}\rho_{M'}$, and $\omega_{\alpha_M\gamma_{M'}}$ is defined in matrix form as,

$$\omega_{\alpha_M\gamma_{M'}}(r) = \rho_M\delta_{MM'}\left\{\delta_{\alpha_M\gamma_M}\delta(\mathbf{r}_{\alpha_M} - \mathbf{r}_{\gamma_M}) + (1-\delta_{\alpha_M\gamma_M})s_{\alpha_M\gamma_M}(\mathbf{r}_{\alpha_M} - \mathbf{r}_{\gamma_M})\right\}, \tag{II-3.13}$$

where $s_{\alpha_M\gamma_M}$ is the intramolecular pair correlation function between distinct sites of the same molecule. The matrices can be reorganized into those with block matrices concerning solute and solvent species. Using the conventional notation devised by H. Friedman in order to distinguish solute (u) and solvent (v), one can rewrite the above equation as a set of equations regarding solvent-solvent (vv), solute-solvent (uv), and solute-solute (uu) correlation functions,

$$\rho^v\mathbf{h}^{vv}\rho^v = \omega^v * \mathbf{c}^{vv} * \omega^v + \omega^v * \mathbf{c}^{vv} * \rho^v\mathbf{h}^{vv}\rho^v + \omega^v * \mathbf{c}^{vu} * \rho^u\mathbf{h}^{uv}\rho^v \tag{II-3.14}$$

$$\rho^u\mathbf{h}^{uv}\rho^v = \omega^u * \mathbf{c}^{uv} * \omega^v + \omega^u * \mathbf{c}^{uv} * \rho^v\mathbf{h}^{vv}\rho^v + \omega^u * \mathbf{c}^{uu} * \rho^u\mathbf{h}^{uv}\rho^v \tag{II-3.15}$$

$$\rho^u\mathbf{h}^{uu}\rho^u = \omega^u * \mathbf{c}^{uu} * \omega^u + \omega^u * \mathbf{c}^{uv} * \rho^v\mathbf{h}^{vu}\rho^u + \omega^u * \mathbf{c}^{uu} * \rho^u\mathbf{h}^{uu}\rho^u \tag{II-3.16}$$

We now consider the cases in which all solute (u) species are at infinite dilution in the solvent mixture, i.e., $\rho^u \rightarrow 0$. We first multiply inverse of the matrix $\boldsymbol{\rho}^v$, or $(\boldsymbol{\rho}^v)^{-1}$, from the left and right sides of Eq. (II-3.13), and then, take the limit, $\boldsymbol{\rho}^u \rightarrow 0$. The operation gives the following equation for solvent.

$$\mathbf{h}^{vv} = \mathbf{w}^v * \mathbf{c}^{vv} * \mathbf{w}^v + \mathbf{w}^v * \mathbf{c}^{vv} * \rho^v \mathbf{h}^{vv}, \qquad \text{(II-3.17)}$$

where $\mathbf{w}^v$ is defined in matrix form by

$$\mathbf{w}^v \equiv \left(\rho^v\right)^{-1} \boldsymbol{\omega}^v \qquad \text{(II-3.18)}$$

The limit of infinite dilution for Eqs. (II-3.15) and (II-3.16) can be taken in similar manner to lead the equations of solute-solvent and solute-solute pair correlation functions,

$$\mathbf{h}^{uv} = \mathbf{w}^u * \mathbf{c}^{uv} * \mathbf{w}^v + \mathbf{w}^u * \mathbf{c}^{uv} * \rho^v \mathbf{h}^{vv} \qquad \text{(II-3.19)}$$

and

$$\mathbf{h}^{uu} = \mathbf{w}^u * \mathbf{c}^{uu} * \mathbf{w}^u + \mathbf{w}^u * \mathbf{c}^{uv} * \rho^v \mathbf{h}^{vu} \qquad \text{(II-3.20)}$$

Structure of water: Here, we take a look at how the XRISM theory is applied to the structure of water. We employ the so-called (12-6-1)-type potential function for the site-site interactions between a pair of water molecules. It is essentially the same potential function developed in the molecular simulation community, such as SPC, TIPS, and so on, with a slight modification concerning the repulsive core of hydrogen atoms [Pettitt and Rossky 1982].

$$u_{\alpha\gamma}(r) = 4\varepsilon_{\alpha\gamma}\left[\left(\frac{\sigma_{\alpha\gamma}}{r}\right)^{12} - \left(\frac{\sigma_{\alpha\gamma}}{r}\right)^{6}\right] + \frac{Z_\alpha Z_\gamma}{r} \qquad \text{(II-3.21)}$$

Here, we use the HNC approximation for the closure of the XRISM equation, or

$$c_{\alpha\gamma} - \phi_{\alpha\gamma} = \exp\left[-\beta u^*_{\alpha\gamma} + Q_{\alpha\gamma} + \tau_{\alpha\gamma}\right] - 1 - \tau_{\alpha\gamma} - Q_{\alpha\gamma} \qquad \text{(II-3.22)}$$

where $\phi_{\alpha\gamma}$, $Q_{\alpha\gamma}$ and $\tau_{\alpha\gamma}$ are given by Eqs. (II-3.5), (II-3.7) and (II-3.10), respectively, and $u^*_{\alpha\gamma}$ is defined by the short-range part of Eq. (II-3.21), namely,

$$u^*_{\alpha\gamma}(r) = 4\varepsilon_{\alpha\gamma}\left[\left(\frac{\sigma_{\alpha\gamma}}{r}\right)^{12} - \left(\frac{\sigma_{\alpha\gamma}}{r}\right)^{6}\right]$$

The algorithm to solve the XRISM equation is illustrated in Fig. II-3.1.

Depicted in Fig. II-3.2 are the pair correlation functions obtained from the XRISM theory. The figure demonstrates the essential features of the pair correlation functions (PCF), exhibited by the experimental studies due to the X-ray and neutron diffraction measurements. (1) The PCF of the O-H pair shows a sharp peak around r = 1.8Å, that represents the hydrogen-bond between a pair of water molecules. (2) A small peak appears around $r = 1.63\sigma$ in the O-O pair, that corresponds to the (O-H---O) tetrahedral coordination characteristic to the structure of ice. That is obviously produced by the hydrogen-bond formation between the O-H pair. (3) The first peak in the O-O pair is eclipsed considerably compared to the simple liquid consisting of spherical molecule such as neon. It is also a manifestation of the tetrahedral coordination typical to ice.

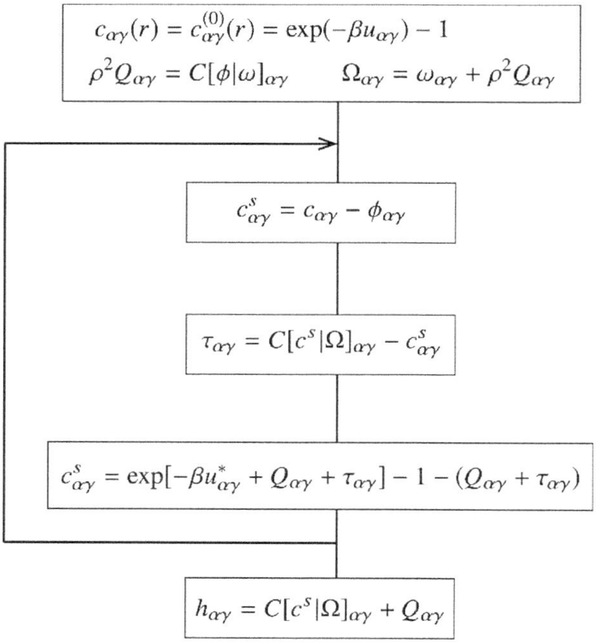

Fig. II-3.1: Algorithm of solving the XRISM equation with HNC closure.

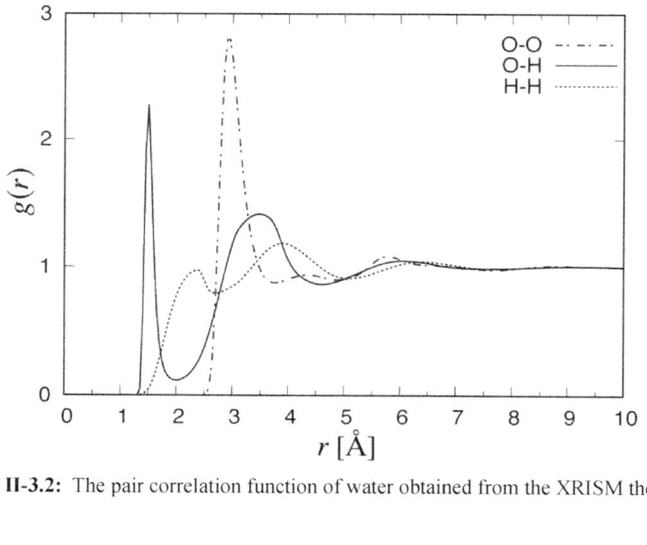

Fig. II-3.2: The pair correlation function of water obtained from the XRISM theory.

***Structure of water-alcohol mixture*:** Water-alcohol mixtures are the most popular solution for human beings. However, understanding of their structure and property is still very primitive even in the 21st century. For example, water and tertiary butanol make a mixture with any concentration, and no phase separation to water and the alcohol is observed [Franks and Ives 1966]. Nevertheless, it is well documented that the solution shows extremely large concentration fluctuation at certain temperature and pressure, which is akin to the phase separation. Such a large concentration fluctuation has been expressed by words such as "micro-cluster" or "quasi phase separation." But, what is the structural entity of "micro cluster" or "quasi phase separation?" Here, we apply the XRISM theory to a water-*t*-butanol mixture to clarify the microscopic structure of solution. In the XRISM calculation, the KH closure Eq. (II-1.31) is employed because it is known that the closure does a better job for a liquid mixture compared to the HNC closure [Yoshida et al. 2002].

Shown in Fig. II-3.3 are the pair correlation functions (or radial distribution function: RDF) between atoms of water and alcohol in the solution. So called "united-atom" model was used for the three methyl-groups of butanol, or $CH_3 \rightarrow C$. Nomenclature of atoms in water and butanol is as follows: methyl group, C; O in OH group, O; H in OH group, H; tertiary carbon, C_c; O in water, O_W; H in water, H_W. The curves drawn by different types of lines distinguish the concentration of solution from the pure water to the pure butanol: for example, the thick solid line in RDFs of the $O_W - O_W$ pair corresponds to RDF in pure water, while the thick dashed line represents that of pure butanol.

The peak position of RDF provides information concerning the distance between a pair of atoms for water-water, water-alcohol, and alcohol-alcohol. Three-dimensional structure of the solution can be deduced from the distance information based on an idea of the trigonometry. Illustrative picture deduced in such a way from RDF is shown in Fig. II-3.4. In case of the infinitely dilute alcohol in water, the hydrogen-bond network of water makes the main frame of the solution structure, and a *t*-butanol molecule is incorporated in the main frame by making

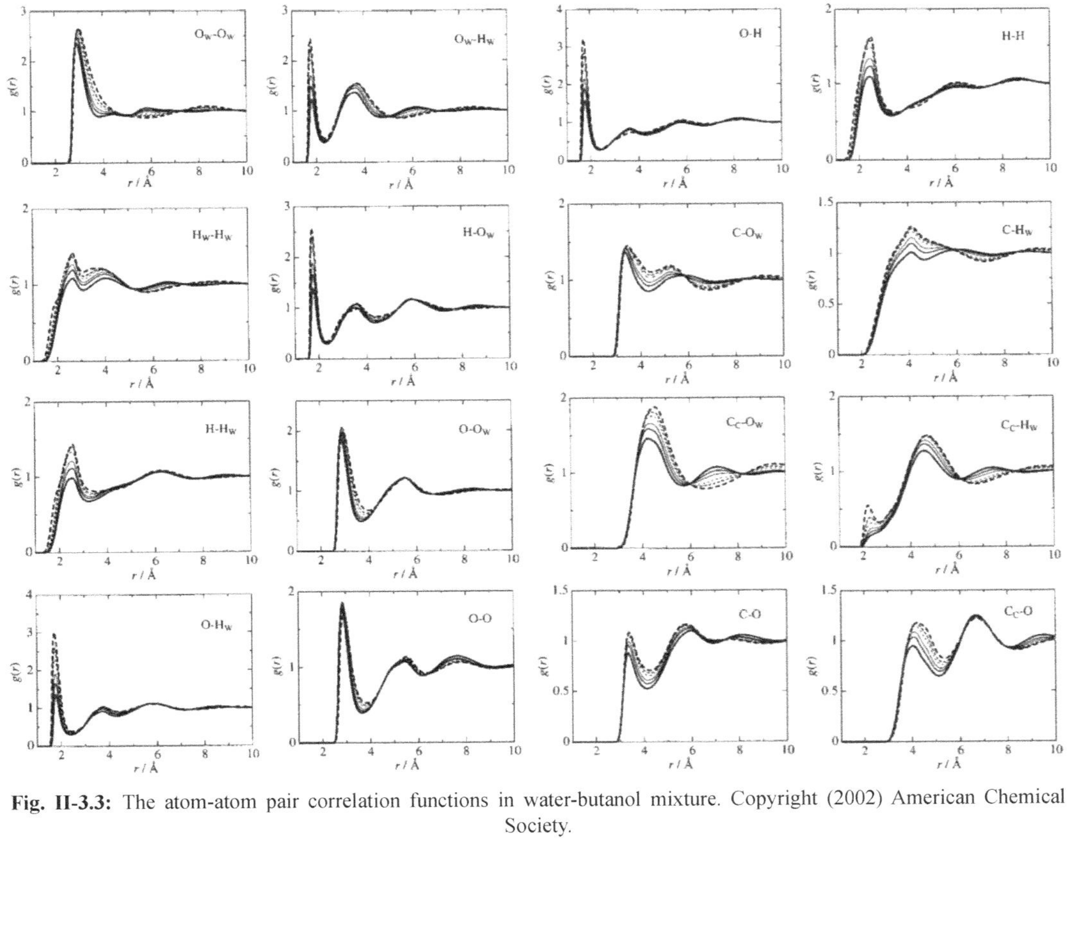

Fig. II-3.3: The atom-atom pair correlation functions in water-butanol mixture. Copyright (2002) American Chemical Society.

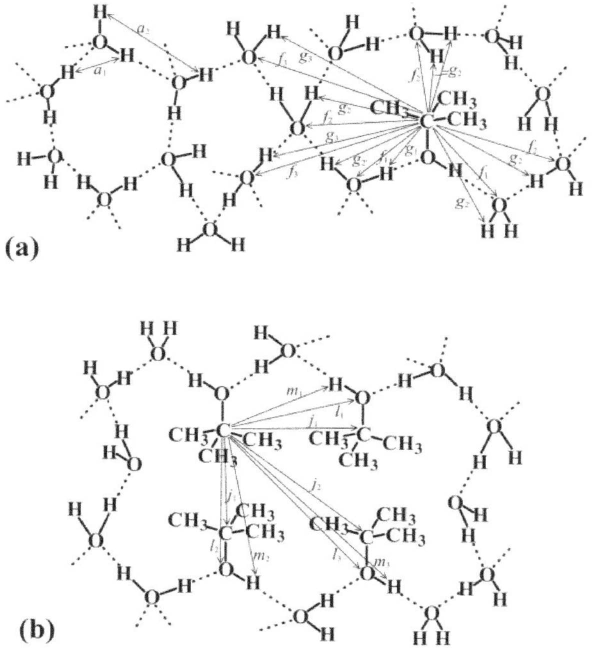

Fig. II-3.4: Solution structure of water-butanol mixture inferred from the correlation function. Copyright (2002) American Chemical Society.

hydrogen-bonds with water molecules. In the case of low concentration of butanol in water, butanol molecules are incorporated in the hydrogen-bond network of water again so that their butyl groups are in contact each other. It looks as if a small "micelle" is formed. We believe this is the identity of "micro-cluster" or "quasi phase separation." On the other hand, in the infinite dilution of water in butanol, a water molecule is incorporated in the hydrogen-bond chain of butanol by making hydrogen–bonds with the alcohol. Those results reveal the microscopic structure of water-alcohol mixture in an unequivocal manner. We will see later (in Section II-4) how the structure of the water-butanol mixture produces the distinct behavior in thermodynamics, or the isothermal compressibility.

II-4 Thermodynamic Functions

II-4.1 Chemical Potential

With Kirkwood's coupling parameter λ, the chemical potential of a solute molecule in solution can be expressed as,

$$\Delta\mu\left(\{\mathbf{r}\}_u\right) = \rho^v \sum_{\alpha_u} \sum_{\gamma_v} \int_0^1 d\lambda \int d\mathbf{r} \frac{\partial u_{\alpha_u\gamma_v}(r;\lambda)}{\partial\lambda} g_{\alpha_u\gamma_v}(r;\lambda) \qquad \text{(II-4.1)}$$

where Greek subscripts α_u and γ_v specify atoms in solute (u) and solvent (v), respectively, and $u_{\alpha_u\gamma_v}$ and $g_{\alpha_u\gamma_v}$ denote the interaction energy and the pair correlation function between solute and solvent atoms, respectively [McQuarrie 1973, Chiles and Rossky]. The expression is general, and it can be applied to any closure relation for the RISM equation by performing the integration over λ numerically. However, the procedure is time consuming, because it requires the converged solution of the RISM equation for each step of the coupling procedure from $\lambda = 0$ to $\lambda = 1$. It has been shown, however, that the integration over λ can be performed analytically. For example, for the HNC closure, Singer and Chandler carried out the integration analytically following the procedure proposed by Morita and Hiroike for the OZ equation of simple liquids [Morita and Hiroike 1960, Singer and Chandler 1985].

Let us write the HNC closure with the coupling parameter λ as follows.

$$g_{\alpha_u\gamma_v}(r;\lambda) = \exp\left[d_{\alpha_u\gamma_v}(r;\lambda)\right] \tag{II-4.2}$$

where $d_{\alpha_u\gamma_v}(r;\lambda)$ is defined by

$$d_{\alpha_u\gamma_v}(r;\lambda) = -\lambda\beta u_{\alpha_u\gamma_v}(r) + h_{\alpha_u\gamma_v}(r;\lambda) - c_{\alpha_u\gamma_v}(r;\lambda) \tag{II-4.3}$$

Taking the derivative of Eq. (II-4.2) with respect to λ gives

$$\frac{dg_{\alpha_u\gamma_v}}{d\lambda} = g_{\alpha_u\gamma_v}\left(-\beta u_{\alpha_u\gamma_v} + \frac{dh_{\alpha_u\gamma_v}}{d\lambda} - \frac{dc_{\alpha_u\gamma_v}}{d\lambda}\right) \tag{II-4.4}$$

where all the arguments of the functions are omitted for simplicity. Rearranging the terms in Eq. (II-4.4) gives,

$$\beta u_{\alpha_u\gamma_v} g_{\alpha_u\gamma_v} = \frac{1}{2}\frac{dh^2_{\alpha_u\gamma_v}}{d\lambda} - \frac{dc_{\alpha_u\gamma_v}}{d\lambda} - h_{\alpha_u\gamma_v}\frac{dc_{\alpha_u\gamma_v}}{d\lambda} \tag{II-4.5}$$

The expression is not appropriate to perform the integration over λ in Eq. (II.A.2-1), since it is not a form of an exact differential, or $d\Delta\mu/d\lambda$. So, we convert it into the exact differential utilizing the solute-solvent RISM equation with the coupling parameter λ.

$$h_{\alpha_u\gamma_v}(r;\lambda) = \sum_{\beta_u}\sum_{\xi_v}\omega_{\alpha_u\beta_u} * c_{\beta_u\xi_v} * \chi_{\xi_v\gamma_v}(r) \tag{II-4.6}$$

where the symbol $*$ indicates the convolution integral. With Eq. (II-4.6), the following equality can be readily proved.

$$\begin{aligned}\frac{\partial}{\partial\lambda}\sum_{\gamma_v}\int d\mathbf{r} h_{\alpha_u\gamma_v}(r;\lambda)c_{\gamma_v\alpha_u}(r;\lambda) &= \frac{\partial}{\partial\lambda}\sum_{\gamma_v}\int d\mathbf{r}\sum_{\beta_u}\sum_{\xi_v}\omega_{\alpha_u\beta_u} * c_{\beta_u\xi_v} * \chi_{\xi_v\gamma_v}(r)c_{\gamma_v\alpha_u}(r;\lambda) \\ &= 2\sum_{\gamma_v}\int d\mathbf{r} h_{\alpha_u\gamma_v}(r;\lambda)\frac{\partial}{\partial\lambda}c_{\gamma_v\alpha_u}(r;\lambda)\end{aligned} \tag{II-4.7}$$

which implies that

$$h_{\alpha_u\gamma_v}(r;\lambda)\frac{dc_{\alpha_u\gamma_v}(r;\lambda)}{d\lambda} = \frac{1}{2}\frac{d}{d\lambda}h_{\alpha_u\gamma_v}(r;\lambda)c_{\alpha_u\gamma_v}(r;\lambda) \tag{II-4.8}$$

Then, Eq. (II-4.5) is converted into the following expression that is an exact differentiation.

$$\begin{aligned}\beta u_{\alpha_u\gamma_v} g_{\alpha_u\gamma_v} &= \frac{1}{2}\frac{dh^2_{\alpha_u\gamma_v}}{d\lambda} - \frac{dc_{\alpha_u\gamma_v}}{d\lambda} - \frac{1}{2}\frac{d(h_{\alpha_u\gamma_v}c_{\alpha_u\gamma_v})}{d\lambda} \\ &= \frac{d}{d\lambda}\left\{\frac{1}{2}h^2_{\alpha_u\gamma_v} - c_{\alpha_u\gamma_v} - \frac{1}{2}h_{\alpha_u\gamma_v}c_{\alpha_u\gamma_v}\right\}\end{aligned} \tag{II-4.9}$$

Substituting the expression into Eq. (II-4.1) and performing the integration over λ yield

$$\Delta\mu^{HNC}\left(\{\mathbf{r}\}_u\right) = \rho^v k_B T \sum_\alpha \sum_\gamma \int d\mathbf{r}\left[\frac{1}{2}h_{\alpha_u\gamma_v}(r)^2 - c_{\alpha_u\gamma_v}(r) - \frac{1}{2}h_{\alpha_u\gamma_v}(r)c_{\alpha_u\gamma_v}(r)\right] \quad \text{(II-4.10)}$$

II-4.2 Partial Molar Volume and Compressibility

The partial molar volume and compressibility play crucial roles in describing the *mechanical stability* and *fluctuation* of solutions. In particular, it has been well clarified recently that the properties have special importance in characterizing the structural stability and fluctuation of protein perturbed by pressure. (The topics will be discussed in the Chapters IV and V.) The site-site expression for the compressibility can be derived readily based on the fluctuation formula. However, the expression for the partial molar volume requires some special care, since it requires a transformation from (V,T) ensemble to (P,T) ensemble. Such a transformation has been originally proposed by Kirkwood and Buff for a solution in which there are no internal degrees of freedom in consisting molecules. Here, we extend the Kirkwood-Buff theory to a liquid-mixture consisting of polyatomic molecules based on the RIMS theory.

Let us define the chemical potential of a molecule specified in a mixture by a sum of atomic contributions, namely,

$$\mu_M = \sum_{\alpha=1}^{n_M} \mu_{\alpha_M} \quad \text{(II-4.11)}$$

where subscripts M and α label a molecular species and constituent atomic species, respectively. n_M denotes the number of atoms in the molecule M. By taking the pressure derivative of Eq. (II-4.11), the partial molar volume of the molecular species M can be expressed as a sum over those of constituent atoms,

$$V_M = \left(\frac{\partial \mu_M}{\partial p}\right)_{T,i\neq M} = \sum_{\alpha=1}^{n_M}\left(\frac{\partial \mu_{\alpha_M}}{\partial p}\right)_{T,i\neq M} = \sum_{\alpha=1}^{n_M} v_{\alpha_M} \quad \text{(II-4.12)}$$

where the subscript i denotes components other than M, and v_{α_M} is defined by

$$v_{\alpha_M} \equiv \left(\frac{\partial \mu_{\alpha_M}}{\partial p}\right)_{T,i\neq M} \quad \text{(II-4.13)}$$

We start the derivation with the well-known relations between the density correlations functions and the fluctuations of the number of atoms,

$$\iint\left[\rho^{(2)}_{\alpha_M\gamma_{M'}}(\mathbf{r}_1,\mathbf{r}_2) - \rho^{(1)}_{\alpha_M}(\mathbf{r}_1)\rho^{(1)}_{\gamma_{M'}}(\mathbf{r}_2)\right]d\mathbf{r}_1 d\mathbf{r}_2 = \left[\langle N_{\alpha_M}N_{\gamma_{M'}}\rangle - \langle N_{\alpha_M}\rangle\langle N_{\gamma_{M'}}\rangle\right] - \delta_{\alpha_M\gamma_{M'}}\langle N_{\alpha_M}\rangle \quad \text{(II-4.14)}$$

where $\rho^{(1)}_{\alpha_M}(\mathbf{r}_1)$ and $\rho^{(2)}_{\alpha_M\gamma_M}(\mathbf{r}_1, \mathbf{r}_2)$ are, respectively, the one particle and pair density correlation functions, and N_{α_M} the number of α atom belonging to the Mth molecular species in the system. $\langle\ \rangle$ signifies the ensemble average.

For an uniform fluid, the one body density function agrees with bulk density ρ_{α_M}, and the pair correlation function can be expressed by the radial distribution function $g_{\alpha_M\gamma_{M'}}(r)$, namely,

$$\rho^{(1)}_{\alpha_M}(\mathbf{r}_1) = \rho_{\alpha_M} \tag{II-4.15}$$

$$\rho^{(2)}_{\alpha_M\gamma_{M'}}(\mathbf{r}_1,\mathbf{r}_2) = \rho_{\alpha_M}\rho_{\gamma_{M'}}g_{\alpha_M\gamma_{M'}}(r) + \rho_{\alpha_M}s_{\alpha\gamma}(r)(1-\delta_{\alpha\gamma})\delta_{MM'} \tag{II-4.16}$$

where $r = |\mathbf{r}_1 - \mathbf{r}_2|$. $s_{\alpha\gamma}(r)$ is an intramolecular correlation function defined by Eq. (II-2.8), or

$$s_{\alpha\gamma}(r) = \frac{1}{4\pi L^2_{\alpha\gamma}}\delta(r - L_{\alpha\gamma}) \tag{II-4.17}$$

where $L_{\alpha\gamma}(r)$ denotes the distance between atoms α and γ. With the use of Eqs. (II-4.16) and (II-4.17), Eq. (II-4.14) reads

$$\begin{aligned}&\int\left[\rho_{\alpha_M}\rho_{\gamma_{M'}}\left(g_{\alpha_M\gamma_{M'}}(r)-1\right)+\rho_{\alpha_M}s_{\alpha\gamma}(r)(1-\delta_{\alpha\gamma})\delta_{MM'}+\rho_{\alpha_M}\delta_{\alpha_M\gamma_{M'}}\right]d\mathbf{r}\\&=\mathrm{V}\left(\left\langle N_{\alpha_M}N_{\gamma_{M'}}\right\rangle-\left\langle N_{\alpha_M}\right\rangle\left\langle N_{\gamma_{M'}}\right\rangle\right)\end{aligned} \tag{II-4.18}$$

Defining the Fourier transform of a function $f(r)$ by $\tilde{f}(k) = \int f(r)e^{-i\mathbf{k}\cdot\mathbf{r}}d\mathbf{r}$, the left-hand side of Eq. (II-4.18) can be written as

$$\begin{aligned}&\int\left[\rho_{\alpha_M}\rho_{\gamma_{M'}}\left(g_{\alpha_M\gamma_{M'}}(r)-1\right)+\rho_{\alpha_M}s_{\alpha\gamma}(r)(1-\delta_{\alpha\gamma})\delta_{MM'}+\rho_{\alpha_M}\delta_{\alpha_M\gamma_{M'}}\right]d\mathbf{r}\\&=\rho_{\alpha_M}\tilde{\omega}_{\alpha_M\gamma_{M'}}(0)+\rho_{\alpha_M}\rho_{\gamma_{M'}}\tilde{h}_{\alpha_M\gamma_{M'}}(0)\\&=\tilde{\chi}_{\alpha_M\gamma_{M'}}(0)\end{aligned} \tag{II-4.19}$$

where $h_{\alpha_M\gamma_{M'}}(r) = g_{\alpha_M\gamma_{M'}}(r) - 1$ and $\omega_{\alpha_M\gamma_{M'}}(r) = \rho_{\alpha_M}s_{\alpha\gamma}(r)(1-\delta_{\alpha\gamma})\delta_{MM'} + \rho_{\alpha_M}\delta_{\alpha_M\gamma_{M'}}$. Using Eqs. (II-4.18) and (II-4.19) and the well-known relation derived from the Gibbs grand canonical ensemble,

$$\begin{aligned}&\left\langle N_{\alpha_M}N_{\gamma_{M'}}\right\rangle-\left\langle N_{\alpha_M}\right\rangle\left\langle N_{\gamma_{M'}}\right\rangle=\left(\mathbf{U}^{-1}\right)_{\alpha_M\gamma_{M'}}\\&U_{\alpha_M\gamma_{M'}}=\frac{1}{k_BT}\left(\frac{\partial\mu_{\alpha_M}}{\partial N_{\gamma_{M'}}}\right)_{V,T,N'}\end{aligned} \tag{II-4.20}$$

we obtain the equation,

$$\frac{1}{k_BT}\left(\frac{\partial\mu_{\alpha_M}}{\partial N_{\gamma_{M'}}}\right)=\frac{1}{V}\left(\tilde{\chi}(0)^{-1}\right)_{\alpha_M\gamma_{M'}} \tag{II-4.21}$$

In order to transform from the (V,T)-ensemble to the (P,T)-ensemble, we obtain,

$$\sum_{M=1}^{x}\sum_{\alpha=1}^{n_M}N_{\alpha_M}\left(\frac{\partial\mu_{\alpha_M}}{\partial N_{\gamma_{M'}}}\right)_{P,T,N'}=0 \tag{II-4.22}$$

and the thermodynamic relation,

$$\left(\frac{\partial\mu_{\alpha_M}}{\partial N_{\gamma_{M'}}}\right)_{V,T,N'}=\left(\frac{\partial\mu_{\alpha_M}}{\partial N_{\gamma_{M'}}}\right)_{P,T,N'}+\frac{v_{\alpha_M}v_{\gamma_{M'}}}{V\chi_T} \tag{II-4.23}$$

Use of Eqs. (II-4.21), (II-4.22), and (II-4.23), we find the expressions for the partial molar volume and the isothermal compressibility of a polyatomic liquid mixture in terms of the site-site density pair correlation functions $\chi(r)$ as

$$v_{\alpha_M} = k_B T \chi_T \sum_{M'=1}^{x} \sum_{\gamma=1}^{n_{M'}} \left(\tilde{\chi}(0)^{-1} \rho \right)_{\alpha_M \gamma_{M'}} \tag{II-4.24}$$

$$k_B T \chi_T = \frac{1}{\sum_{M=1}^{x} \sum_{M'=1}^{x} \sum_{\alpha=1}^{n_M} \sum_{\gamma=1}^{n_{M'}} \left(\rho \tilde{\chi}(0)^{-1} \rho \right)_{\alpha_M \gamma_{M'}}} \tag{II-4.25}$$

The expressions involve a singular functions as can be readily verified. Let us define a matrix V (hereafter, we omit "(0)," which represents $k = 0$, from the matrix expression) as,

$$\mathbf{V} = \tilde{\boldsymbol{\chi}}^{-1} \boldsymbol{\rho} \tag{II-4.26}$$

From Eq. (A8) with use of the RISM equation, we obtain

$$\tilde{\boldsymbol{\chi}}^{-1} = \left(\boldsymbol{\rho} \tilde{\boldsymbol{\omega}} \right)^{-1} - \tilde{\mathbf{c}} \tag{II-4.27}$$

Therefore,

$$\mathbf{V} = \tilde{\boldsymbol{\omega}}^{-1} - \tilde{\mathbf{c}} \boldsymbol{\rho} \tag{II-4.28}$$

All the elements in the matrix $\tilde{\omega}$ in Eq. (II-4.27) is unity, thereby an inverse of the matrix diverges. In order to bypass the singularity, we use the following trick. Multiplying Eq. (II-4.28) by ω on the both sides, we obtain

$$\tilde{\boldsymbol{\omega}} \mathbf{V} \tilde{\boldsymbol{\omega}} = \tilde{\boldsymbol{\omega}} - \tilde{\boldsymbol{\omega}} \mathbf{c} \boldsymbol{\rho} \tilde{\boldsymbol{\omega}} \tag{II-4.29}$$

Hereafter, for simplicity, we consider a two-component system, that is, solute (n sites) and solvents (m sites). In that case, Eq. (II-4.29) is expressed by using the block matrices as

$$\begin{pmatrix} \tilde{\omega}_{uu} \mathbf{V}_{uu} \tilde{\omega}_{uu} & \tilde{\omega}_{uu} \mathbf{V}_{uv} \tilde{\omega}_{vv} \\ \tilde{\omega}_{vv} \mathbf{V}_{vu} \tilde{\omega}_{uu} & \tilde{\omega}_{vv} \mathbf{V}_{vv} \tilde{\omega}_{vv} \end{pmatrix} = \begin{pmatrix} \tilde{\omega}_{uu} - \tilde{\omega}_{uu} \tilde{\mathbf{c}}_{uu} \rho_u \tilde{\omega}_{uu} & -\tilde{\omega}_{uu} \tilde{\mathbf{c}}_{uv} \rho_v \tilde{\omega}_{vv} \\ -\tilde{\omega}_{vv} \tilde{\mathbf{c}}_{vu} \rho_u \tilde{\omega}_{uu} & \tilde{\omega}_{vv} - \tilde{\omega}_{vv} \tilde{\mathbf{c}}_{vv} \rho_v \tilde{\omega}_{vv} \end{pmatrix} \tag{II-4.30}$$

All the elements of $\tilde{\omega}_{MM}$ are unity so that an each element of $\tilde{\omega}_{MM} \mathbf{A} \tilde{\omega}_{M'M'}$ results in a sum of all the elements of **A**, where M and M' represents u and v, and **A** represents an arbitrary matrix. For example,

$$\sum_{\alpha=1}^{n} \sum_{\alpha'=1}^{n} V_{\alpha_u \alpha'_u} = 1 - \rho_u \sum_{\alpha=1}^{n} \sum_{\alpha'=1}^{n} \tilde{c}_{\alpha_u \alpha'_u}(0) \tag{II-4.31}$$

$$\sum_{\alpha=1}^{n} \sum_{\gamma=1}^{m} V_{\alpha_u \gamma_v} = -\rho_v \sum_{\alpha=1}^{n} \sum_{\gamma=1}^{m} \tilde{c}_{\alpha_u \gamma_v}(0) \tag{II-4.32}$$

$$V_u = k_B T \chi_T \sum_{\alpha=1}^{n} \left(\sum_{\alpha'=1}^{n} V_{\alpha_u \alpha'_u} + \sum_{\gamma=1}^{m} V_{\alpha_u \gamma_v} \right) \tag{II-4.33}$$

$$V_u = k_B T \chi_T \left(1 - \rho_u \sum_{\alpha=1}^{n} \sum_{\alpha'=1}^{n} \tilde{c}_{\alpha_u \alpha'_u}(0) - \rho_v \sum_{\alpha=1}^{n} \sum_{\gamma=1}^{m} \tilde{c}_{\alpha_u \gamma_v}(0) \right) \tag{II-4.34}$$

The expression for the partial molar volume at the infinite dilution, V_u^0 is obtained by taking the limit $\rho_u \to 0$,

$$V_u^0 = k_B T \chi_T \left(1 - \rho_v \sum_{\alpha=1}^{n} \sum_{\gamma=1}^{m} \tilde{c}_{\alpha_u \gamma_v}(0) \right) \tag{II-4.35}$$

In the same way, the isothermal compressibility of a two-component system can be expressed as follows:

$$k_B T \chi_T = \frac{1}{\rho_u + \rho_v - \rho_u^2 \sum_{\alpha=1}^{n} \sum_{\alpha'=1}^{n} \tilde{c}_{\alpha_u \alpha_u'}(0) - \rho_v^2 \sum_{\gamma=1}^{m} \sum_{\gamma'=1}^{m} \tilde{c}_{\alpha_v \gamma_v'}(0) - 2\rho_u \rho_v \sum_{\alpha=1}^{n} \sum_{\gamma=1}^{m} \tilde{c}_{\alpha_u \gamma_v}(0)} \tag{II-4.36}$$

At the limit of the infinite dilution,

$$k_B T \chi_T^0 = \frac{1}{\rho_v \left(1 - \rho_u \sum_{\gamma=1}^{m} \sum_{\gamma'=1}^{m} \tilde{c}_{\gamma_v \gamma_v'}(0) \right)} \tag{II-4.37}$$

The expression for a multicomponent system can be readily obtained after straight forward extension as follows,

$$V_M = k_B T \chi_T \left(1 - \sum_{M'=1}^{x} \rho_M \sum_{\alpha=1}^{n_M} \sum_{\gamma=1}^{m_{M'}} \tilde{c}_{\alpha_M \gamma_{M'}}(0) \right) \tag{II-4.38}$$

$$k_B T \chi_T^0 = \frac{1}{\sum_{M=1}^{x} \rho_M - \sum_{M=1}^{x} \sum_{M'=1}^{x} \rho_M \rho_{M'} \sum_{\gamma=1}^{n_M} \sum_{\gamma'=1}^{n_{M'}} \tilde{c}_{\alpha_M \gamma_{M'}}(0)} \tag{II-4.39}$$

In the following, we investigate the isothermal compressibility of water-alcohol mixture as an example of the study based on the theory just described. The study of the partial molar volume is presented later in Chapter IV in connection with the solvation of biomolecules.

Isothermal compressibility of water-t-butanol mixture: In the Section II-3, We have revealed the microscopic structure of the water-*t*-butanol mixture based on the XRISM theory. The most important conclusion concerning the liquid structure in low concentration of butanol is that the butanol molecules are incorporated in the hydrogen-bond network of water so that their butyl groups are in contact each other to form a micelle-like structure (Fig. II-3.4). So, the question to be asked is "how the structure is manifested in thermodynamic properties?" Since the words such as "micro-cluster" and/or "quasi phase separation" are terminologies expressing the density and/or concentration fluctuations in solution, it is natural to analyze the compressibility of the solution. It has been well documented that the compressibility is a measure of the density and/or concentration fluctuation. The isothermal compressibility of a liquid mixture can be evaluated using Eq. (II-4.35), provided the pair correlation functions calculated based on the XRISM theory [Omelyan et al. 2003]. The results are compared with the experimental data in Fig. II-4.1 [Kubota et al. 1987, Tabellout et al. 1990].

Two distinct features are observed in the experimental data, Fig. II-4.1(a). The first is the turn over observed in temperature dependence at $x \sim 0.02$, that make an *isosbestic* point. The

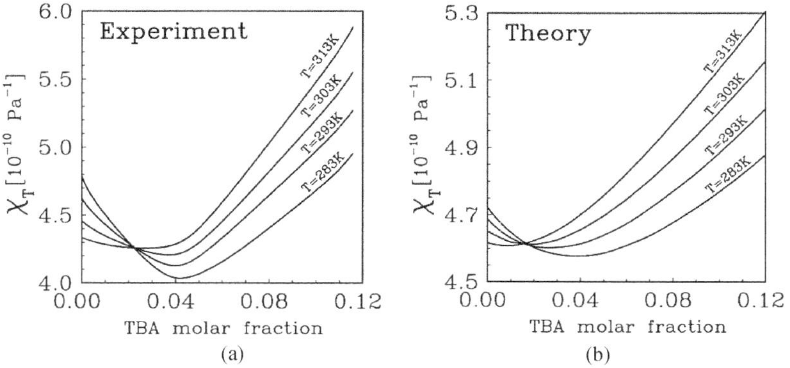

Fig. II-4.1: Isothermal compressibility of water-t-butanol mixture: (a) experiment, (b) XRISM. Copyright (2003) World Scientific Publishing Comany.

second feature is the minima seen in the concentration dependence at all temperatures. It is apparent that the XRISM theory reproduces faithfully the two features in the experimental results. From the results, the following picture concerning the anomalous behavior of the water-alcohol mixture can be drawn.

A) In pure water, the compressibility reduces with increasing temperature in this temperature range. The behavior can be explained in terms of the ice-like structure or hydrogen-bond network which is bulkier and more compressible than the non-hydrogen-bonded structure. The ice-like structure "melts" with rising temperature, which makes water less compressible.

B) The compressibility of pure butanol increases with rising temperature. This behavior is a common feature shared by many other liquids but water and gallium. The behavior can be explained in terms of the thermal expansion that increases with rising temperature, that is, the compressible space increases as temperature increases.

C) Concentration dependence of the compressibility shows characteristic minima in the temperature range investigated. The existence of mina demonstrates that there are two competing effects in the concentration dependence which balances at some concentration. One of those effects is the increase of compressibility with increasing concentration of butanol, which can be explained in terms of a model of ideal mixing. Suppose the compressibility of the mixture is expressed by a simple superposition of the compressibility of the component liquids, weighted by respective concentration or mol fraction in such a way that $\chi_T = (x-1)\chi_{\text{water}} + x\chi_{\text{butanol}}$. Then, the contribution of butanol to the compressibility, which is greater than water, becomes more dominative as the concentration increases, thereby the compressibility increases with increasing concentration at the butanol-rich region. On the other hand, such a simple picture does not apply any more at the lower concentration or the water rich region. The structure of water-butanol mixture, deduced from the pair correlation functions, exhibited in Fig. II-3.4, is quite useful and instructive for explaining the anomalous behavior. Butanol molecules in the low concentration regime are incorporated in the hydrogen-bond network of water through their OH-group, and their alkyl groups fill the void space of the hydrogen-bond network, as is shown in Fig. II-3.4. That is, the alkyl groups reduces the compressibility by filling the *compressible space* of water.

II-4.3 Interaction-site Model of the Dielectric Constant

The theory of dielectric constant by Chandler

Here, the theory of dielectric constant based on the interaction-site model, formulated by D. Chandler is reviewed [Chandler 1977]. Although Chandler's theory treats a general case of a multi-component system consisting of flexible as well as polarizable molecules, we confine ourselves to a single component liquid consisting of rigid and non-polarizable molecules for simplicity. For the general case, one has to consult with the paper by Chandler.

The derivation is based on the test charge hypothesis as follows. Suppose a test charge Z is placed at a point in a polar liquid. The electrostatic potential created by the test charge at the position far distant from the test charge can be expressed as

$$V(\mathbf{r}) : Z/\varepsilon r, \quad \text{for large } r \tag{II-4.40}$$

where r is the distance from the charge Z placed at the origin. The test charge hypothesis identifies the screening constant ε as the phenomenological dielectric constant of the polar liquid. So, if one is able to derive the expression for the electrostatic potential around a test charge in terms of the site-site pair correlation functions, one can find an ISM expression for ε in the limit of large distance r, by comparing the two equations.

The electrostatic potential created by a test charge is written in terms of the density field of site α of a molecule as

$$V_{mol}(\mathbf{r}) = \sum_{\alpha} Z_{\alpha} \int d\mathbf{r}' \frac{v_{\alpha}(\mathbf{r}')}{|\mathbf{r} - \mathbf{r}'|} \tag{II-4.41}$$

where $v_{\alpha}(\mathbf{r}')$ is the density field of site α at $\mathbf{r}'$ defined by

$$v_{\alpha}(\mathbf{r}') = \sum_{i=1} \delta(\mathbf{r}' - \mathbf{r}_i^{(\alpha)}) \tag{II-4.42}$$

The electrostatic potential V_{mol} created by Z can be divided into two contributions: the direct and solvent induced parts.

$$\begin{aligned} V(\mathbf{r}) &= (Z/r) + \langle V_{mol}(\mathbf{r}) \rangle_Z \\ &= (Z/r) + \sum_{\alpha} Z_{\alpha} \int d\mathbf{r}' \frac{\rho_{\alpha}(\mathbf{r}'; \psi^Z)}{|\mathbf{r} - \mathbf{r}'|} \end{aligned} \tag{II-4.43}$$

where $\rho_{\alpha}(\mathbf{r}'; \psi^Z)$ denotes the density distribution function of solvent-site α at the position $\mathbf{r}'$, which is influenced by the test charge Z placed at the origin. Therefore, the second term in the above equation represents the average electrostatic potential created by solvent sites. At a position which is far distant from the origin, the influence of the test charge is small, thereby one can expand $\rho_{\alpha}(\mathbf{r}'; \psi^Z)$ in the functional Taylor series, namely,

$$\rho_{\alpha}(\mathbf{r}'; \psi^Z) = \rho_{\alpha}(\mathbf{r}'; \psi^Z = 0) + \sum_{\gamma} \int d\mathbf{r}'' \frac{Z_{\gamma} Z}{r''} \left[\frac{\delta \rho_{\alpha}(\mathbf{r}')}{\delta \psi_{\gamma}^Z(\mathbf{r}'')} \right]_{\psi^Z = 0} + O(Z^2) \tag{II-4.44}$$

The functional derivative is

$$\frac{\delta \rho_{\alpha}(\mathbf{r}')}{\delta \psi_{\gamma}^Z(\mathbf{r}'')} = -\beta \chi_{\alpha\gamma}(|\mathbf{r}' - \mathbf{r}''|) \tag{II-4.45}$$

where $\chi_{\alpha\gamma}(r)$ is the site-site pair correlation function defined by

$$\chi_{\alpha\gamma}\left(|\mathbf{r}-\mathbf{r}'|\right)=\rho\omega_{\alpha\gamma}\left(|\mathbf{r}-\mathbf{r}'|\right)+\rho^2 h_{\alpha\gamma}\left(|\mathbf{r}-\mathbf{r}'|\right) \tag{II-4.46}$$

$$\omega_{\alpha\gamma}\left(|\mathbf{r}-\mathbf{r}'|\right)=\delta_{\alpha\gamma}\delta\left(|\mathbf{r}-\mathbf{r}'|\right)+s_{\alpha\gamma}\left(|\mathbf{r}-\mathbf{r}'|\right) \tag{II-4.47}$$

$h_{\alpha\gamma}(r)$ are the site-site pair correlation function, and $\omega_{\alpha\gamma}(r)$ is the intramolecular correlation function which includes the Dirac delta-function as the normalized distribution functions of the distance among atoms in a molecule.

$$s_{\alpha\gamma}(r)=\frac{1}{4\pi L_{\alpha\gamma}^2}\delta(r-L_{\alpha\gamma}) \tag{II-4.48}$$

Putting Eqs. (II-4.44) ~ (II-4.48) into (II-4.43), one finds the following equation that relates the site-site pair correlation function to the electrostatic potential created at position **r** by the test charge Z placed at the origin.

$$V(\mathbf{r})=Z\left\{\frac{1}{r}-\beta\sum_{\alpha,\gamma}Z_\alpha Z_\gamma\int d\mathbf{r}'\int d\mathbf{r}''\frac{1}{r''}\frac{\chi_{\alpha\gamma}\left(|\mathbf{r}''-\mathbf{r}'|\right)}{|\mathbf{r}''-\mathbf{r}|}\right\}+O\left(Z^2\right) \tag{II-4.49}$$

The further manipulation is easier in the Fourier space, since we can use the theorem of the convolution integrals. We define the Fourier transform of $V(\mathbf{r})$ by

$$\tilde{V}(\mathbf{k})=\int d\mathbf{r}\exp(-i\mathbf{k}\cdot\mathbf{r})V(\mathbf{r})$$

The phenomenological electrostatic potential (II-4.40) and the microscopic counter-part (II-4.49) are respectively expressed in the Fourier space as

$$\tilde{V}(k):\frac{4\pi Z}{\varepsilon k^2} \tag{II-4.50}$$

$$\tilde{V}(\mathbf{k})=\frac{4\pi Z}{k^2}\left\{1-\frac{4\pi\beta}{k^2}\sum_{\alpha,\gamma}Z_\alpha Z_\gamma\tilde{\chi}_{\alpha\gamma}(k)\right\}+O\left(Z^2\right) \tag{II-4.51}$$

Equating the two expressions to the first order of Z, one finds the following expression for the macroscopic dielectric constant.

$$\left(\frac{1}{\varepsilon}-1\right)=\lim_{k\to 0}\left\{-\frac{4\pi\beta}{k^2}\sum_{\alpha,\gamma}Z_\alpha Z_\gamma\tilde{\chi}_{\alpha\gamma}(k)\right\} \tag{II-4.52}$$

Now, let us consider the small k expansion of $\tilde{\chi}_{\alpha\gamma}(k)$. The first term becomes in the limit of $k\to 0$,

$$\lim_{k\to 0}\tilde{\chi}_{\alpha\gamma}(k)=\left(\frac{\partial\rho_M}{\partial\beta\mu_M}\right) \tag{II-4.53}$$

So, it does not depend on a pair of atoms due to the compressibility theorem. Therefore, its contribution to the dielectric constant vanishes due to the charge neutrality condition. The non-zero term in the expansion starts from the second order in k as,

$$\sum_{\alpha,\gamma} Z_\alpha Z_\gamma \tilde{\chi}_{\alpha\gamma}(k) = k^2 \sum_{\alpha,\gamma} Z_\alpha Z_\gamma \tilde{\chi}^{(2)}_{\alpha\gamma}(k) + \cdots \tag{II-4.54}$$

where

$$\tilde{\chi}^{(2)}_{\alpha\gamma}(k) = -\frac{1}{6}\int d\mathbf{r} r^2 \chi_{\alpha\gamma}(r) \tag{II-4.55}$$

The second moment $\tilde{\chi}^{(2)}_{\alpha\gamma}(k)$ is split into two contributions; the intramolecular and intermolecular parts (Eq. II-4.46). The intramolecular part is expanded into the power of k as

$$\tilde{\omega}_{\alpha\gamma}(k) = 1 + k^2\omega^{(2)}_{\alpha\gamma} + O(k^4) \tag{II-4.56}$$

that gives rise to the contribution from the dipole moment in gas phase,

$$\sum_{\alpha,\gamma} Z_\alpha Z_\gamma \omega^{(2)}_{\alpha\gamma} = \frac{1}{3}\mu_M^2 \tag{II-4.57}$$

where μ_M is the dipole moment.

$$\mu_M = \left|\sum_\alpha Z_\alpha \mathbf{L}_1^{(\alpha)}\right| \tag{II-4.58}$$

The second term in Eq. (II-4.46) is expanded into the power of k as

$$\tilde{h}_{\alpha\gamma}(k) = h^{(0)} + k^2 h^{(2)}_{\alpha\gamma} + \cdots \tag{II-4.59}$$

The contribution of the intermolecular term, or, $\tilde{h}_{\alpha\gamma}(k)$, to the dielectric constant becomes,

$$\sum_{\alpha,\gamma} Z_\alpha Z_\gamma h^{(2)}_{\alpha\gamma} = -\frac{1}{6}\int d\mathbf{r} r^2 \left[\sum_{\alpha,\gamma} Z_\alpha Z_\gamma h_{\alpha\gamma}(r)\right] \tag{II-4.60}$$

Inserting Eqs. (II-4.57) and (II-4.60) into (II-4.52), one finds

$$1 - \frac{1}{\varepsilon} = \frac{4\pi}{3}\beta\rho\mu^2 + 4\pi\beta\rho^2 \sum_{\alpha,\gamma} Z_\alpha Z_\gamma h^{(2)}_\alpha \tag{II-4.61}$$

A remark on the dielectric constant obtained from XRISM

The dielectric constant can be calculated in principle based on Eq. (II-4.61) from the XRISM theory. However, the XRISM with conventional closure relations such as HNC and PY gives a trivial dielectric constant corresponding to non-interacting dipolar gas, or the ideal gas with dipole moments.

$$\varepsilon = 1 + 3y \tag{II-4.62}$$

where y is defined by $y \equiv 4\pi\beta\mu^2\rho/9$ with the number density ρ of the liquid and dipole moment μ. The reason why the XRISM/HNC theory produces the dielectric constant for ideal gas has been *mathematically* proved by Sullivan and Gray unambiguously [Sullivan and Gray 1981, Hirata et al. 1982, Cummings and Stell 1982]. However, a *diagnosis* to treat the disease has not been clarified yet. Perkyns and Pettitt has proposed an empirical correction to XRISM, called DRISM, that introduces the phenomenological value of the dielectric constant into the theory to screen the Coulomb interaction between charges in solution [Perkyns and Pettitt 1992a, b]. The theory has been applied to explain the potential of mean force between two ions in solution. The theory has scored a great success in explaining the structural as well as thermodynamic properties of electrolyte solutions. The theory, however, is essentially an empirical theory, and it cannot predict the dielectric constant. In the following, we revisit the problem to explore a possibility of reproducing the dielectric constant without using phenomenological parameters, but with a parameter which reflects the effect of short-range correlation onto the dipole-dipole interaction between the two molecules. In order to clarify the physical reason why the XRISM/HNC theory gives rise to a trivial dielectric constant, let's summarize the theory below.

Summary of the XRISM theory: The renormalized RISM theory, or the XRISM theory, with the HNC closure is summarized in a compact manner using the chain sum convention.

$$\tau = \mathcal{C}\left[c - \phi|\omega + \rho \mathbf{Q} \rho\right] \tag{II-4.63}$$

$$c_{\alpha\gamma} - \phi_{\alpha\gamma} = \exp\left[-\beta u^*_{\alpha\gamma} + \tau_{\alpha\gamma} + Q_{\alpha\gamma}\right] - 1 - \tau_{\alpha\gamma} - Q_{\alpha\gamma} \tag{II-4.64}$$

where $u^*_{\alpha\gamma}(r)$ denote the short-range interaction between a pair of intermolecular interaction-sites α and γ. $\phi_{\alpha\gamma}$ is the Coulomb interaction between the partial charges of the sites α and γ reduced by the thermal energy (kT)

$$\phi_{\alpha\gamma}(r) = -\beta \frac{Z_\alpha Z_\gamma}{r} \tag{II-4.65}$$

$Q_{\alpha\gamma}(r)$ is the renormalized Coulomb interaction defined using the chain sum notation by

$$\mathbf{Q} = \mathcal{C}\left[\phi|\omega\right] \tag{II-4.66}$$

where $\omega_{\alpha\gamma}(r)$ is the intra-molecular correlation function defined by Eq. (II-4.47).

Two points are worthwhile to be noted. The XRISM theory treats the correlation between sites or atoms of molecules. Therefore, it concerns essentially the interactions between a pair of charged particles, or *ions* with *partial charges*, not the *dipole-dipole* interaction directly. The other point to be made is related to the HNC closure employed in the XRISM theory that neglects the *bridge* diagrams (see the Section II-1.5 for the bridge diagrams). For examples, in the site-site viewpoint, the first few bridge diagrams for a diatomic fluid can be listed as

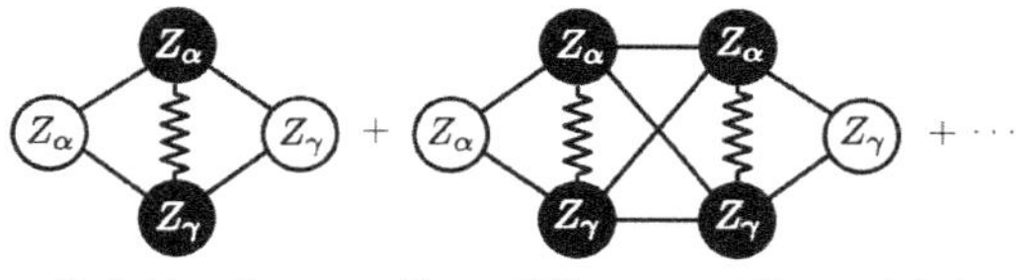

Fig. II-4.2(a): Illustration of site-site bridge diagrams. Zα and Zβ represent the partial charges of a molecule. The solid and zig-zag lines connecting the two circles denote, respectively, the φ-bond defined by Eq. II-4.65 and the s-bond defined by Eq. II-4.47. Copyright (2019) Elsevier.

However, such diagrams can be replaced, in the molecular viewpoint, by the following diagrams,

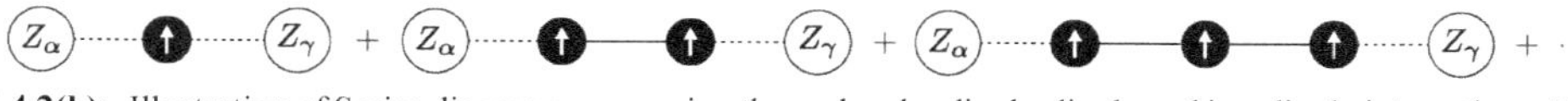

Fig. II-4.2(b): Illustration of Series diagrams concerning the molecular dipole-dipole and ion-dipole interactions. Zα and Zγ denote the partial charges of molecules. The arrow represents an ideal dipole, the dipole moment of which is equivalent to $\mu = \sum_\alpha Z_\alpha l_\alpha$. The black circles with arrow represent integrals over the Cartesian as well as angular coordinates. The solid line between the two ideal dipoles denotes φDD-bond defined by Eq. (C7) in the Appendix of this chapter, while the dotted line between the partial charges and the ideal dipole represents that of the ion-dipole interaction. Copyright (2019) Elsevier.

The diagrams in the molecular viewpoint are nothing but the *series diagrams* that can be treated by means of the convolution integrals as been done by Hoye and Stell (see Appendix C). The identification of the *series* diagrams with respect to the dipole-dipole interaction in the molecular picture with the *bridge* diagrams concerning charge-charge interaction in the site-site picture is the main point made here. The identification implies that the XRISM/HNC or XRISM/PY theory, which ignores the bridge diagrams for the dipole-dipole interactions, lacks the most important ingredient contributing to the dielectric constant. That is the physics behind the reason why the XRISM/HNC theory gives rise to a trivial dielectric constant for gas phase, which entirely ignores the dipole-dipole interaction between the dipoles [Ikuta and Hirata 2019].

XRISM with dipole-bridge correction (XRISM-DB): Now, we develop a new XRISM theory that incorporates the chain-sum of the series diagrams concerning the dipole-dipole interaction as the bridge diagrams with respect to the charge-charge interaction [Ikuta and Hirata 2019]. With the bridge diagrams, the site-site interaction between the partial charges can be expressed as

$$\phi_{\alpha\gamma}^{DB}(r) = \phi_{\alpha\gamma}(r) + b_{\alpha\gamma}(r) \tag{II-4.67}$$

where $b_{\alpha\gamma}(r)$ is the sum of the site-site bridge diagrams in Fig. II-4.2(a), and $\phi_{\alpha\gamma}$ is the Coulomb interaction between the two sites, defined by Eq. (II-4.65). Since $b_{\alpha\gamma}(r)$ in Eq. (II-4.67) is identified as the series diagrams in Fig. II-4.2(b), $\phi_{\alpha\gamma}^{DB}$ can be written using the screening constant defined by Eq. (C.12) in Appendix C,

$$\phi_{\alpha\gamma}^{DB}(r) = \phi_{\alpha\gamma}(r) / \varepsilon_{CC} \tag{II-4.68}$$

where ε_{CC} is defined by

$$\varepsilon_{CC} \equiv (1-z)/(1+2z) \tag{II-4.69}$$

and z is defined as,

$$z = y\left(1 + \frac{1}{3}B\right) \tag{II-4.70}$$

In the equations, $y \equiv 4\pi\beta\mu^2\rho/9$, and B is a molecular parameter reflecting the contribution from the short-range interaction to the correlation between the two ideal dipoles (see Appendix C). Then, the RISM equation renormalized for the Coulomb interaction, or the XRISM equation, can be written in the chain sum notation as,

$$\tau = \mathcal{C}\left[c - \boldsymbol{\phi}^{DB} \middle| \boldsymbol{\omega} + \boldsymbol{\rho}\mathbf{Q}^{DB}\boldsymbol{\rho}\right] \tag{II-4.71}$$

$$c_{\alpha\gamma} - \phi_{\alpha\gamma}^{DB} = \exp\left[-\beta u_{\alpha\gamma}^{*} + \tau_{\alpha\gamma} + Q_{\alpha\gamma}^{DB}\right] - 1 - \tau_{\alpha\gamma} - Q_{\alpha\gamma}^{DB} \tag{II-4.72}$$

where $\mathbf{Q}^{DB}$ is the chain sum of $\phi_{\alpha\gamma}^{DB}$ defined by Eq. (II-4.68).

The new equation looks almost identical to the XRISM/HNC equation, except for the superscript "DB" which indicates that the chain sum for the Coulomb interaction includes the bridge diagrams concerning the dipole-dipole interaction. We refer to the new equation as

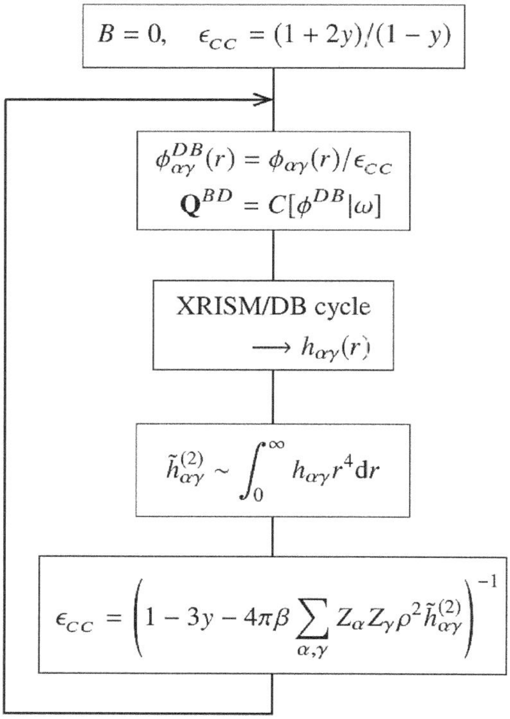

Fig. II-4.3: Iterative scheme to calculate the dielectric constant. Copyright (2019) Elsevier.

XRISM/DB. The XRISM-DB equation can be solved numerically by the standard procedure performed for the XRISM/HNC equation, just dividing the Coulomb interaction by ε_{CC} defined by Eq. (II-4.69). The numerical solution, however, requires a single parameter B in Eq. (II-4.70), that reflects the effect of the short-range correlations, $c_{\alpha\gamma} - \phi_{\alpha\gamma}^{DB}$, originated from the molecular geometry and charge distribution, upon the dielectric screening of the charge-charge interaction. The parameter could in principle be determined in a self-consistent manner in the iterative procedure for solving Eqs. (II-4.71) and (II-4.72), since the effect of the short-range correlation represented by $c_{\alpha\gamma} - \phi_{\alpha\gamma}^{DB}$ in Eq. (II-4.71) is a converged result of the procedure, while ε_{CC} in Eq. (II-4.68) depends on the short-range correlation. A possible iteration scheme is illustrated in Fig. II-4.3.

However, some difficulty is anticipated for the numerical procedure, which is concerned with the calculation of $\hat{h}_{\alpha\gamma}^{(2)}$. The calculation of $\hat{h}_{\alpha\gamma}^{(2)}$ involves the integration of $r^4 h_{\alpha\gamma}(r)$ as is indicated in Fig. II-4.3. So, the calculation requires an integration of very slowly converging function over the infinite range from $r = 0$ to $r = \infty$. Nevertheless, there is a possibility to find the constant B in Eq. (II-4.70) without performing such an intensive calculation. If the constant does not depend so much on the characteristics of polar fluids, we can regard the parameter as a universal constant. In the following section, we determine the parameter B in Eq. (II-4.70) for several polar fluids to see if the parameter in fact is *universal*.

***Determining B in Eq.* Eq. (II-4.70):** Here, a simple way to determine the dielectric constant of polar fluids using a pair of ions as "test charges" is proposed. The potential of mean force

$W_{ij}(r)$ between the two ions, *i* and *j*, in polar fluid is defined from the pair correlation function between the two ions as,

$$W_{ij}(r) \equiv -\frac{1}{\beta}\log g_{ij}(r) = -\frac{1}{\beta}\log\left[h_{ij}(r)+1\right] \tag{II-4.73}$$

where $g_{ij}(r) \equiv h_{ij}(r) + 1$ is the pair correlation function. The pair correlation function between two ions at infinite dilution can be readily calculated using Eq. (II-3.20) in Section II-3 and the XRISM-DB Eq. (II-4.71) with some closure relations such as HNC or KH. The quantity can be also calculated based on the Singer-Chandler formula for the excess chemical potential, for examples Eq. (II-4.10) for the HNC closure, described earlier in this chapter, because the solvent induced part of the potential of mean force. $W_{ij}(r)$ between two ions separated by *distance r* is nothing but the excess chemical potential $\Delta\mu(r)$ of a *diatomic molecule* with a *bond length* fixed to *r*, that is $W_{ij}(r) = Z_iZ_j/r + \Delta\mu(r)$. So, the potential of men force $W_{ij}(r)$ for a distance *r* can be obtained by changing the *bond length r* as an *order parameter*. At large distance, or $r \to \infty$, the potential of mean force should approach *asymptotically* to the Coulomb interaction screened by the dielectric constant ε_{CC} defined by Eq. (II-4.69), or

$$W_{ij}(r) = \frac{Z_iZ_j}{\varepsilon_{cc}r} \quad \text{for} \quad r \to \infty \tag{II-4.74}$$

On the other hand, the phenomenological Coulomb interaction $W_{ij}^{obs}(r)$ between the two ions at $r = \infty$ should become

$$W_{ij}^{obs}(r) = \frac{Z_iZ_j}{\varepsilon_{obs}r} \quad \text{for} \quad r \to \infty \tag{II-4.75}$$

where ε_{obs} is the phenomenological dielectric constant. Therefore, $W_{ij}(r)$ should coincide with $W_{ab}^{obs}(r)$ at $r \to \infty$, which results in $\varepsilon_{CC} = \varepsilon_{obs}$. Then, using Eqs. (II-4.69) and (II.4.70), one can find the parameter *z* or *B* which gives rise to the phenomenological dielectric constant.

Depicted in Fig. II-4.4 is the potential of mean force (PMF) between Na^+ and Cl^- in water, calculated by Eq. (II-4.74) with the DB correction, and by Eq. (II-4.75) with the empirical dielectric constant. The PMF from the XRISM equation is also shown for comparison. The bridge diagrams included in the XRISM-DB equation are just concerned with the dipole-dipole interactions, not for entire interactions. Therefore, the equation should be supplemented by some closure relations such as HNC and KH [Kovalenko and Hirata 1999, Hirata 2003]. Although we have employed the HNC and KH closures to calculate PMF, only those from the KH closure are shown. As can be readily seen, the PMF from XRISM/DB with the parameter $B \sim -3.3$ converges to the screened Coulomb interaction with the phenomenological dielectric constant ε_{obs} at large distance.

In Fig. II-4.5, plotted are the dielectric constants of water calculated by the procedure described above, varying *B* as the test parameter. The empirical dielectric constant at 25 C is the horizontal line. The intersection gives the value of *B* predicted by the theory, which agrees with the phenomenological dielectric constant. Ikuta and Hirata have carried out the same analysis for several different dipolar liquids that are employed popularly in experiments. The results for the *B*-value are listed in Table II-4.1. (Note that the HNC results for ε_D of water and methanol are not converged results.) All the *B*-parameters for the different dipolar liquids are clustered around –3.2. The values do not depend on the closure relations employed either.

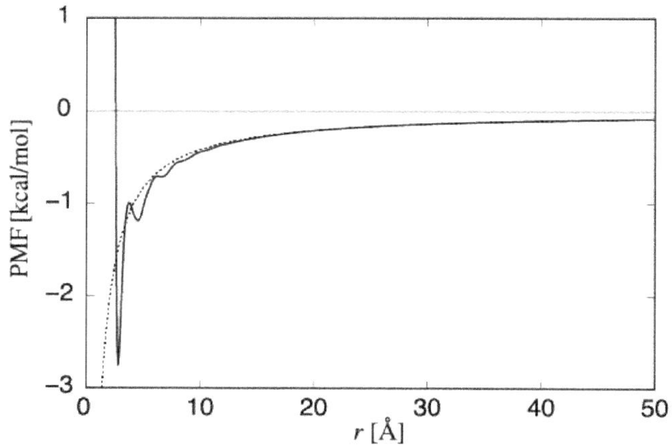

Fig. II-4.4: Potential of mean force between Na^+ and Cl^- in water: dotted line, $W_{ij}^{obs}(r)$; solid line, $W_{ij}(r)$ from XRISM/DB. Copyright (2019) Elsevier.

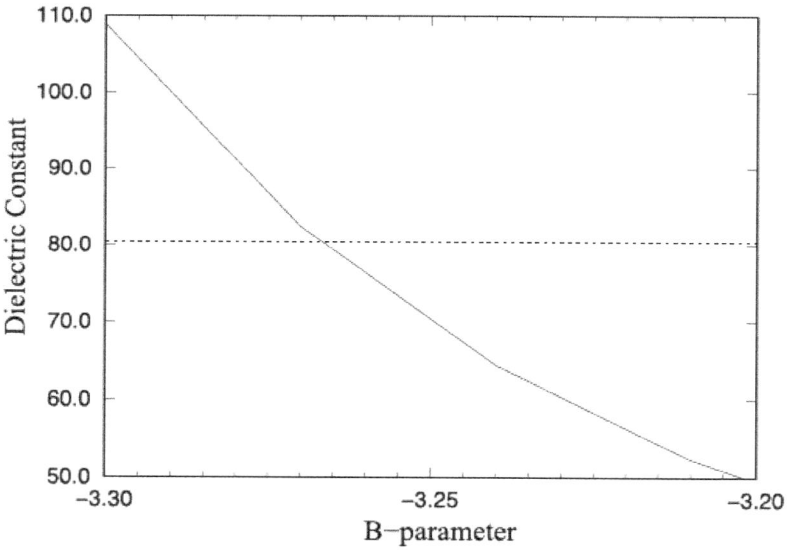

Fig. II-4.5: Dielectric constant calculated from XRISM-DB with changing parameter B. Dotted line indicates the phenomenological dielectric constant.

Table II-4.1: Dielectric Constants (ϵ_D) and B-parameter. Copyright (2019) Elsevier.

		HNC		KH	
entry	$\epsilon_D^{exp.}$	$\epsilon_D^{calc. (HNC)}$	B_{HNC}-parameter	$\epsilon_D^{calc. (KH)}$	B_{KH}-parameter
H_2O	80.4	52.3	−3.210	80.9	−3.268
MeCN	37.5	37.4	−3.055	37.4	−3.055
MeOH	32.7	28.4	−3.386	32.7	−3.412
DMSO	46.7	45.8	−3.026	46.7	−3.041

Those results indicate that the B-parameter is essentially a *universal* constant, although there is a minor variation reflecting the specificity of the polar liquids.

Two questions may be naturally raised. Why should the sign of B-parameter be negative, and what is the physics behind the universality of the parameter? Answer to the first question is simple. It is because the parameter reflects the sign of $\langle \hat{\mathbf{s}}_1 \cdot \hat{\mathbf{s}}_2 \rangle$ or $\langle \cos\theta \rangle$, in which θ denotes the angle between the two ideal dipoles in dipolar liquids, and $\langle \cdots \rangle$ represents the ensemble average [Hpye and Stell 1974, Friedman 1985]. Due to the dipole-dipole interactions, two dipole moments in the fluid tend to orient toward opposite directions in the average. That makes the B-parameter negative. So, in the gas phase in which the dipole-dipole interaction can be ignored, the B-parameter vanishes.

The second question may be non-trivial to be answered, but Ikuta and Hirata speculated that the parameter reflects the short-range correlation which is shared by any molecules with different geometry and charge distribution. As is clarified in Section II-1, the B-parameter in the Hoye-Stell theory reflects the short-range correlation between two ideal dipoles, which originates essentially from the repulsive core placed at the center of each molecule. The main role of such a correlation is to prevent the ideal dipoles from collapsing together. In the case of XRISM-DB, such a short-range correlation corresponding to the dipole-dipole interaction is not included explicitly in the equation, since it does not include any bridge diagrams corresponding to the core repulsion. Instead, the theory involves the core-repulsion among the interaction-sites or atoms. The core-repulsion among atoms may be working as a *surrogate* of the short-range correlation between ideal dipoles. There are some deviations from the universality, depending on the molecular species. The deviation may be reflecting the specificity in the geometry and the higher order moment of the charge distribution of molecular species, that is not counted by the correlation between the ideal dipoles.

Effect of the dipole bridge correction to the pair correlation functions (g(r)): Depicted in Fig. II-4.6 are the pair correlations functions (PCF) of water calculated using the XRISM/DB equation, compared with those from the XRISM/KH equation with the same molecular parameters. The results from the MD simulations are also plotted in the figure.

It has been well documented that the PCF of water shows three distinct features, reflecting the hydrogen-bond network in the liquid [Eisenberg and Kauzman 1969, Thissen and Narten 1982]. The first one is the first peak appearing in the PCF of the O-H pair, that probes the hydrogen-bond between two neighboring water molecules. The second feature is the position of the second peak in the O-O PCF at $r \sim 4.5$A, that is a manifestation of the tetrahedral coordination characteristic to ice. The third feature is the narrowing of the first peak in the O-O PCF, compared to simple liquids, which also indicates the tetrahedral coordination, or less coordinated molecules compared to the simple liquids. The results from XRISM-DB reproduce qualitatively such distinct features of water structure as well as the other methods. The new method apparently exaggerates the features of water structure manifested in $g_{\alpha\gamma}(r)$ compared to the other methods. However, those quantitative disagreements may be reconciled by tuning the potential energy parameters, which are characteristic to the method, for examples, the Lennard-Jones parameters for hydrogen atoms, and the partial charges which are concerned with not only the Coulomb interaction among atoms but also the ideal dipole-moment.

The result is also suggestive of a molecular picture behind the large dielectric constant exhibited by water. The dielectric constant of water is 80, which is unusually large for its dipole moment, ~ 1.8 Debye, compared to other polar liquids, for examples, acetonitrile: acetonitrile

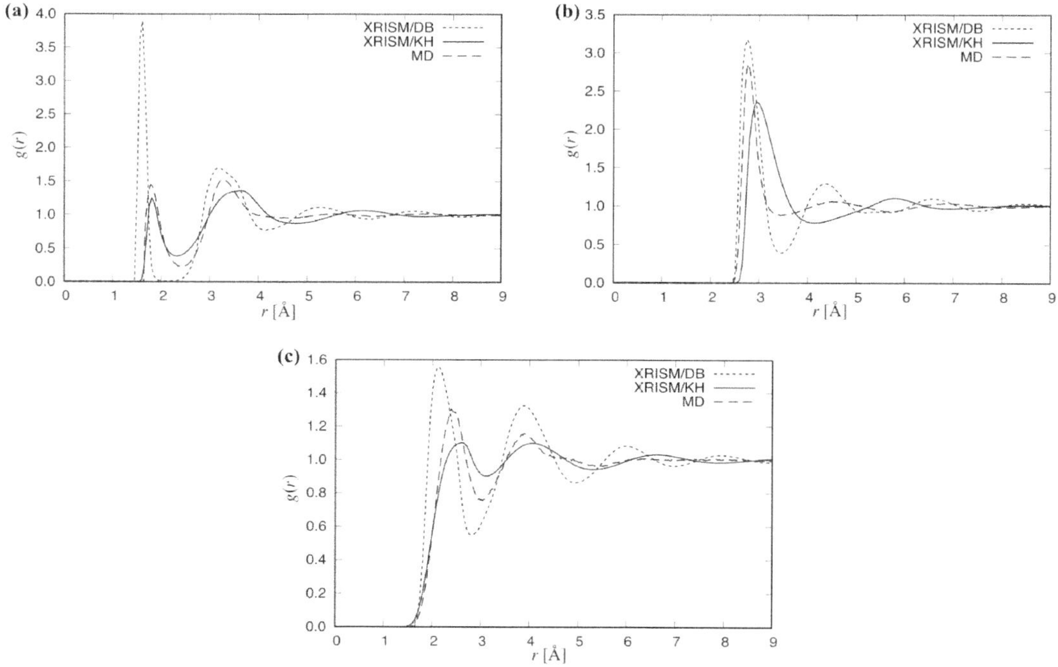

Fig. II-4.6: The pair correlation function of water: (a) O-H pair, (b) O-O pair, (c) H-H pair. Solid line, XRISM/KH; dotted line, XRISM-DB; dashed line, MD. Copyright (2019) Elsevier.

has rather large dipole moment, ~ 3.5 Debye, but its dielectric constant is around 38. Since main difference between the two liquids is whether they make the hydrogen-bonded structure or not, it will be reasonable to attribute the large dielectric constant exhibited by water to the hydrogen-bond network which is manifested in $g_{\alpha\gamma}(r)$ of water. The molecular picture for the dielectric constant of water, abstracted from the XRISM/DB theory, has some similarity with the model drawn naively by many authors in the past. The naive model attributes the large dielectric constant of water to the effective dipole moment of hydrogen-bonded clusters having the ice-like structure.

II-5 Electronic Structure of a Molecule in Solutions: RISM-SCF Theory

The electronic structure plays a crucial role in chemical reactions. No wonder so many theoretical studies have been devoted to clarify the electronic structure of a molecule based on the quantum chemistry. Most of those studies have been concerned with the electronic structure of a molecule isolated in *vacuum*. However, it is a common understanding in the field of synthetic organic chemistry that the reactivity is largely controlled by *solvent*. The experience by organic chemists suggests strongly that the electronic structure of a molecule in the real world should be evaluated in liquids or solutions. The statement applies also to chemical reactions due to enzyme taking place in living systems, although the reactions in that case are governed not only by the electronic structure but also by other factors such as the molecular recognition and structural fluctuation of protein. In the present section, we briefly touch a method to evaluate the electronic structure of a molecule in solutions (For detail of the method, consult the book edited by the author, Hirata 2003).

II-5.1 Hartree-Fock Equation

The RISM-SCF theory was developed in the framework of the so called *molecular orbital theory*. In the conventional molecular orbital theory for an isolated molecule in gas phase, the wave function of electrons are represented by the linear combination of the atomic orbitals (LCAO), satisfying Pauli's principle. Such representation of the electronic wave function converts the Schrodinger equation into

$$\mathcal{F}\psi_i(\mathbf{r}) = \varepsilon_i \psi_i(\mathbf{r}) \quad \text{(II-5.1)}$$

where $\psi_i(\mathbf{r})$ is the one particle orbital, and F, called the Fock operator, takes the following form

$$\mathcal{F} = -\frac{h^2}{2m_e}\nabla_i^2 + \sum_{\alpha=1}^{M}\frac{eZ_\alpha}{|\mathbf{r}_i - \mathbf{R}_\alpha|} + \sum_j^N\left(\mathcal{J}_j - \mathcal{K}_j\right) \quad \text{(II-5.2)}$$

where e denotes the electronic charge, m_e the mass of an electron, h the Planck constant, Z_α the charge of nucleus α, $\mathbf{r}_i$ position of the electron, $\mathbf{R}_\alpha$ and the position of α-th nucleus in the molecule, respectively. The last term represents the Coulombic and exchange operators defined by,

$$\mathcal{J}_j\psi_i(\mathbf{r}_1) = \int d\mathbf{r}_2\psi_j^*(\mathbf{r}_2)\frac{e^2}{|\mathbf{r}_1 - \mathbf{r}_2|}\psi_j(\mathbf{r}_2)\psi_i(\mathbf{r}_1)d\mathbf{r}_2$$

$$\mathcal{K}_j\psi_i(\mathbf{r}_1) = \int d\mathbf{r}_2\psi_j^*(\mathbf{r}_2)\frac{e^2}{|\mathbf{r}_1 - \mathbf{r}_2|}\psi_i(\mathbf{r}_2)\psi_j(\mathbf{r}_1)d\mathbf{r}_2 \quad \text{(II-5.3)}$$

If one expresses $\psi_i(\mathbf{r})$ by the linear combination of atomic orbital (LCAO), or,

$$\psi_i(\mathbf{r}) = \sum_\mu C_{i\mu}\chi_\mu(\mathbf{r}) \quad \text{(II-5.4)}$$

Equation (II-5.2) is converted into the eigen-value equation called the Rootharn equation.

$$\mathbf{FC} = \mathbf{SC}\,\varepsilon \quad \text{(II-5.5)}$$

where **F** and **S** are the matrices, elements of which are defined by,

$$F_{\mu\nu} = \int d\mathbf{x}\chi_\mu^*(\mathbf{r})\mathcal{F}^{(0)}\chi_\nu(\mathbf{r}) \quad \text{(Fock matrix)} \quad \text{(II-5.6)}$$

$$S_{\mu\nu} = \int d\mathbf{x}\chi_\mu^*(\mathbf{r})\chi_\nu(\mathbf{r}) \quad \text{(Overlap matrix)} \quad \text{(II-5.7)}$$

II-5.2 Solvated Fock Operator

In order to treat a molecule in solution, we define the solvated Fock operator by

$$\mathcal{F}_i^{solv} = \mathcal{F}_i - f_i \sum_{\lambda \in solute} V_\lambda b_\lambda \tag{II-5.8}$$

where $\mathcal{F}_i$ is the Fock operator of an isolated molecule defined by Eq. (II-5.2), and the second term represents the solvent effect on the electronic structure. V_λ represents the electrostatic potential produced by solvent molecules around the solute, or the *reaction field*, and b_λ denotes the so called *population operator* of the solute atoms, that determines the electronic charge density at each atom in the solute molecule.

In the historical development of the RISM-SCF theory, the reaction field V_λ was originally given by a heuristic consideration concerning the physics of the quantity as

$$V_{\lambda \in solute} = \rho \sum_{\alpha \in solvent} \int_V \frac{Z_\alpha}{r} g_{\lambda\alpha}(r) d\mathbf{r}\, d\mathbf{V} \tag{II-5.9}$$

where $r = |\mathbf{r}_\alpha - \mathbf{R}_\lambda|$, and ρ denotes the density of solvent [Tenno et al. 1993, 1994]. In the equation, $\rho g_{\alpha\lambda}(|\mathbf{r}_\alpha - \mathbf{R}_\lambda|)$ signifies the *local* density of solvent atom α at the position $\mathbf{r}_\alpha$ that is separated by r from the position of the solute atom λ, or $\mathbf{R}_\lambda$ (see Fig. II-5.1). Therefore, $\rho g_{\lambda\alpha}(|\mathbf{r}_\alpha - \mathbf{R}_\lambda|)d\mathbf{V}$ means the number of solvent atoms in the *volume element* $d\mathbf{V}$ at $\mathbf{r}_\alpha$. So, $(Z_\alpha/r)g_{\lambda\alpha}(r)d\mathbf{V}$ in the integrand is the electrostatic potential produced by the charge Z_α of the solvent atoms in the volume element $d\mathbf{r}$. Thus, $\int_V (Z_\alpha/r)g_{\lambda\alpha}(r)d\mathbf{V}$ signifies the electrostatic potential produced by all the solvent atoms in the system. Finally, by summing up the contribution from the different atoms in solvent, one gets the expression for the reaction field, or Eq. (II-5.9).

Few years later, the solvated Fock operator was derived by Sato et al. based on the variational principle as follows [Sato et al. 1996].

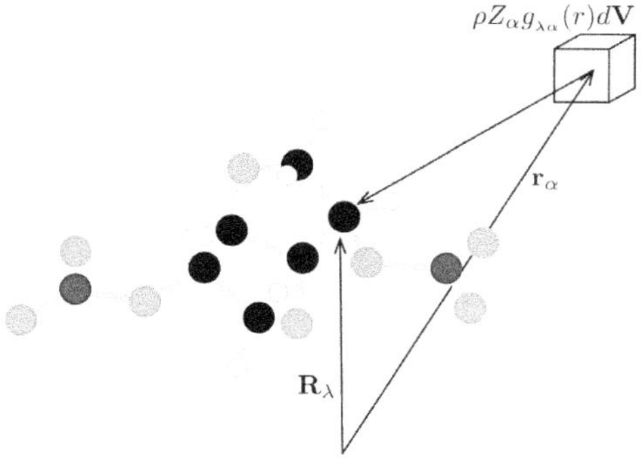

Fig. II-5.1: Illustration to show the reaction field produced by solvent molecules on an atom of solute.

II-5.3 The RISM-SCF Equation based on the Variational Principle

We define the Helmholtz free energy A of a solute molecule in solution as a sum of the electronic energy of the solute and the excess chemical potential as

$$\mathrm{A} = E_{solute} + \langle \Psi | \Delta\mu | \Psi \rangle \tag{II-5.10}$$

E_{solute} can be calculated by the *ab initio* electronic structure theory such as MCSCF.

$$\begin{aligned} E_{solute} &= E_{nuc} + E_{el} \\ &= \sum_{\lambda\eta} \frac{Z_\lambda Z_\eta}{R_{\lambda\eta}} + \sum_{ij} \gamma_{ij} h_{ij} + \frac{1}{2} \Gamma_{ijkl} \left(\phi_i \phi_j \middle| \phi_k \phi_l \right) \\ &= \sum_{\lambda\eta} \frac{Z_\lambda Z_\eta}{R_{\lambda\eta}} + \sum_{ij} \gamma_{ij} h_{ij} + \frac{1}{2} \Gamma_{ijkl} \left(\phi_i \middle| g_{kl} \middle| \phi_j \right) \end{aligned} \tag{II-5.11}$$

In the equation, the first term on the right hand side is the electrostatic interaction among nuclei of solute atoms, γ_{ij} and Γ_{ijkl} are the vector coupling coefficients,

$$\gamma_{ij} = \sum_{I,J} C_I C_J \gamma_{ij}^{IJ} \tag{II-5.12}$$

$$\Gamma_{ijkl} = \sum_{I,J} C_I C_J \Gamma_{ijkl}^{IJ} \tag{II-5.13}$$

with the CI coefficient, C_I. In Eq. (II-5.10), $\Delta\mu$ is the excess chemical potential of the solute molecule given in terms of the site-site pair correlation function, or Eq. (II-4.10), or

$$\Delta\mu = \rho k_B T \sum_{\lambda\in u} \sum_{\alpha\in v} \int d\mathbf{r} \left[\frac{1}{2} h_{\lambda\alpha}(r)^2 - c_{\lambda\alpha}(r) - \frac{1}{2} h_{\lambda\alpha}(r) c_{\lambda\alpha}(r) \right],$$

where $\lambda \in u$ and $\alpha \in v$ indicate that λ is an atom in solute, while α belongs to a solvent molecule. Therefore, $\langle \psi | \Delta\mu | \psi \rangle$ in Eq. (II-5.10) is a functional not only of the correlation functions $h_{\lambda\alpha}$ and $c_{\lambda\alpha}$, but also of the electronic wave function $|\psi\rangle$. The solvated Fock operator is obtained by solving the following variational principle with constraints for the orthonomality of wave functions or the configuration state functions (C) and one particle orbital (φ_i) (see Chapter I for the variational principle).

$$\delta(\mathbf{A}[\mathbf{c}, \mathbf{h}, \mathbf{t}, \mathbf{V}, \mathbf{C}] - [\text{constraints to orthonomarity}]) = 0 \tag{II-5.14}$$

where $\mathbf{t} = \mathbf{h} - \mathbf{c}$, and $\mathbf{V}$ is defined by Eq. (II-5.9). The variational principle leads,

$$\mathcal{F}_i^{solv} = \mathcal{F}_{ij} - \gamma_{ij} \sum_{\lambda\in solute} b_\lambda \frac{\partial}{\partial Z_\lambda} \left(-\frac{\rho}{\beta} \sum_{\alpha\lambda} \int d\mathbf{r} \exp\left[-\beta u_{\lambda\alpha}(r) + h_{\lambda\alpha}(r) - c_{\lambda\alpha}(r) \right] \right) \tag{II-5.15}$$

If classical Coulomb interactions are assumed for the electrostatic interactions among point charges between solute and solvent molecules, and the term concerning the CI coefficient (**C**) are omitted, the solvated Fock operator is reduced to Eq. (II-5.8).

The theory has been applied to a variety of chemical processes in solution, including solvatochromism [Tenno et al. 1994, Sato et al. 1998, Nishiyama et al. 2012], chemical reactions [Ishida et al. 1998, 1999a, Naka et al. 1999, Sato and Hirata 1999a, Harano et al. 2000a, b, Sato et al. 2003, 2004, Lee et al. 2006, Yoshida et al. 2008], acid-base equilibrium [Kawata et al. 1995, 1996, Sato and Hirata 1999b, c], solvation dynamics of excited state [Ishida et al. 1999b], and so on. Later in Chapter IV, we briefly review one of those works devoted to the hydrolysis reaction of ATP and its analogue as an example of applications of the RISM-SCF equation. (The actual calculation was carried out using the spatial distribution function obtained from the 3D-RISM equation, which will be explained in Chapter IV.)

The Literature

The formulation of the simple liquid in Section I-1 is based on a note of lectures by K. Hiroike delivered at Hokkaido University in 1971, which was given based on his paper [Morita and Hiroike 1961]. Although the lecture and his paper are complete in the sense that it includes both the functional differentiation and graphical approaches, the latter is entirely skipped here. For those who are interested in developing the liquid state theory, for example, beyond the HNC approximation, the book by Hansen and McDonald (1986) is recommended, since it includes the entire theory included in the paper by Morita and Hiroike. The definition of the density field in terms of the Dirac delta-function, and its application to deriving the statistical-mechanical formula of the density distribution functions were taken from the book by Harashima (1954). The definition makes the physical meaning of the distribution functions transparent.

APPENDICES

Appendix A Algebra of the Chain Sum

In the chapter, we have introduced a mathematical device called "chain sum," that facilitates the solution of the Ornstein-Zernike-type equation. The chain sum is defined by the sum of series diagrams

$$C\left[A\middle|\rho\right] \equiv$$ 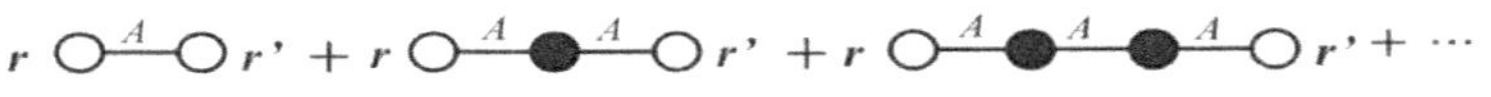 (A.1)

where $A(\mathbf{r}, \mathbf{r}'')$, $B(\mathbf{r}'', \mathbf{r}')$ are functions to be integrated, and the black circles represent the convolution integral defined as

 $$= \rho \int A(\mathbf{r},\mathbf{r}'')B(\mathbf{r}'',\mathbf{r}')d\mathbf{r}'' \tag{A.2}$$

The Ornstein-Zernike equation can be expressed in terms of the chain-sum notation as

$$h(r) = C\left[c\middle|\rho\right] \tag{A.3}$$

where c denotes the direct correlation function.

When one is trying to solve the Ornstein-Zernike equation for some liquids or solutions, there is a case in which the integrand includes two different functions, say A and B: for instance, in the electrolyte solution case, it includes functions representing a short-range and the Coulomb interactions. The renormalization of such composite functions is usually quite tedious. However, such renormalization is facilitated by the following theorems given by Dale and Rossky [Rossky and Dale 1980].

[Theorem 1]

$$C\left[A+B\middle|\rho\right] = C\left[A\middle|\rho\right] + C\left[B\middle|\overline{C\left[A\middle|\rho\right]}\right] \tag{A.4}$$

where $\overline{C\left[A\middle|\rho\right]}$ is called "hyper vertex" function defined as

$$\overline{C\left[A\middle|\rho\right]} \equiv \rho\delta(\mathbf{r}-\mathbf{r}') + \rho^2 C\left[A\middle|\rho\right] \tag{A-5}$$

and $\delta(\mathbf{r} - \mathbf{r}')$ is the Dirac delta function.

[Proof]

It is easy to see that the left-hand-side of Eq. (4) can be expanded into

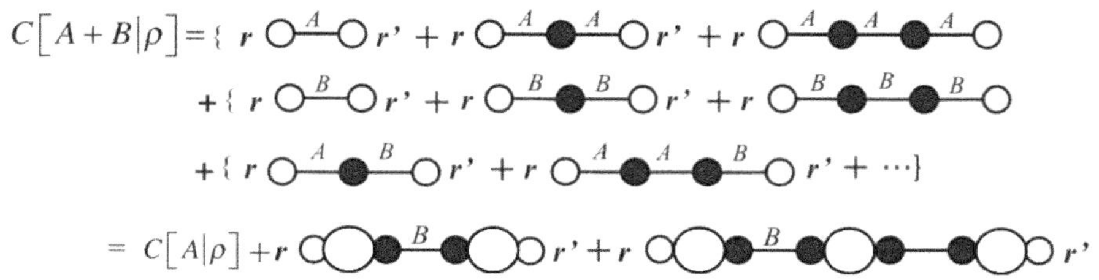

(A.6)

where

 $=\overline{C[A|\rho]}$

The diagrams in the first equality correspond to all the series diagrams produced by A and B with arbitrary combination. In the second equality, $C[A|\rho]$ corresponds to all the diagrams in the first parenthesis in the right hand side. The diagrams with the hyper-vertex in the second equality correspond to the sum of the all diagrams in the second and third parenthesis in the first equality. That can be readily verified by substituting (A.2) into (A.3), and apply the theorem of the Dirac delta-function. The chain sum of B with the hypervertex function $C[A|\rho]$ is nothing but the second term of Eq. (A.1). That proves Theorem 1.

As a corollary of Theorem 1, the following relation can be readily verified.

$$
\begin{aligned}
C'[A+B|\rho] &= C[A+B|\rho]-(A+B) \\
&= C'[A|\rho]+C'\left[B\middle|\overline{C[A|\rho]}\right]
\end{aligned}
\tag{A.7}
$$

[Theorem 2]

$$
C\left[A+B\middle|\bar{D}\right]=C\left[A\middle|\bar{D}\right]+C\left[B\middle|\overline{D+C\left[A\middle|\bar{D}\right]}\right] \tag{A.8}
$$

[Proof]

To prove the theorem, first expand the left-hand side of Eq. (A.5) into the series diagrams with hyper vertex $\bar{D}$ as

$C\left[A+B\middle|\bar{D}\right]$

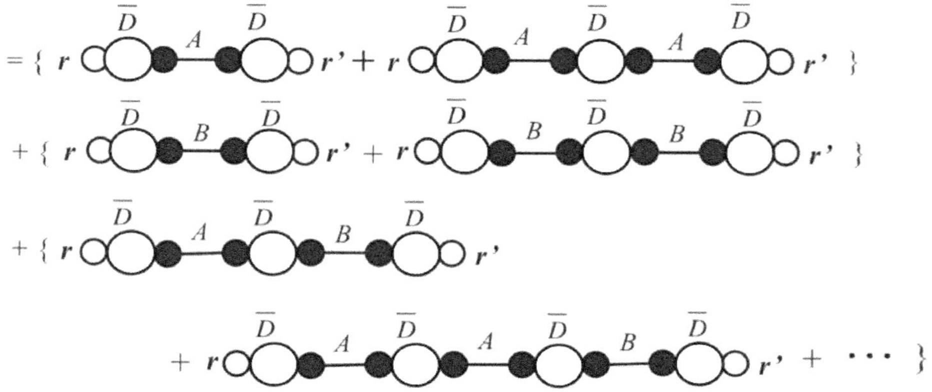

(A.9)

The diagrams in the first parenthesis in the right hand side correspond to $C\left[A\middle|\overline{D}\right]$ as can be readily verified. The sum of the diagrams in the second and third parenthesis is equal to $C\left[B\middle|\overline{D+C\left[A\middle|\overline{D}\right]}\right]$, namely

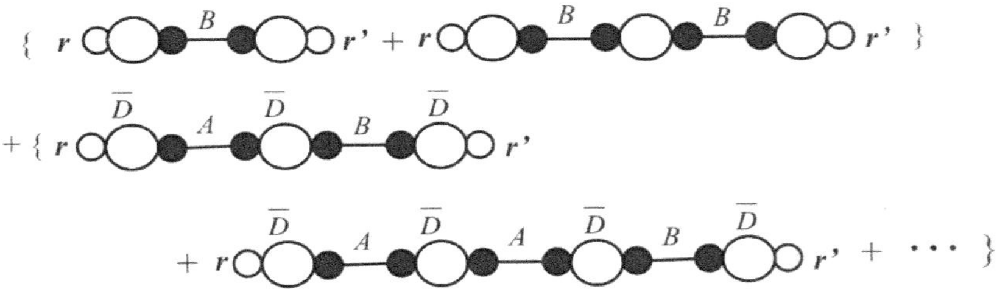

$$= C\left[B\middle|\overline{D+C\left[A\middle|\overline{D}\right]}\right]$$

Therefore, Theorem 2 was proved. As a corollary of the theorem, the following relation can be readily verified.

$$C'\left[A+B\middle|\overline{D}\right] = C'\left[A\middle|\overline{D}\right] + C'\left[B\middle|\overline{D+C\left[A\middle|\overline{D}\right]}\right] \qquad \text{(A.10)}$$

Appendix B Derivation of the RISM Equation

In this appendix, we apply the Chandler-Andersen transformation to Eq. (II-2.2)

$$h(1,2) = c(1,2) + \left(\frac{\rho}{\Omega}\right)\int c(1,3)h(3,2)d(3) \qquad \text{(B.1)}$$

with the *RISM approximation*, or Eq. (II-2.3),

$$c(1,2) = \sum_{\alpha=1}^{n}\sum_{\gamma=1}^{n} c\left(\left|\mathbf{r}_1^\alpha - \mathbf{r}_2^\gamma\right|\right) \qquad \text{(B.2)}$$

in order to derive the RISM equation [Chandler and Andersen 1972].

The left-hand-side (l.h.s) of Eq. (B.2) can be expanded into an infinite series of (ρ/Ω) by convoluting h in (l.h.s) into h in (r.h.s), iteratively, starting from $h(3, 2) = c(3, 2)$ as

$$\begin{aligned} h(1,2) &= c(1,2) + \left(\frac{\rho}{\Omega}\right)\int d(3)c(1,3)h(3,2) \\ &= c(1,2) + \left(\frac{\rho}{\Omega}\right)\int d(3)c(1,3)c(3,2) + \left(\frac{\rho}{\Omega}\right)^2 \int d(3)\int d(4)c(1,3)c(3,4)c(3,2) + \cdots \end{aligned} \qquad \text{(B.3)}$$

Now, we apply the Chandler-Andersen transformation,

$$f_{\alpha\gamma}(r) = \frac{1}{\Omega^2}\int f(1,2)\delta\left(\mathbf{R}_1 + \mathbf{l}_1^\alpha\right)\delta\left(\mathbf{R}_2 + \mathbf{l}_2^\gamma - \mathbf{r}\right)d(1)d(2),$$

to Eq. (B.3).

The (l.h.s) of (B.3) become obviously,

$$h_{\alpha\gamma}(r) = \frac{1}{\Omega^2}\int h(1,2)\delta\left(\mathbf{R}_1 + \mathbf{l}_1^{\alpha}\right)\delta\left(\mathbf{R}_2 + \mathbf{l}_2^{\gamma} - \mathbf{r}\right)d(1)d(2) \tag{B.4}$$

We express the (r.h.s) of the transformation as

$$h_{\alpha\gamma}(r) = \sum_{i=1} I_i \tag{B.5}$$

where I_i are written as

$$I_1 = \Omega^{-2}\int d(1)d(2)\delta(\mathbf{R}_1 + \mathbf{l}_1^{\alpha})\delta(\mathbf{R}_2 + \mathbf{l}_2^{\gamma} - \mathbf{r})c(1,2) \tag{B.6}$$

$$I_2 = \Omega^{-2}\int d(1)d(2)\delta(\mathbf{R}_1 + \mathbf{l}_1^{\alpha})\delta(\mathbf{R}_2 + \mathbf{l}_2^{\gamma} - \mathbf{r})\left(\frac{\rho}{\Omega}\right)\int d(3)c(1,3)c(3,2) \tag{B.7}$$

$$I_3 = \Omega^{-2}\int d(1)d(2)\delta(\mathbf{R}_1 + \mathbf{l}_1^{\alpha})\delta(\mathbf{R}_2 + \mathbf{l}_2^{\gamma} - \mathbf{r})\left(\frac{\rho}{\Omega}\right)^2\int d(3)d(4)c(1,3)c(3,4)c(4,2) \tag{B.8}$$

....

With the RISM approximation, Eq. (B.6) is written as

$$I_1 = \Omega^{-2}\int d(1)d(2)\delta(\mathbf{R}_1 + \mathbf{l}_1^{\alpha})\delta(\mathbf{R}_2 + \mathbf{l}_2^{\gamma} - \mathbf{r})\sum_{\eta,\nu} c_{\eta\nu}\left(\left|\mathbf{r}_1^{\eta} - \mathbf{r}_2^{\nu}\right|\right) \tag{B.9}$$

Let us express $c_{\eta\nu}(r)$ in terms of its Fourier transform as,

$$c_{\eta\nu}\left(\left|\mathbf{r}_1^{\eta} - \mathbf{r}_2^{\nu}\right|\right) = (2\pi)^{-3}\int d\mathbf{k}\tilde{c}_{\eta\nu}(\mathbf{k})\exp\left[i\mathbf{k}\cdot(\mathbf{r}_1^{\eta} - \mathbf{r}_2^{\nu})\right] \tag{B.10}$$

Then, (B.9) can be manipulated as follows

$$\begin{aligned}
I_1 =\ & \Omega^{-2}\int d\mathbf{R}_1 d\Omega_1 \int d\mathbf{R}_2 d\Omega_2 \delta(\mathbf{R}_1 + \mathbf{l}_1^{\alpha})\delta(\mathbf{R}_2 + \mathbf{l}_2^{\gamma} - \mathbf{r})\sum_{\eta,\nu}(2\pi)^{-3}\int \tilde{c}_{\eta\nu}(\mathbf{k})\exp\left[i\mathbf{k}\cdot(\mathbf{r}_1^{\eta} - \mathbf{r}_2^{\nu})\right] \\
=& \sum_{\eta,\nu}(2\pi)^{-3}\int d\mathbf{k}\tilde{c}_{\eta\nu}(\mathbf{k})\Omega^{-2}\int d\Omega_1 d\Omega_2 \exp\left[i\mathbf{k}\cdot(\mathbf{l}_1^{\eta} - \mathbf{l}_2^{\nu})\right]\int d\mathbf{R}_1\delta(\mathbf{R}_1 + \mathbf{l}_1^{\alpha})\exp\left[i\mathbf{k}\cdot\mathbf{R}_1\right]\times \\
& \int d\mathbf{R}_2\delta(\mathbf{R}_2 + \mathbf{l}_2^{\gamma} - \mathbf{r})\exp\left[-i\mathbf{k}\cdot\mathbf{R}_2\right] \\
=& \sum_{\eta,\nu}(2\pi)^{-3}\int d\mathbf{k}\tilde{c}_{\eta\nu}(\mathbf{k})\Omega^{-2}\int d\Omega_1 d\Omega_2 \exp\left[i\mathbf{k}\cdot(\mathbf{l}_1^{\eta} - \mathbf{l}_2^{\nu})\right]\exp\left[-i\mathbf{k}\cdot\mathbf{l}_1^{\alpha}\right]\exp\left[-i\mathbf{k}\cdot(-\mathbf{l}_2^{\gamma} + \mathbf{r})\right] \\
=& (2\pi)^{-3}\int d\mathbf{k}\sum_{\eta,\nu}\tilde{c}_{\eta\nu}(\mathbf{k})\Omega^{-2}\int d\Omega_1 d\Omega_2 \exp\left[-i\mathbf{k}\cdot(\mathbf{l}_1^{\alpha} - \mathbf{l}_1^{\eta})\right]\exp\left[-i\mathbf{k}\cdot(\mathbf{l}_2^{\nu} - \mathbf{l}_2^{\gamma})\right]\exp\left[-i\mathbf{k}\cdot\mathbf{r}\right] \\
=& (2\pi)^{-3}\int d\mathbf{k}\sum_{\eta,\nu}\tilde{c}_{\eta\nu}(\mathbf{k})\Omega^{-1}\left\{\int d\Omega_1 \exp\left[-i\mathbf{k}\cdot(\mathbf{l}_1^{\alpha} - \mathbf{l}_1^{\eta})\right]\right\}\left\{\Omega^{-1}\int d\Omega_2 \exp\left[-i\mathbf{k}\cdot(\mathbf{l}_2^{\nu} - \mathbf{l}_2^{\gamma})\right]\right\}\exp\left[-i\mathbf{k}\cdot\mathbf{r}\right] \\
=& (2\pi)^{-3}\int d\mathbf{k}\sum_{\eta,\nu}\tilde{\omega}_{\alpha\eta}(\mathbf{k})\tilde{c}_{\eta\nu}(\mathbf{k})\tilde{\omega}_{\nu\gamma}(\mathbf{k})\exp\left[-i\mathbf{k}\cdot\mathbf{r}\right]
\end{aligned}$$

In the last step of manipulation, we have defined $\tilde{\omega}_{\alpha\gamma}(\mathbf{k})$ as

$$\tilde{\omega}_{\alpha\gamma}(\mathbf{k}) \equiv \Omega^{-1}\int d\Omega_1 \exp\left[-i\mathbf{k}\cdot(\mathbf{l}_1^{\alpha} - \mathbf{l}_1^{\eta})\right] \tag{B.11}$$

This is a quite important consequence of the Chandler-Andersen transformation. We will look at the physical meaning of the function $\omega_{\alpha\gamma}$ later. In any case, we find the first term of the transformation in matrix form as

$$I_1 = (2\pi)^{-3} \int d\mathbf{k} \left[\tilde{\mathbf{w}}(\mathbf{k}) \tilde{\mathbf{c}}(\mathbf{k}) \tilde{\mathbf{w}}(\mathbf{k}) \right]_{\alpha\gamma} \exp[-i\mathbf{k} \cdot \mathbf{r}] \tag{B.12}$$

The transformation of the other terms in the series can be made basically in the same way after quite tedious manipulations. For example, the manipulation of the second term becomes,

$$\begin{aligned}
I_2 &= \Omega^{-2} \int d(1)d(2)\delta(\mathbf{R}_1 + \mathbf{l}_1^{\alpha})\delta(\mathbf{R}_2 + \mathbf{l}_2^{\gamma} - \mathbf{r}) \frac{\rho}{\Omega} \int d(3)c(1,3)c(3,2) \\
&= \Omega^{-2} \int d(1)d(2)\delta(\mathbf{R}_1 + \mathbf{l}_1^{\alpha})\delta(\mathbf{R}_2 + \mathbf{l}_2^{\gamma} - \mathbf{r}) \frac{\rho}{\Omega} \int d(3) \sum_{\eta,\nu} c_{\eta\nu}\left(\left|\mathbf{r}_1^{\eta} - \mathbf{r}_3^{\nu}\right|\right) \sum_{\xi,\varsigma} c_{\xi\varsigma}\left(\left|\mathbf{r}_3^{\xi} - \mathbf{r}_2^{\varsigma}\right|\right) \\
&= \frac{\rho}{\Omega} \int d(3)\Omega^{-2} \int d\mathbf{R}_1 d\Omega_1 \int d\mathbf{R}_2 d\Omega_2 \delta(\mathbf{R}_1 + \mathbf{l}_1^{\alpha})\delta(\mathbf{R}_2 + \mathbf{l}_2^{\gamma} - \mathbf{r}) \\
&\qquad \times \sum_{\eta,\nu} (2\pi)^{-3} \int \tilde{c}_{\eta\nu}(\mathbf{k}) \exp\left[i\mathbf{k} \cdot (\mathbf{r}_1^{\eta} - \mathbf{r}_3^{\nu})\right] d\mathbf{k} \sum_{\xi,\varsigma} (2\pi)^{-3} \int \tilde{c}_{\eta\nu}(\mathbf{k}') \exp\left[i\mathbf{k}' \cdot (\mathbf{r}_3^{\xi} - \mathbf{r}_2^{\varsigma})\right] d\mathbf{k}' \\
&= \sum_{\eta,\nu} \sum_{\xi,\varsigma} \frac{\rho}{\Omega} \int d(3)\Omega^{-2} \int d\Omega_1 \int d\Omega_2 \int d\mathbf{R}_1 \int d\mathbf{R}_2 \delta(\mathbf{R}_1 + \mathbf{l}_1^{\alpha})\delta(\mathbf{R}_2 + \mathbf{l}_2^{\gamma} - \mathbf{r}) \\
&\qquad \times (2\pi)^{-3} \int \hat{c}_{\eta\nu}(\mathbf{k}) \exp\left[i\mathbf{k} \cdot (\mathbf{r}_1^{\eta} - \mathbf{r}_3^{\nu})\right] d\mathbf{k} (2\pi)^{-3} \int \hat{c}_{\xi\varsigma}(\mathbf{k}') \exp\left[i\mathbf{k}' \cdot (\mathbf{r}_3^{\xi} - \mathbf{r}_2^{\varsigma})\right] d\mathbf{k}' \\
&= \sum_{\eta,\nu} \sum_{\xi,\varsigma} \frac{\rho}{\Omega} \int d(3)\Omega^{-2} \int d\Omega_1 \int d\Omega_2 \int d\mathbf{R}_1 \int d\mathbf{R}_2 \delta(\mathbf{R}_1 + \mathbf{l}_1^{\alpha})\delta(\mathbf{R}_2 + \mathbf{l}_2^{\gamma} - \mathbf{r}) \\
&\qquad \times (2\pi)^{-3} \int \tilde{c}_{\eta\nu}(\mathbf{k}) \exp\left[i\mathbf{k} \cdot \left\{(\mathbf{R}_1 + \mathbf{l}_1^{\eta} - (\mathbf{R}_3 + \mathbf{l}_3^{\nu})\right\}\right] d\mathbf{k} \\
&\qquad \times (2\pi)^{-3} \int \tilde{c}_{\xi\varsigma}(\mathbf{k}') \exp\left[i\mathbf{k}' \cdot \left\{(\mathbf{R}_3 + \mathbf{l}_3^{\xi} - (\mathbf{R}_2 + \mathbf{l}_2^{\varsigma})\right\}\right] d\mathbf{k}' \\
&= \sum_{\eta,\nu} \sum_{\xi,\varsigma} \frac{\rho}{\Omega} \int d(3)\Omega^{-2} \int d\Omega_1 \int d\Omega_2 (2\pi)^{-3} \int \tilde{c}_{\eta\nu}(\mathbf{k}) \exp\left[i\mathbf{k} \cdot \left\{(-\mathbf{l}_1^{\alpha} + \mathbf{l}_1^{\eta} - (\mathbf{R}_3 + \mathbf{l}_3^{\nu})\right\}\right] d\mathbf{k} \\
&\qquad \times (2\pi)^{-3} \int \tilde{c}_{\xi\varsigma}(\mathbf{k}') \exp\left[i\mathbf{k}' \cdot \left\{(\mathbf{R}_3 + \mathbf{l}_3^{\xi} - (-\mathbf{l}_2^{\gamma} + \mathbf{r} + \mathbf{l}_2^{\varsigma})\right\}\right] d\mathbf{k}' \\
&= \sum_{\eta,\nu} \sum_{\xi,\varsigma} \frac{\rho}{\Omega} \int d(3)(2\pi)^{-3} \int d\mathbf{k}\tilde{c}_{\eta\nu}(\mathbf{k})(2\pi)^{-3} \int d\mathbf{k}'\tilde{c}_{\xi\varsigma}(\mathbf{k}') \exp(-i\mathbf{k} \cdot \mathbf{R}_3) \exp(i\mathbf{k}' \cdot \mathbf{R}_3) \\
&\qquad \times \Omega^{-1} \int d\Omega_1 \exp\left[-i\mathbf{k} \cdot \left\{(\mathbf{l}_1^{\alpha} - \mathbf{l}_1^{\eta}) + \mathbf{l}_3^{\nu}\right\}\right] \Omega^{-1} \int d\Omega_2 \left[i\mathbf{k}' \cdot \left\{(\mathbf{l}_3^{\xi} - (\mathbf{l}_2^{\varsigma} - \mathbf{l}_2^{\gamma})\right\}\right] \exp(i\mathbf{k}' \cdot \mathbf{r}) \\
&= \sum_{\eta,\nu} \sum_{\xi,\varsigma} \frac{\rho}{\Omega} \int d(3)(2\pi)^{-3} \int d\mathbf{k}\tilde{c}_{\eta\nu}(\mathbf{k})(2\pi)^{-3} \int d\mathbf{k}'\tilde{c}_{\xi\varsigma}(\mathbf{k}') \exp(-i\mathbf{k} \cdot \mathbf{R}_3) \exp(i\mathbf{k}' \cdot \mathbf{R}_3) \\
&\qquad \times \exp\left(-i\mathbf{k} \cdot \mathbf{l}_3^{\nu} + i\mathbf{k}' \cdot \mathbf{l}_3^{\xi}\right) \Omega^{-1} \int d\Omega_1 \exp\left[-i\mathbf{k} \cdot (\mathbf{l}_1^{\alpha} - \mathbf{l}_1^{\eta})\right] \Omega^{-1} \int d\Omega_2 \left[-i\mathbf{k}' \cdot (\mathbf{l}_2^{\varsigma} - \mathbf{l}_2^{\gamma})\right] \\
&= \sum_{\eta,\nu} \sum_{\xi,\varsigma} \frac{\rho}{\Omega} \int d(3)(2\pi)^{-3} \int d\mathbf{k}\tilde{c}_{\eta\nu}(\mathbf{k})(2\pi)^{-3} \int d\mathbf{k}'\tilde{c}_{\xi\varsigma}(\mathbf{k}') \exp(-i\mathbf{k} \cdot \mathbf{R}_3) \exp(i\mathbf{k}' \cdot \mathbf{R}_3) \\
&\qquad \times \exp\left(-i\mathbf{k} \cdot \mathbf{l}_3^{\nu} + i\mathbf{k}' \cdot \mathbf{l}_3^{\xi}\right) \omega_{\alpha\eta}(\mathbf{k}) \omega_{\varsigma\gamma}(\mathbf{k}') \exp(i\mathbf{k}' \cdot \mathbf{r}) \\
&= \sum_{\eta,\nu} \sum_{\xi,\varsigma} (2\pi)^{-3} \int d\mathbf{k}\tilde{c}_{\eta\nu}(\mathbf{k})(2\pi)^{-3} \int d\mathbf{k}'\tilde{c}_{\xi\varsigma}(\mathbf{k}') \frac{\rho}{\Omega} \int d\Omega_3 \int d\mathbf{R}_3 \exp(-i\mathbf{R}_3 \cdot (\mathbf{k} - \mathbf{k}')) \\
&\qquad \times \exp\left(-i\mathbf{k} \cdot \mathbf{l}_3^{\nu} + i\mathbf{k}' \cdot \mathbf{l}_3^{\xi}\right) \tilde{\omega}_{\alpha\eta}(\mathbf{k}) \tilde{\omega}_{\varsigma\gamma}(\mathbf{k}') \exp(i\mathbf{k}' \cdot \mathbf{r}) \\
&= \sum_{\eta,\nu} \sum_{\xi,\varsigma} (2\pi)^{-3} \int d\mathbf{k}\tilde{c}_{\eta\nu}(\mathbf{k})(2\pi)^{-3} \int d\mathbf{k}'\tilde{c}_{\xi\varsigma}(\mathbf{k}') \frac{\rho}{\Omega} \int d\Omega_3 \delta(\mathbf{k} - \mathbf{k}') \exp\left(-i\mathbf{k} \cdot \mathbf{l}_3^{\nu} + i\mathbf{k}' \cdot \mathbf{l}_3^{\xi}\right) \\
&\qquad \times \tilde{\omega}_{\alpha\eta}(\mathbf{k}) \tilde{\omega}_{\varsigma\gamma}(\mathbf{k}') \exp(i\mathbf{k}' \cdot \mathbf{r}) \\
&= \sum_{\eta,\nu} \sum_{\xi,\varsigma} (2\pi)^{-3} \int d\mathbf{k}\tilde{c}_{\eta\nu}(\mathbf{k})\tilde{c}_{\xi\varsigma}(\mathbf{k}) \frac{\rho}{\Omega} \int d\Omega_3 \exp\left(-i\mathbf{k} \cdot (\mathbf{l}_3^{\nu} - \mathbf{l}_3^{\xi}\right) \tilde{\omega}_{\alpha\eta}(\mathbf{k}) \tilde{\omega}_{\varsigma\gamma}(\mathbf{k}) \exp(i\mathbf{k} \cdot \mathbf{r}) \\
&= \sum_{\eta,\nu} \sum_{\xi,\varsigma} (2\pi)^{-3} \rho \int d\mathbf{k}\tilde{c}_{\eta\nu}(\mathbf{k})\tilde{c}_{\xi\varsigma}(\mathbf{k}) \tilde{\omega}_{\nu\xi}(\mathbf{k}) \tilde{\omega}_{\alpha\eta}(\mathbf{k}) \tilde{\omega}_{\varsigma\gamma}(\mathbf{k}) \exp(i\mathbf{k} \cdot \mathbf{r})
\end{aligned}$$

In the matrix form, one finds,

$$I_2 = (2\pi)^{-3} \int d\mathbf{k} \left[\rho \mathbf{w}(k)\mathbf{c}(k)\mathbf{w}(k)\mathbf{c}(k)\mathbf{w}(k) \right]_{\alpha\gamma} \exp[-i\mathbf{k}\cdot\mathbf{r}] \tag{B.13}$$

Summing up the infinite series of $I_i (i = 1, 2, \mathrm{L})$, one gets,

$$h_{\alpha\gamma}(r) = \frac{\rho}{(2\pi)^3} \int \left[\frac{\mathbf{w(k)c(k)w(k)}}{1 - \rho\mathbf{c(k)w(k)}} \right]_{\alpha\gamma} \exp(i\mathbf{k}\cdot\mathbf{r})\, d\mathbf{k} \tag{B.14}$$

This is an expression of the RISM equation.

Here, we come back to the physical meaning of $\tilde{\omega}_{\alpha\gamma}(\mathbf{k})$ in Eq. (B.11). By taking the inverse Fourier transform of Eq. (B.11), one finds,

$$\omega_{\alpha\gamma}(r) = \delta_{\alpha\gamma}\delta(r) + (1 - \delta_{\alpha\gamma})s_{\alpha\gamma}(r) \tag{B.15}$$

where $\delta_{\alpha\gamma}$ and $\delta(r)$ denote, respectively, the Cronekker and the Dirac delta-functions.
The $s_{\alpha\gamma}(r)$ is defined by

$$s_{\alpha\beta}(r) = \frac{1}{4\pi L_{\alpha\beta}^2} \delta(r - L_{\alpha\beta}) \tag{B.16}$$

where $L_{\alpha\gamma} (= |\mathbf{l}_i^{\alpha} - \mathbf{l}_i^{\gamma}|)$ is the distance between the atoms (or sites) α and γ in the *same* molecule, *i*. So, (B.15) indicates that $s_{\alpha\gamma}(r)$ has an infinitely high peak at $r = L_{\alpha\gamma}$. In other words, the distance between the two atoms in a molecule is fixed in the distance $L_{\alpha\gamma}$. Given the distance (or *length*) between all pairs of atoms in a molecule, the geometry or structure of the molecule is defined.

Appendix C Brief Review of the Hoye-Stell Theory of the Dielectric Constant

Let us first review briefly the theory of dielectric constant for dipolar fluid presented by Hoye and Stell [Hoye and Stell 1974] in terms of the renormalization of molecular Ornstein-Zernike (MOZ) equation.

The MOZ equation is written as

$$h(1,2) = c(1,2) + \eta \int c(1,2)\rho(3)h(3,2)d(3) \tag{C.1}$$

where $h(1,2)$ and $c(1,2)$ are the total and direct correlation functions, respectively, and $\eta = \rho / \Omega$. The equation can be expanded in terms of diagrammatic representation as

$h(1,2) =$  (C.2)

$$\equiv \mathcal{C}[c|\eta]$$

where the white circles (○) denote the coordinates of terminus molecules 1 and 2, and the black circles (●) represent the coordinates of molecules, over which the integration over the position and angle should be carried out. The diagrams with black circles are referred to as the *series diagrams*, and the sum of the infinite series is called "chain sum." [Morita-Hiroike] $\mathrm{C}[c|\eta]$ is a book-keeping device to represent the chain sum. The diagrams feature the convolution integrals at the black circles. By virtue of the convolution integral, the Fourier transform of each term converts the series into a simple power series of $\tilde{c}(\mathbf{k})$, provided that the integral over angular coordinates is possible. Actually, the integration for the case of the dipolar liquids has been carried out using the method of spherical invariant [Friedman].

Let us apply the equation to dipolar liquid in which a pair of molecules is interacting with the following type of potential function.

$$u(1,2) = u^*(1,2) + u_{DD}(1,2) \tag{C.3}$$

where $u^*(1,2)$ denotes the short-range interaction between molecules 1 and 2, and $u_{DD}(1,2)$ represents the dipole-dipole interaction defined by the following equation

$$u_{DD}(1,2) = \frac{1}{r^3} D(1,2) \tag{C.4}$$

where $D(1,2)$ is defined by the

$$D(1,2) = 3(\hat{\mathbf{s}}_1 \cdot \hat{\mathbf{r}})(\hat{\mathbf{s}}_2 \cdot \hat{\mathbf{r}}) - (\hat{\mathbf{s}}_1 \cdot \hat{\mathbf{s}}_2) \tag{C.5}$$

In the equation, $\hat{\mathbf{r}}$ denotes the unit vector of $\mathbf{r}$ separating the two ideal dipoles, and $\hat{\mathbf{s}}_1$ and $\hat{\mathbf{s}}_2$ represent the unit vectors defining the orientation of the two dipoles.

Now, let us renormalize Eq. (C.1) or Eq. (C.2) applying the technique developed by Dale-Rossky-Friedman (see Appendix A in this chapter]. First, we split the direct correlation function $c(1,2)$ into short-range part and that of the dipole-dipole interaction.

$$c(1,2) = c_s(1,2) + \phi_{DD}(1,2) \tag{C.6}$$

where ϕ_{DD} is defined by

$$\phi_{DD}(1,2) = -\beta u_{DD}(1,2) \tag{C.7}$$

The equation can be renormalized to give the following equation

$$\eta^2 h(1,2) = c_s(1,2) + q_{DD}(1,2) + \tau_{DD}(1,2) \tag{C.8}$$

where $q_{DD}(1,2)$ and $\tau_{DD}(1,2)$ are defined, respectively, by

$$q_{DD}(1,2) \equiv \mathrm{C}\left[\phi_{DD}\middle|\eta\right] \tag{C.9}$$

and

$$\tau_{DD}(1,2) \equiv \mathrm{C}\left[c_s\middle|\overline{q}_{DD}\right] - c_s(1,2) \tag{C.10}$$

The equation just derived is essentially the same equation given by Hoye and Stell, or Eq. (31) in [Hoye-Stell], except for the notation and the level of renormalization. ($c_s(1,2)$ and $\phi_{DD}(1,2)$ correspond, respectively, to $W(1,2)$ and $V(1,2)$ in [Hoye-Stell]. $V(1,2)$ or $\phi_{DD}(1,2)$ is further renormalized here to produce $q_{DD}(1,2)$). Using q_{DD}, the renormalized interaction between two molecules can be expressed by

$-\beta w(1,2) =$  (C.11)

In the diagrams, the black circles represent positional as well as orientational coordiates of the ideal dipoles to be integrated, and the white circles denote the coordinates of either ions or dipoles. The solid line connecting the circles represent either ion-ion, or ion-dipole, or dipole-

dipole interaction depending on the kind of circles at terminus, ion or dipole, and ~~~ denotes $h(1,2)$. Hoye and Stell carried out the convolution integral in q_{DD} and τ_{DD} in Eq. (C.11) using the method of spherical invariance [Friedman], and obtained a general expression for the screening constant, or dielectric constant, which are common to the ion-ion, ion-dipole, and dipole-dipole interactions, that reads

$$\varepsilon_{CC} = \varepsilon_{CD} = \varepsilon_{DD} = \frac{1+2z}{1-z} \tag{C.12}$$

In Eq. (C.12), z is defined by

$$z = y\left(1 + \frac{1}{3}B\right) \tag{C.13}$$

$y \equiv 4\pi\beta\rho\mu^2/9$, and B is a constant reflecting the contribution from the short-range interaction to the correlation between the two ideal dipoles [Hoye-Stell]. The constant is related to $\langle \hat{\mathbf{s}}_1 \cdot \hat{\mathbf{s}}_2 \rangle$ in Eq. (C.5). If one ignores the short-range correlation to the dipole-dipole interaction, the contributions form $c_s(1,2)$ and $\tau_{DD}(1,2)$ vanish, and it makes $B = 0$, that gives the following expression for the dielectric constant of dipolar liquids in which the interaction between a pair molecules is expressed by Eq. (C.4).

$$\varepsilon_{CC} = \varepsilon_{CD} = \varepsilon_{DD} = \frac{1+2y}{1-y} \tag{C.14}$$

It is important to note that the infinite sum of the *series* diagrams concerning the dipole-dipole interaction, or q_{DD}, makes the most essential contribution to the dielectric constant.

CHAPTER III

Dynamics of Liquids and Solutions

In the previous chapter, we have studied theories to describe equilibrium properties of molecular liquids and solutions. There, the density pair correlation functions play a central role in describing the structure and thermodynamic properties of liquids and solutions. In the present chapter, we derive equations to describe dynamics of molecular liquids and solutions. The main theme of the chapter is to formulate a variety of time correlation functions based on the first principle, or the intermolecular interaction, from which all the phenomenological properties including transport coefficients and the time resolved spectroscopy. But, before doing so, we briefly review the phenomenological theory of Brownian motion.

III-1 Phenomenological Langevin Theory

A particle in a liquid system is in a random motion due to fluctuating force exerted by surrounding molecules. Such a motion is first observed by R. Brown in 1827 for a pollen particle in fluid. In order to describe the phenomenon, Langevin proposed an equation of motion, which bears his name, or the "Langevin equation." The equation is a conceptual extension of the Newtonian equation to a Brownian particle that reads,

$$m\frac{d\mathbf{v}}{dt} = -\zeta'\mathbf{v} + \mathbf{f}'(t) \tag{III-1.1}$$

The force acting on the particle consists of two parts: a random force due to the thermal motion of surrounding particles, $\mathbf{f}'(t)$, and the frictional force proportional to the velocity of the particle, $\zeta'\mathbf{v}$. Since the random force bears stochastic characteristics, the equation is called "stochastic differential equation." It is convenient to simplify the equation with $\zeta' = m\zeta$ and $\mathbf{f}'(t) = m\mathbf{f}(t)$, that gives,

$$\frac{d\mathbf{v}}{dt} = -\zeta\mathbf{v} + \mathbf{f}(t) \tag{III-1.2}$$

We are now ready to solve the equation. But, what does it mean to "solve" such an equation that is *stochastic*, not *deterministic*, in nature? Since $\mathbf{f}(t)$ in the equation is stochastic, you never know *for sure* the velocity of the particle after some time t. Nevertheless, you may be able to find probability distribution of the particle having some velocity $\mathbf{v}(t)$, provided an initial velocity at time $t = 0$. So, the problem is to find the *probability distribution* of velocity of a Brownian particle at time t. But, how can it be done? Remember, to find the probability distribution is equivalent to find the *moments* of the probability distribution (see Chapter I). So, our task is to find the moments of the probability distribution of velocity, $P(\mathbf{v}, t; \mathbf{v}_0)$.

In order to solve the equation, we have to keep two boundary conditions in mind, at $t \to 0$ and $t \to \infty$. First of all, it is a premise that the velocity of the particle at $t = 0$ is known, that is the initial condition. In other words, the probability distribution should be sharply peaked at some initial velocity $\mathbf{v}_0$. The condition can be written in terms of the Dirac delta-function as,

$$P(\mathbf{v},t;\mathbf{v}_0) \to \delta(\mathbf{v}-\mathbf{v}_0) \text{ at } t \to 0 \qquad \text{(III-1.3)}$$

where $P(\mathbf{v}, t; \mathbf{v}_0)$ denotes the probability distribution of the velocity at time t. On the other hand, after long time has elapsed, the velocity of the particle is thermalized to satisfy the equilibrium condition, that is a Maxwellian distribution, or,

$$P(\mathbf{v},t;\mathbf{v}_0) \to \left(\frac{m}{2\pi kT}\right)^{3/2} \exp\left(-\frac{mv^2}{2kT}\right) \quad t \to \infty \qquad \text{(III-1.4)}$$

The formal solution of the equation can be obtained by integrating both sides of the equation with respect to time as

$$\mathbf{V} \equiv \mathbf{v}(t) - \mathbf{v}_0 e^{-\zeta t} = e^{-\zeta t}\int_0^t e^{\zeta s}\mathbf{f}(s)ds \qquad \text{(III-1.5)}$$

Now, let's find the nth moments of the distribution function $P(\mathbf{v}, t; \mathbf{v}_0)$. If one takes the ensemble average of the both sides of Eq. (III-1.5), the right hand side vanishes, since $\langle \mathbf{f}(s)\rangle = 0$ by definition. So, it gives the first moment of the velocity distribution as

$$\langle \mathbf{v}(t)\rangle = \mathbf{v}_0 e^{-\zeta t} \qquad \text{(III-1.6)}$$

Namely, the average velocity of the particle decays exponentially from the initial velocity, and the rate of decay is characterized by the friction constant. The second moment is obtained by taking the ensemble average of $\mathbf{V}^2$ as follows

$$\langle V^2\rangle = \langle v^2\rangle - v_0^2 e^{-2\zeta t} = e^{-2\zeta t}\int_0^t\int_0^t e^{\zeta(t'+t'')}\langle \mathbf{f}(t')\cdot\mathbf{f}(t'')\rangle dt'dt'' \qquad \text{(III-1.7)}$$

where the correlation function of the random forces at time t' and t'' can be written in terms of the time difference $|t' - t''|$ as

$$\langle \mathbf{f}(t')\mathbf{f}(t'')\rangle = \phi_t\left(|t'-t''|\right) \qquad \text{(III-1.8)}$$

For a Brownian particle, $\phi(t)$ is assumed to decay in infinitesimally short time, and it can be expressed by a Dirac delta-function as, $\phi(t) = (\tau/2)\delta(t)$, where $\delta(t)$ is the Dirac delta-function. In order to perform the integral in Eq. (III-1.7), we change the variable as $t' + t'' = x$ and $t' + t'' = y$. Then, we get,

$$\langle V^2\rangle = \frac{1}{2}e^{-2\zeta t}\int_0^{2t} e^{\zeta x}dx\int_{-\infty}^{\infty}\phi_t(y)dy \qquad \text{(III-1.9)}$$

The integration concerning $\phi(t) = (\tau/2)\delta(t)$ gives a constant τ, and the integration with respect to x can be readily carried out to gives rise to

$$\langle V^2 \rangle = \frac{\tau}{2\zeta}(1 - e^{-2\zeta\tau}) \tag{III-1.10}$$

We can evaluate τ by realizing that equipartition $(1/2)m\langle V^2\rangle = (3/2)kT$ must apply when $t \rightarrow \infty$, or in equilibrium.

$$\frac{3kT}{m} = \frac{\tau}{2\zeta} \tag{III-1.11}$$

That gives,

$$\langle V^2 \rangle = \frac{3kT}{m}\left(1 - e^{-2\zeta t}\right) \tag{III-1.12}$$

Using Eq. (III-1.7), we may write the equation as follows

$$\langle v^2 \rangle = \frac{3kT}{m} + \left(v_0^2 - \frac{3kT}{m}\right)e^{-2\zeta t} \tag{III-1.13}$$

The equation shows how the average kinetic energy of a Brownian particle approaches its equilibrium value, or the equipartition law. In the same way, all the higher order moment of the probability distribution of V can be produced to give

$$\langle V^{2n+1} \rangle = 0 \tag{III-1.14}$$

$$\langle V^{2n} \rangle = 1 \cdot 3 \cdot 5 \cdots (2n-1)\langle V^2 \rangle^n \tag{III-1.15}$$

The results indicate that all the odd moments of V disappear, and all the even moments are factorized into the second moment. Such behaviors of the moments of the probability distribution is that characteristic to a Gaussian distribution. So, it can be written as

$$P(\mathbf{v},t;\mathbf{v}_0) = \left[\frac{m}{2\pi kT(1-e^{-2\zeta t})}\right]^{3/2} \exp\left[\frac{m\left|\mathbf{v} - \mathbf{v}_0 e^{-\zeta t}\right|^2}{2\pi kT(1-e^{-2\zeta t})}\right] \tag{III-1.16}$$

The probability distribution function of the displacement vector $\mathbf{r}$ can be obtained in a similar way. The displacement of the Brownian particle can be obtained from the velocity $\mathbf{v}$ by an integration with respect to time, or,

$$\mathbf{r} - \mathbf{r}_0 = \int_0^t \mathbf{v}(t')dt' \tag{III-1.17}$$

Substituting Eq. (III-1.5) in to the equation,

$$\mathbf{r} - \mathbf{r}_0 = \int_0^t dt'\left[\mathbf{v}_0 e^{-\zeta t'} + \int_0^{t'} dt'' e^{\zeta t''}\mathbf{f}(t'')\right],$$

or

$$\mathbf{r} - \mathbf{r}_0 - \zeta^{-1}\mathbf{v}_0\left(1 - e^{-\zeta t}\right) = \int_0^t dt' e^{-\zeta t'} \int_0^{t'} dt'' e^{\zeta t''}\mathbf{f}(t''). \tag{III-1.18}$$

The integration on the right hand side, with respect to t', can be made by parts, to give,

$$\mathbf{r}-\mathbf{r}_0-\zeta^{-1}\mathbf{v}_0\left(1-e^{-\zeta t}\right)=\int_0^t\zeta^{-1}\left[1-e^{\zeta(t'-t)}\right]\mathbf{f}(t')dt'. \qquad \text{(III-1.19)}$$

Taking the ensemble average of both sides, one finds for the first moment of the displacement vector as,

$$\left\langle\mathbf{r}-\mathbf{r}_0\right\rangle=\zeta^{-1}\mathbf{v}_0\left(1-e^{-\zeta t}\right), \qquad \text{(III-1.20)}$$

where $\langle\mathbf{f}(t)\rangle=0$ is taken into account. The second moment, or the variance, can be obtained in a similar manner as

$$\left\langle\left|\mathbf{r}-\mathbf{r}_0\right|^2\right\rangle=\zeta^{-2}v_0^2\left(1-e^{-\zeta t}\right)^2+\frac{3kT}{m\zeta^2}\left(2\zeta t-3+4e^{-\zeta t}-e^{-2\zeta t}\right) \qquad \text{(III-1.21)}$$

The higher order moments can be found in a similar manner, and are proven to have characteristics of the Gaussian distribution. Therefore, we may write,

$$P\left(\mathbf{r},t;\mathbf{r}_0,\mathbf{v}_0\right)=\left[\frac{m\zeta^2}{2\pi kT(2\zeta t-3+4e^{-\zeta t}-e^{-2\zeta t})}\right]^{3/2}\times\exp\left\{-\frac{m\zeta^2\left|\mathbf{r}-\mathbf{r}_0-\zeta^{-1}\mathbf{v}_0\left(1-e^{-\zeta t}\right)^2\right|}{2kT(2\zeta t-3+4e^{-\zeta t}-e^{-2\zeta t})}\right\} \qquad \text{(III-1.22)}$$

It is interesting to look at the limiting behavior of the second moment, or the mean square displacement, at $t\longrightarrow 0$ and $t\longrightarrow\infty$. At $t\longrightarrow 0$,

$$\left\langle\left|\mathbf{r}-\mathbf{r}_0\right|^2\right\rangle\rightarrow\left|v_0^2\right|t^2 \qquad \text{(III-1.23)}$$

This behavior is expected when the particle is in a free motion with the velocity v_0. In a very short time after a Brownian particle has started to move, it will keep its initial velocity before it collides with surrounding molecules. On the other hand, at $t\longrightarrow\infty$, the mean square displacement behaves as,

$$\left\langle\left|\mathbf{r}-\mathbf{r}_0\right|^2\right\rangle=\frac{6kT}{m\zeta}t=6Dt \qquad \text{(III-1.24)}$$

Namely, the mean square displacement becomes proportional to time t that is characteristic of the diffusive motion, and the proportional constant D is the diffusion constant defined by

$$D\equiv\frac{kT}{m\zeta} \qquad \text{(III-1.25)}$$

The equation is an expression of the *fluctuation-dissipation* theorem found by A. Einstein.

III-2 Phase Space and Liouville Equation

In the previous section, we have treated the dynamics of a fluid system in terms of a Brownian motion, in which media surrounding a Brownian particle are regarded as continuum. The description, however, is too crude to explain life phenomena that involve water and biomolecules. Here, we review the fundamental equation to describe time evolution of dynamic variables in phase space, or the Liouville equation.

Let's think of a classical system consisting of N molecules. N is a large number, as many as $\sim 10^{23}$. Since the system is classical, there is no correlation between spatial and momentum coordinates of each molecule. Therefore, a dynamic state of the system can be uniquely defined by a point in the space spanned by $6N$ independent variables, $3N$ for the spatial coordinates, q_x, q_y, q_z for each molecule, and $3N$ for x, y, z-components of the momentum, p_x, p_y, p_z. (It should be noted that in a quantum system, the two variables of a molecule are not independent due to the Heisenberg uncertainty principle, or $\Delta x \Delta p_x > h$). The $6N$ dimensional space is called the "phase space," and a point in the space is referred to as "phase point," and denoted as $\Gamma_N = \{q_1, p_1, \cdots, q_{3N}, p_{3N}\}$.

We define the Hamiltonian of the system by

$$H = \sum_i^{3N} \frac{p_i^2}{2m_i} + U_N(\mathbf{q}_1, \mathbf{q}_2, \cdots, \mathbf{q}_N) \tag{III-2.1}$$

By taking the derivative of Eq. (III-2.1) with respect to the momentum and spatial coordinates, one gets the Hamilton equation of motion as (See Chapter I for the Hamilton equation of motion.)

$$\frac{\partial H}{\partial p_i} = \dot{q}_i, \qquad \frac{\partial H}{\partial q_i} = -\dot{p}_i \tag{III-2.2}$$

It is important to realize that all the mechanical properties, dynamic as well as static, depend just on the phase point Γ_N. Such properties in phase space are referred to as "dynamic variables." The Hamiltonian and its components, the kinetic and potential energies, the density fields described in Chapter II, are typical examples of the dynamic variables.

Now let us consider the derivative of a dynamic variable, A, with respect to time, that is,

$$\frac{dA}{dt} = \left(\frac{\partial A}{\partial t}\right)_{\Gamma_N} + \sum_i \left(\frac{\partial A}{\partial q_i}\dot{q}_i + \frac{\partial A}{\partial p_i}\dot{p}_i\right). \tag{III-2.3}$$

The first term on the right-hand-side vanishes, or

$$\left(\frac{\partial A}{\partial t}\right)_{\Gamma_N} = 0, \tag{III-2.4}$$

because the dynamic variable depends only on the phase Γ_N by definition. Therefore, one gets

$$\frac{dA}{dt} = -\sum_i \left(\frac{\partial H}{\partial q_i}\frac{\partial A}{\partial p_i} - \frac{\partial H}{\partial p_i}\frac{\partial A}{\partial q_i}\right) \equiv H^{\times} A, \tag{III-2.5}$$

where the Liouville operator $H^{\times}$ is defined as

$$H^{\times} \equiv -\sum_i^{3N} \left(\frac{\partial H}{\partial q_i}\frac{\partial}{\partial p_i} - \frac{\partial H}{\partial p_i}\frac{\partial}{\partial q_i}\right), \tag{III-2.6}$$

or in a more explicit form,

$$H^{\times} = \sum_{i}^{3N} \left(\frac{p_i}{m_i} \frac{\partial}{\partial q_i} + F_i \frac{\partial}{\partial p_i} \right), \tag{III-2.7}$$

where $F_i = -dH/dq_i = -dU/dq_i$. We refer Eq. (III-2.5) to as "Liouville equation of motion."

The Liouville equation of motion just derived is essentially the same as the Hamilton equation of motion or the Newton equation of motion. Therefore, a trajectory obtained from the equation is deterministic. Such equations have been employed by the molecular dynamics simulation to calculate the trajectory of a system typically consisting of $\sim 10^5$ to $\sim 10^6$ molecules. However, the system is far less than a real system that consists of $\sim 10^{23}$ molecules. The concern is not limited to the system size, but also to the *sampling* of the phase space. It is the probability distribution function that replaces the deterministic description by the stochastic one.

III-2.1 Liouville Theorem and Liouville Equation

Consider an ensemble of replicas of a system consisting of N molecules. Each replica of the ensemble can be represented by a point in the phase space. So, replicas make a distribution in the phase space, which varies with time. Let us denote the distribution of phase points by f_N. The time derivative of f_N is

$$\begin{aligned} \frac{df_N}{dt} &= \left(\frac{\partial f_N}{\partial t} \right)_{\Gamma_N} + \sum_{i=1}^{3N} \left(\frac{\partial f_N}{\partial q_i} \dot{q}_i + \frac{\partial f_N}{\partial p_i} \dot{p}_i \right) \\ &= \left(\frac{\partial f_N}{\partial t} \right)_{\Gamma_N} + H^{\times} f_N \end{aligned}. \tag{III-2.8}$$

It should be noted that the first term on the right-hand-side does not vanish because the population of replicas at a phase point varies with time. On the other hand, it can be proven that the left-hand-side vanishes, or,

$$\frac{df_N}{dt} = 0. \tag{III-2.9}$$

The relation is referred to as "Liouville theorem." The proof of the theorem is provided in the following.

The distribution of replicas in the phase space can be viewed as the density of fluid in which particles are not interacting with each other. For such a fluid system, the equation of continuity holds as

$$\left(\frac{\partial f_N}{\partial t} \right)_{\Gamma_N} = \nabla_{\Gamma_N} \cdot \mathbf{j}_{\Gamma_N}, \tag{III-2.10}$$

where ∇_{Γ_N} and $\mathbf{j}_{\Gamma_N}$ denote vectors of the differential operator in the phase space and the *flux* of phase points defined as,

$$\nabla_{\Gamma_N} \equiv \left(\frac{\partial}{\partial \mathbf{q}_1}, \frac{\partial}{\partial \mathbf{q}_2}, \cdots, \frac{\partial}{\partial \mathbf{q}_N}; \frac{\partial}{\partial \mathbf{p}_1}, \frac{\partial}{\partial \mathbf{p}_2}, \cdots, \frac{\partial}{\partial \mathbf{p}_N} \right), \tag{III-2.11}$$

and

$$\mathbf{j}_{\Gamma_N} \equiv \left(f_N\dot{\mathbf{q}}_1, f_N\dot{\mathbf{q}}_2, \cdots, f_N\dot{\mathbf{q}}_N; f_N\dot{\mathbf{p}}_1, f_N\dot{\mathbf{p}}_2, \cdots, f_N\dot{\mathbf{p}}_N\right). \tag{III-2.12}$$

Let us rewrite the equation of continuity (III-2.10) as

$$\left(\frac{\partial f_N}{\partial t}\right)_{\Gamma_N} = -\sum_i \left[\frac{\partial}{\partial q_i}(f_N\dot{q}_i) + \frac{\partial}{\partial p_i}(f_N\dot{p}_i)\right]. \tag{III-2.13}$$

Due to the property of the partial derivative, one finds from Eq. (III-2.2),

$$\frac{\partial \dot{q}_i}{\partial q_i} = \frac{\partial^2 H}{\partial q_i \partial p_i} = -\frac{\partial \dot{p}_i}{\partial p_i}. \tag{III-2.14}$$

Substituting the relation into Eq. (III-2.13) leads to

$$\left(\frac{\partial f_N}{\partial t}\right)_{\Gamma_N} = -\sum_i \left[\frac{\partial f_N}{\partial q_i}\dot{q}_i + \frac{\partial f_N}{\partial p_i}\dot{p}_i\right]. \tag{III-2.15}$$

This proves the Liouville theorem, or $df_N/dt = 0$.
Using the Liouville operator defined by Eq. (III-2.8), Eq. (III-2.15) can be written as

$$\left(\frac{\partial f_N}{\partial t}\right)_{\Gamma_N} = -H^\times f_N. \tag{III-2.16}$$

The equation is called "Liouville equation," which describes the time evolution of the distribution of phase points.

III-2.2 Ensemble Average of Dynamic Variables

In equilibrium, the distribution function in the phase space can be described by the Boltzmann distribution,

$$f_N^{eq}(\Gamma_N) = \frac{\exp[-\beta H(\Gamma_N)]}{\int \exp[-\beta H(\Gamma_N)] d\Gamma_N}. \tag{III-2.17}$$

The distribution function does not depend on time in the equilibrium as can be readily proved as,

$$\dot{f}_N^{eq}(\Gamma_N) = -H^\times f_N^{eq}(\Gamma_N) \propto -H^\times e^{-\beta H} = \beta e^{-\beta H} H^\times H = 0. \tag{III-2.18}$$

The formal solutions of the Liouville equation of motion and the Liouville equation are obtained by integrating the equations, (III-2.5) and (III-2.16) as

$$A(t) = e^{H^\times t} A(0), \tag{III-2.19}$$

and

$$f_N(\Gamma_N, t) = e^{-H^\times t} f_N(\Gamma_N, 0), \tag{III-2.20}$$

where $A(0)$ and $f_N(\Gamma_N, 0)$ denote the initial value of the dynamic variable and the distribution, respectively. The ensemble average of a dynamic variable D is defined as,

$$\langle D \rangle \equiv \int D(\Gamma_N) f_N^{eq}(\Gamma_N) d\Gamma_N \qquad \text{(III-2.21)}$$

III-2.3 Time Correlation Functions

Now, let us consider the ensemble average of a product of two dynamic variables A and B at different times, $t = 0$ and $t = t$, that is, $D \equiv A(0)B(t)$.

$$\begin{aligned} \langle D \rangle &= \langle A(0)B(t) \rangle \\ &= \left\langle A e^{H^{\times} t} B \right\rangle \\ &= \int f_N^{eq}(\Gamma_N) A(\Gamma_N) e^{H^{\times} t} B(\Gamma_N) d\Gamma_N \end{aligned} \quad . \qquad \text{(III-2.22)}$$

The ensemble average is referred to as the "time correlation function," that plays a central role in characterizing dynamic properties of solution, including biomolecular systems. The two dynamic variables can be either same or different. If $A = B$, the time correlation function is called the *auto*- or *self*-correlation function. On the other hand, in the case of $A \neq B$, it is called the *distinct*- or *joint*- or *cross*-correlation function.

The time correlation function has an important feature, that is, *stationary*, meaning that for any time t', one has

$$\langle A(t')B(t+t') \rangle = \langle A(0)B(t) \rangle = \langle A(-t)B(0) \rangle \qquad \text{(III-2.23)}$$

The equality can be written also in terms of the Liouville operator as, (Eq. (III-2.19)),

$$\left\langle A e^{tH^{\times}} B \right\rangle = \left\langle (e^{-tH^{\times}} A) B \right\rangle . \qquad \text{(III-2.24)}$$

In order to prove the theorem, let us derive the following relation first

$$\left\langle A H^{\times} B \right\rangle = -\left\langle (H^{\times} A) B \right\rangle \qquad \text{(III-2.25)}$$

The relation is derived as follows.

$$\begin{aligned} \left\langle A H^{\times} B \right\rangle &= \int d\Gamma_N f_N^{eq}(\Gamma_N) A H^{\times} B \\ &= -\int d\Gamma_N f_N^{eq}(\Gamma_N) A \dot{\Gamma}_N \cdot \frac{\partial}{\partial \Gamma_N} B \\ &= -\left\{ \left[f_N^{eq}(\Gamma_N) A \dot{\Gamma}_N B \right]_{-\infty}^{\infty} - \int d\Gamma_N \dot{\Gamma}_N \cdot \frac{\partial \left\{ f_N^{eq}(\Gamma_N) A \right\}}{\partial \Gamma_N} B \right\} \\ &= \int d\Gamma_N f_N^{eq}(\Gamma_N) \left\{ \dot{\Gamma}_N \cdot \frac{\partial A}{\partial \Gamma_N} \right\} B \\ &= -\int d\Gamma_N f_N^{eq}(\Gamma_N) (H^{\times} A) B \\ &= -\left\langle (H^{\times} A) B \right\rangle \end{aligned} \qquad \text{(III-2.26)}$$

In the second equality above, the Livouville operator $H^{\times}$ is expressed in its explicit form in terms of the coordinates of phase space, that is,

$$H^{\times} = \dot{\Gamma}_N \cdot \frac{\partial}{\partial \Gamma_N} \tag{III-2.27}$$

In the third equality, the integration over Γ_N is performed by parts, where the first term disappears due to the boundary condition at $\Gamma = \pm\infty$. In the fourth equality, we have used the obvious relation that implies the time independence of the equilibrium distribution function, that is,

$$H^{\times} f_N^{eq}(\Gamma_N) = \dot{\Gamma}_N \cdot \frac{\partial f_N^{eq}(\Gamma_N)}{\partial \Gamma_N} = 0. \tag{III-2.28}$$

Finally, Eq. (III-2.24) is derived as follows,

$$\left\langle A e^{tH^{\times}} B \right\rangle = \sum_n \frac{t^n}{n!} \left\langle A H^{\times n} B \right\rangle = \sum_n \frac{t^n}{n!} (-1)^n \left\langle (H^{\times n} A) B \right\rangle = \left\langle (e^{-tH^{\times}} A) B \right\rangle. \tag{III-2.29}$$

From Eq. (III-2.29), one finds,

$$\left\langle \dot{B}(t) A \right\rangle = -\left\langle B \dot{A}(-t) \right\rangle = -\left\langle B(t) \dot{A} \right\rangle, \tag{III-2.30}$$

and

$$\left\langle \dot{B} A \right\rangle = -\left\langle B \dot{A} \right\rangle. \tag{III-2.31}$$

These equations hold for the case of $B = A$, so we have,

$$\left\langle A \dot{A} \right\rangle = 0. \tag{III-2.32}$$

It implies that a dynamic variable and its time derivative does not correlate, or "orthogonal" in the phase space.

At this point, it will be noteworthy to mention some important properties of the Liouville operator $H^{\times}$ defined by Eqs. (III-2.6) or (III-2.7). The operator is "anti-Hermitian". It will be desirable to convert it into Hermitian in order to use good properties of the Hermitian operator; all the eigen values of a Hermitian operator are *real*. Let us define the new Liouville operator by

$$\mathcal{L} = -iH^{\times} = i \sum_{i=1}^{3N} \left[\frac{\partial H}{\partial q_i} \frac{\partial}{\partial p_i} - \frac{\partial H}{\partial p_i} \frac{\partial}{\partial q_i} \right]. \tag{III-2.33}$$

The Liouville equation of motion and the Liouville equation are rewritten in terms of the new operator as,

$$\frac{dA}{dt} = i\mathcal{L}A, \quad \text{and} \quad \frac{df_N}{dt} = -i\mathcal{L}f_N. \tag{III-2.34}$$

The new operator is Hermitian as is proven in the following.

$$\left\langle -i\mathcal{L}AB^* \right\rangle = -\left\langle A(-i\mathcal{L})^* B^* \right\rangle$$

$$-i\left\langle \mathcal{L}AB^* \right\rangle = -i\left\langle A\mathcal{L}^* B^* \right\rangle$$

$$\left\langle \mathcal{L}AB^* \right\rangle = \left\langle A(\mathcal{L}B)^* \right\rangle$$

III-3 Generalized Langevin Theory

Here, we introduce the generalized Langevin equation (GLE), which enables atomistic description of the dynamics of liquids and solutions. A generalized Langevin equation can be derived by defining a set of dynamic variables as a function of the representative point ($\Gamma(t)$) in the phase space (Hilbert space), and by projecting all other degrees of freedom onto those variables. Let a column vector **A** be the dynamic variable. Time evolution of a variable $\mathbf{A}(t)$ in the phase space can be described by the Liouville equation, derived in the previous section, or Eq. (III-2.34).

$$\frac{d\mathbf{A}(t)}{dt} = i\mathcal{L}\mathbf{A}(t), \tag{III-3.1}$$

where L is the Liouville operator defined by the following equation.

$$\mathcal{L} \equiv i\sum_i \left[\left(\frac{\partial H}{\partial \mathbf{r}_i}\right) \cdot \left(\frac{\partial}{\partial \mathbf{p}_i}\right) - \left(\frac{\partial H}{\partial \mathbf{p}_i}\right) \cdot \left(\frac{\partial}{\partial \mathbf{r}_i}\right) \right]. \tag{III-3.2}$$

GLE is obtained by projecting $\mathbf{A}(t)$ onto a vector **A**. An operator (P) for the projection is defined by the following equation,

$$\mathcal{P}\mathbf{A}(t) = \frac{\mathbf{A} \cdot (\mathbf{A}, \mathbf{A}(t))}{(\mathbf{A}, \mathbf{A})}, \tag{III-3.3}$$

or

$$\mathcal{P} = \frac{\mathbf{A} \cdot (\mathbf{A}, \ )}{(\mathbf{A}, \mathbf{A})}. \tag{III-3.4}$$

The projection operator satisfies the following relation as can be readily proved by operating P onto the both sides of Eq. (III-3.4).

$$\mathcal{P}\,\mathcal{P} = \mathcal{P} \tag{III-3.5}$$

In Eq. (III-3.3), $\mathbf{A}(t)$ is a variable in the phase space, and "•" means a matrix product, $(\cdots, \cdots)$ signifies the inner product of two vectors in the Hilbert space, defined by,

$$(\mathbf{A}, \mathbf{B}) = \langle \mathbf{A}\tilde{\mathbf{B}} \rangle = \int d\Gamma f^{eq}(\Gamma)\mathbf{A}(\Gamma)\tilde{\mathbf{B}}(\Gamma). \tag{III-3.6}$$

In the equation, $f^{eq}(\Gamma)$ denotes a probability distribution function in equilibrium, $\tilde{\mathbf{B}}$ indicates the adjoining vector of **B**. The generalized Langevin equation can be derived by applying the projection operator P to the Liouville equation for a variable $\mathbf{A}(t)$ to read

$$\frac{d\mathbf{A}(t)}{dt} = \mathrm{i}\Omega\mathbf{A}(t) + \int_0^t \mathbf{K}(t-\tau)\mathbf{A}(t)d\tau + \mathbf{f}(t) \tag{III-3.7}$$

where $i\Omega$ is defined by

$$i\Omega \equiv \frac{(\dot{\mathbf{A}}, \mathbf{A})}{(\mathbf{A}, \mathbf{A})} \tag{III-3.8}$$

The quantity is sometimes referred to as "Collective Frequency" because it is related to a *frequency* of the collective fluctuation of the dynamic variables. (The meaning of the quantity may become clearer later when the theory is applied to water and protein.) $\mathbf{f}(t)$ is a random force defined by

$$\mathbf{f}(t) \equiv \exp(i\mathcal{Q}\mathcal{L}t)\mathcal{Q}\dot{\mathbf{A}} \tag{III-3.9}$$

where $Q \equiv 1 - P$ is an operator to project the variables onto the space orthogonal to the space spanned by the dynamic variables **A**. Due to the projection, $\mathbf{f}(t)$ does not have correlation with the dynamic variables. It is the reason why the quantity is called "random force." $\mathbf{K}(t)$ in Eq. (III-3.7) is defined by

$$\mathbf{K}(t) \equiv \frac{(\mathbf{f}(t), \mathbf{f}(0))}{(\mathbf{A}, \mathbf{A})} \tag{III-3.10}$$

The quantity is called the friction kernel because it corresponds to the friction coefficient in the phenomenological Langevin equation, described in the previous section. However, it is not just an instantaneous friction, but it has a past memory as is indicated by the integral with respect to time in Eq. (III-3.7). Therefore, the friction kernel is sometimes called a memory kernel. The relation expressed by Eq. (III-3.10) corresponds to the fluctuation-dissipation theorem described by a phenomenological Langevin theory, or Eq. (III-1.25). Derivation of the Eq. (III-3.10) is outlined briefly in the following.

For the derivation, we start from the general solution of the Liouville equation (III-3.1).

$$\mathbf{A}(t) = \exp(i\mathcal{L}t)\mathbf{A}(0) = \exp(i\mathcal{L}t)\mathbf{A}. \tag{III-3.11}$$

Taking the derivative of Eq. (III-3.11) with respect to time t, one finds

$$\frac{d\mathbf{A}(t)}{dt} = \exp(i\mathcal{L}t)\, i\mathcal{L}\,\mathbf{A}. \tag{III-3.12}$$

We split the right-hand-side of the equation into two parts with the operator $\mathcal{Q} \equiv 1 - \mathcal{P}$,

$$\frac{d\mathbf{A}(t)}{dt} = \exp(i\mathcal{L}t)\, i(\mathcal{P} + \mathcal{Q})\mathcal{L}\,\mathbf{A}. \tag{III-3.13}$$

The first term in the right-hand-side of Eq. (III-3.13) can be rewritten as

$$\exp(i\mathcal{L}t)i\mathcal{P}\mathcal{L}\mathbf{A} = i\exp(i\mathcal{L}t)\frac{(\mathcal{L}\mathbf{A}, \mathbf{A})}{(\mathbf{A}, \mathbf{A})}\mathbf{A} = \frac{(\dot{\mathbf{A}}, \mathbf{A})}{(\mathbf{A}, \mathbf{A})}\mathbf{A}(t) \equiv \mathrm{i}\Omega\mathbf{A}(t), \tag{III-3.14}$$

where $i\Omega$ is defined by Eq. (III-3.8). This is the first term in Eq. (III-3.13). In order to manipulate the second term, or $\exp(i\mathcal{L}t)i\mathcal{Q}\mathcal{L}\,\mathbf{A}$, we use the following identity,

$$\exp\left[(\mathcal{F} + \mathcal{G})t\right] = \exp(\mathcal{F}\,t) + \int_0^t d\tau \exp\left[(\mathcal{F} + \mathcal{G})(t - \tau)\right]\mathcal{G}\exp(\mathcal{F}\,\tau), \tag{III-3.15}$$

that can be readily proved by multiplying both sides by $\exp[-t\mathcal{F}]$ and taking the derivative with respect to time t. With the identifications, $\mathcal{F} = i\mathcal{Q}\mathcal{L}$ and $\mathcal{G} = i\mathcal{P}\mathcal{L}$,

$$\exp(i\mathcal{L}t) = \exp(i\mathcal{Q}\mathcal{L}\,t) + \int_0^t d\tau \exp\left[i\mathcal{L}(t - \tau)\right]i\mathcal{P}\mathcal{L}\exp(i\mathcal{Q}\mathcal{L}\tau) \tag{III-3.16}$$

Then, the second term in Eq. (III-3.13) can be rewritten as

$$\begin{aligned} \exp(i\mathcal{L}t)i\mathcal{Q}\mathcal{L}\mathbf{A} &= \exp\left[i(\mathcal{Q} + \mathcal{P})\mathcal{L}t\right]\mathcal{Q}\dot{\mathbf{A}} \\ &= \exp(i\mathcal{Q}\mathcal{L}t)\mathcal{Q}\dot{\mathbf{A}} + \int_0^t d\tau \exp\left[i\mathcal{Q}(t - \tau)\right](i\mathcal{P}\mathcal{L})\exp(i\mathcal{Q}\mathcal{L}\tau)\mathcal{Q}\dot{\mathbf{A}} \end{aligned} \tag{III-3.17}$$

The first term of the equation is the same as the random force term defined by Eq. (III-3.9), while the second term is identified as the friction term defined by Eq. (III-3.7) as follows,

$$\begin{aligned}\int_0^t d\tau \exp\left[i\mathcal{L}(t-\tau)\right](\mathcal{P}\,i\mathcal{L})\exp(i\mathcal{QL}\tau)\mathcal{Q}\dot{\mathbf{A}} &= \int_0^t d\tau \exp\left[i\mathcal{L}(t-\tau)\right](\mathcal{P}\,i\mathcal{L})\mathbf{f}(\tau) \\ &= \int_0^t d\tau \exp\left[i\mathcal{L}(t-\tau)\right]\frac{\left(i\mathcal{L}\,\mathbf{f}(\tau),\mathbf{A}\right)}{\left(\mathbf{A},\mathbf{A}\right)}\mathbf{A} \\ &= \int_0^t d\tau \exp\left[i\mathcal{L}(t-\tau)\right]\frac{\left(\mathbf{f}(\tau),i\mathcal{L}\mathbf{A}\right)}{\left(\mathbf{A},\mathbf{A}\right)}\mathbf{A} \\ &= \int_0^t d\tau \exp\left[i\mathcal{L}(t-\tau)\right]\frac{\left(\mathbf{f}(\tau),\dot{\mathbf{A}}\right)}{\left(\mathbf{A},\mathbf{A}\right)}\mathbf{A} \\ &= \int_0^t d\tau \frac{\left(\mathbf{f}(\tau),\dot{\mathbf{A}}\right)}{\left(\mathbf{A},\mathbf{A}\right)}\dot{\mathbf{A}}(t-\tau) \\ &\equiv \int_0^t d\tau \mathbf{K}(\tau)\dot{\mathbf{A}}(t-\tau)\end{aligned} \quad \text{(III-3.18)}$$

where $\mathbf{K}(\tau)$ is the friction (or memory) kernel defined by Eq. (III-3.10). In the manipulation, we have used the following obvious relations,

$$\left(\mathbf{f}(t),(1-\mathcal{Q})\dot{\mathbf{A}}\right)=0, \quad \text{(III-3.19)}$$

and

$$\begin{aligned}\left(\mathbf{f}(t),\dot{\mathbf{A}}\right) &= \left(\mathbf{f}(t),\mathcal{Q}\dot{\mathbf{A}}\right) \\ &= \left(\mathbf{f}(t),\mathbf{f}(0)\right)\end{aligned} \quad \text{(III-3.20)}$$

That completes the derivation of GLE.

The GLE shares many physical characteristics with the phenomenological Langevin equation described in the preceding section: the random force, the friction term, and so on. Especially, the fluctuation dissipation theorem, that features the old theory, is inherited by Eq. (III-3.10) which bridges the fluctuational force $\mathbf{f}(t)$ and the friction kernel $\mathbf{K}(t)$: notice that the "friction" and the "dissipation" have the same physical origin. The GLE is yet to have distinct features, which the phenomenological Langein equation misses. The primary feature of GLE is that it is a first principle theory, meaning that all the terms of the equation can be evaluated in principle from a Hamiltonian of the system under consideration. There are two important factors in the equation that should be evaluated from a Hamiltonian: the collective frequency $i\Omega$ and the friction kernel $\mathbf{K}(t)$. In the following sections, we take the structural dynamics of water as an example of applying GLE to liquids.

III-4 Generalized Langevin Equation of Molecular Liquids

In the previous sections, we have derived the generalized Langevin equation (GLE) for a set of dynamical variables defined in the phase space. In this section, we apply the theory to the structural dynamics of molecular liquids. Remember the equilibrium structure of liquids is characterized by the density pair correlation function (Chapter I). So, by the structural

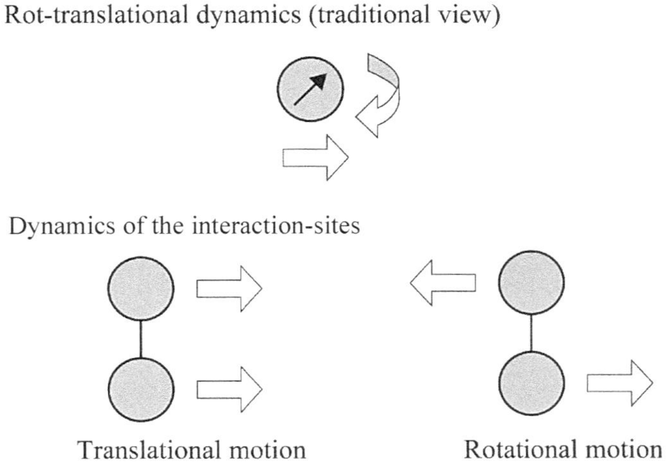

Fig. III-4.1: Two pictures for the dynamics of a molecule in liquids.

dynamics of liquids, we mean the time evolution of the density pair correlation function. Such a correlation function is called the space-time correlation function, or the "van Hove function" taking the name of the person who invented the function [Egelstaff 1967]. It is crucial for theory to choose a set of dynamic variables which can describe properly the time evolution of the space time correlation function.

When water is concerned with, the most important requirement for the variables is how successful they are to account for chemical specificity of water such as the hydrogen-bonds. The chemical specificity is reflected in the dynamics through the structure of liquids. There are two pictures conceivable for the dynamics of molecular liquids. One of those pictures has been practiced conventionally in most of the theories of liquid dynamics, which splits the dynamics into the translational and rotational degrees of freedom [Abragam 1961]. A drawback of such a model is slow convergence of the spherical-harmonic expansion in describing the rotational motion. Hirata has proposed an entirely different picture based on the interaction site model of liquids to describe the liquid dynamics [Hirata 1992]. According to the new model, the dynamics of a molecule in liquid is viewed as a correlated translational-diffusion of each atom in the molecule. For example, let us think the dynamics of a diatomic molecule in the liquid, illustrated in Fig. III-4.1. If the two atoms (or sites) in a molecule move in the *same* direction, the entire molecular motion is a *translational* motion. On the other hand, if the two atoms move in the *opposite* direction, the entire motion becomes a *rotational* motion. The actual motion is a combination of those two. The advantage of the model is to make a connection with the RISM theory for the liquid structure, described in Chapter II.

Let's look at the dynamics of molecular liquids as an illustrative example of application of GLE. We employ the interaction-site-model for the interaction potential among molecules, described in the preceding chapters. Then, the Hamiltonian can be written as

$$H = \sum_{i=1}^{N} \sum_{\alpha=1}^{n} \left[\frac{\mathbf{p}_i^{\alpha} \cdot \mathbf{p}_i^{\alpha}}{2m_{\alpha}} + \sum_{j \neq i} \sum_{\beta \neq \alpha} u\left(\left| \mathbf{r}_i^{\alpha} - \mathbf{r}_j^{\beta} \right| \right) \right], \tag{III-4.1}$$

where $\mathbf{p}_i^{\alpha}$ and m_{α} are the momentum and the mass of atom α of the i-th molecule, $u(|\mathbf{r}_i^{\alpha} - \mathbf{r}_j^{\beta}|)$ is the interaction energy between atom α in the i-th molecule and atom β in the j-th molecule:

N and n denote the number of molecules in the system, and the number of atoms in a water molecule. The Liouville operator (Eq. III-3.2) is then defined as

$$i\mathcal{L} \equiv \sum_{i=1}^{N}\sum_{\alpha=1}^{n}\left[\frac{1}{2m_\alpha}\mathbf{p}_i^\alpha \cdot \frac{\partial}{\partial \mathbf{r}_i^\alpha} - \sum_{j\neq i}\sum_{\beta\neq\alpha}\frac{\partial u\left(\left|\mathbf{r}_i^\alpha - \mathbf{r}_j^\beta\right|\right)}{\partial \mathbf{r}_i^\alpha}\cdot\frac{\partial}{\partial \mathbf{p}_i^\alpha}\right] \qquad \text{(III-4.2)}$$

We choose the fluctuation of the number density ($\delta\rho^\alpha$) of atoms and its conjugated momentum (J^α) as the dynamic variable as,

$$\delta\rho^\alpha(\mathbf{r},t) \equiv \sum_i \delta\left(\mathbf{r} - \mathbf{r}_i^\alpha(t)\right) - \rho_\alpha \qquad \text{(III-4.3)}$$

and

$$\mathbf{J}^\alpha(t) = \sum_i \mathbf{p}_i^\alpha \delta\left(\mathbf{r} - \mathbf{r}_i^\alpha(t)\right). \qquad \text{(III-4.4)}$$

In the actual derivation of the GLE, we work on the expression in the Fourier space rather than the real space,

$$\mathbf{A}(t) = \begin{pmatrix} \delta\rho_\mathbf{k}^\alpha(t) \\ J_\mathbf{k}^\alpha(t) \end{pmatrix}, \qquad \text{(III-4.5)}$$

where $\delta\rho_k^\alpha$ and $\mathbf{J}_k^\alpha$ are defined by

$$\delta\rho_\mathbf{k}^\alpha(t) \equiv \int d\mathbf{r} e^{i\mathbf{k}\cdot\mathbf{r}}\delta\rho^\alpha(\mathbf{r},t) = \sum_i e^{i\mathbf{k}\cdot\mathbf{r}_i^\alpha(t)} - (2\pi)^3\rho_\alpha\delta(\mathbf{k}), \qquad \text{(III-4.6)}$$

and

$$\mathbf{J}_\mathbf{k}^\alpha(t) \equiv \int d\mathbf{r} e^{i\mathbf{k}\cdot\mathbf{r}}\sum_i \mathbf{p}_i^\alpha \delta\left(\mathbf{r} - \mathbf{r}_i^\alpha(t)\right) = \sum_i \frac{\mathbf{p}_i^\alpha(t)}{m_\alpha} e^{i\mathbf{k}\cdot\mathbf{r}_i^\alpha(t)}. \qquad \text{(III-4.7)}$$

Now, let's derive each term in the GLE (Eq.III-3.7).

The collective frequency term: The collective frequency term $i\Omega = (\dot{\mathbf{A}}, \mathbf{A})/(\mathbf{A}, \mathbf{A})$ for molecules can be written in terms of the definition for $\delta\rho_k^\alpha$ and $\mathbf{J}_k^\alpha$ as

$$(\mathbf{A},\mathbf{A}) = \begin{pmatrix} \left(\delta\rho_\mathbf{k}^\alpha, \delta\rho_\mathbf{k}^\beta\right) & \left(\mathbf{J}_\mathbf{k}^\alpha, \delta\rho_\mathbf{k}^\beta\right) \\ \left(\delta\rho_\mathbf{k}^\alpha, \mathbf{J}_\mathbf{k}^\beta\right) & \left(\mathbf{J}_\mathbf{k}^\alpha, \mathbf{J}_\mathbf{k}^\beta\right) \end{pmatrix}. \qquad \text{(III-4.8)}$$

The off diagonal elements in the matrix are zero by definition of the dynamic variables, namely

$$\left(\delta\rho_\mathbf{k}^\alpha, \mathbf{J}_\mathbf{k}^\beta\right) = 0 \quad \text{and} \quad \left(\mathbf{J}_\mathbf{k}^\alpha, \delta\rho_\mathbf{k}^\beta\right) = 0.$$

We define the momentum correlation function by

$$J_{\alpha\beta}(k) \equiv \frac{1}{N}\left\langle \mathbf{J}_{-\mathbf{k}}^\alpha, \mathbf{J}_\mathbf{k}^\beta \right\rangle \qquad \text{(III-4.9)}$$

Substituting Eq. (III-4.7) into (III-4.9), one gets,

$$
\begin{aligned}
J_{\alpha\beta}(k) &= \frac{1}{N}\sum_i\sum_j \frac{1}{m_\alpha}\frac{1}{m_\beta}\left\langle \mathbf{p}_i^\alpha \cdot \mathbf{p}_j^\beta \exp\left\{-i\mathbf{k}\cdot\left(\mathbf{r}_i^\alpha - \mathbf{r}_j^\beta\right)\right\}\right\rangle \\
&= \frac{1}{N}\sum_i\sum_j \frac{1}{m_\alpha}\frac{1}{m_\beta}\left\langle \mathbf{p}_i^\alpha \cdot \mathbf{p}_j^\beta\right\rangle\left\langle \exp\left\{-i\mathbf{k}\cdot\left(\mathbf{r}_i^\alpha - \mathbf{r}_j^\beta\right)\right\}\right\rangle \\
&= \frac{1}{N}\sum_i\sum_j \left\langle \mathbf{v}_i^\alpha \cdot \mathbf{v}_j^\beta\right\rangle \delta_{ij}\left\langle \exp\left\{-i\mathbf{k}\cdot\left(\mathbf{r}_i^\alpha - \mathbf{r}_j^\beta\right)\right\}\right\rangle \\
&= \frac{1}{N}\sum_i \left\langle \mathbf{v}_i^\alpha \cdot \mathbf{v}_i^\beta\right\rangle\left\langle \exp\left\{-i\mathbf{k}\cdot\left(\mathbf{r}_i^\alpha - \mathbf{r}_i^\beta\right)\right\}\right\rangle
\end{aligned}, \tag{III-4.10}
$$

where we have used the relation $\langle \mathbf{v}_i^\alpha \cdot \mathbf{v}_j^\beta \rangle = \langle \mathbf{v}_i^\alpha \rangle \langle \mathbf{v}_j^\beta \rangle = 0$ if $i \neq j$. Then, the correlation matrix $(\mathbf{A},\mathbf{A})$ and its inverse become,

$$
(\mathbf{A},\mathbf{A}) = \begin{pmatrix} N\chi_{\alpha\beta}(\mathbf{k}) & 0 \\ 0 & J_{\alpha\beta}(k) \end{pmatrix} \tag{III-4.11}
$$

and

$$
(\mathbf{A},\mathbf{A})^{-1} = \begin{pmatrix} N^{-1}\chi_{\alpha\beta}(\mathbf{k})^{-1} & 0 \\ 0 & J_{\alpha\beta}(k)^{-1} \end{pmatrix}, \tag{III-4.12}
$$

where the matrix inverse is defined by

$$
\sum_{\gamma=1}^{n} \chi_{\alpha\gamma}(k)\chi_{\gamma\beta}^{-1}(k) = \delta_{\alpha\beta} \quad \text{and} \quad \sum_{\gamma=1}^{n} J_{\alpha\gamma}(k)J_{\gamma\beta}^{-1}(k) = \delta_{\alpha\beta}.
$$

The correlation matrix $(\dot{\mathbf{A}}, \mathbf{A})$ reads

$$
(\mathbf{A},\dot{\mathbf{A}}) = \begin{pmatrix} \left(\delta\rho_{\mathbf{k}}^\alpha, \dot{\rho}_{\mathbf{k}}^\beta\right) & \left(\mathbf{J}_{\mathbf{k}}^\alpha, \dot{\rho}_{\mathbf{k}}^\beta\right) \\ \left(\delta\rho_{\mathbf{k}}^\alpha, \dot{\mathbf{J}}_{\mathbf{k}}^\beta\right) & \left(\mathbf{J}_{\mathbf{k}}^\alpha, \dot{\mathbf{J}}_{\mathbf{k}}^\beta\right) \end{pmatrix}, \tag{III-4.13}
$$

where $\dot{\rho}_k^\alpha = iL\,\rho_k^\alpha$, and $\dot{\mathbf{J}}_k^\alpha$ is

$$
\dot{\mathbf{J}}_{\mathbf{k}}^\alpha = \sum_i \frac{1}{m_\alpha}\left(\dot{\mathbf{p}}_i^\alpha + \mathbf{p}_i^\alpha i\mathbf{k}\cdot\frac{\mathbf{p}_i^\alpha}{m_\alpha}\right)\exp\left(i\mathbf{k}\cdot\mathbf{r}_i^\alpha\right). \tag{III-4.14}
$$

Considering that $(\delta\rho_{\mathbf{k}}^\alpha, \dot{\rho}_{\mathbf{k}}^\beta) = \mathbf{0}$, one finds the following expression for the collective frequency term

$$
i\Omega \equiv \frac{(\mathbf{A},\dot{\mathbf{A}})}{(\mathbf{A},\mathbf{A})} = \begin{pmatrix} \mathbf{0} & i\mathbf{k}\delta_{\alpha\beta} \\ i\mathbf{k}\sum_{\gamma=1}^{n} J_{\alpha\gamma}(k)\chi_{\gamma\beta}^{-1}(k) & \mathbf{0} \end{pmatrix}. \tag{III-4.15}
$$

Then, the reversible part of the generalized Langevin equation (III-3.7) is given by,

$$i\Omega\cdot\mathbf{A}(t)=\begin{pmatrix} i\mathbf{k}\cdot J_{\mathbf{k}}^{\alpha}(t) \\ i\mathbf{k}\sum_{\beta\gamma}J_{\alpha\gamma}(k)\chi_{\gamma\beta}^{-1}(k)\delta\rho_{\mathbf{k}}^{\beta}(t)\end{pmatrix}. \tag{III-4.16}$$

The fluctuating force (random force): From Eq. (III-3.9), the fluctuating force at $t = 0$ is given by

$$\mathbf{f}=(1-\mathcal{P})\dot{\mathbf{A}}=\dot{\mathbf{A}}-i\Omega\cdot\mathbf{A}, \tag{III-4.17}$$

where $P\dot{\mathbf{A}} = (\mathbf{A},\dot{\mathbf{A}})\cdot(\mathbf{A},\mathbf{A})^{-1}\mathbf{A} = i\Omega\cdot\mathbf{A}$ is taken into account. $\dot{\mathbf{A}}$ and $i\Omega\cdot\mathbf{A}$ are

$$\dot{\mathbf{A}}=\begin{pmatrix}\dot{\rho}_{\mathbf{k}}^{\alpha} \\ \dot{\mathbf{J}}_{\mathbf{k}}^{\alpha}\end{pmatrix}=\begin{pmatrix} i\mathbf{k}\cdot\mathbf{J}_{\mathbf{k}}^{\alpha} \\ \dot{\mathbf{J}}_{\mathbf{k}}^{\alpha}\end{pmatrix} \tag{III-4.18}$$

and

$$i\Omega\cdot\mathbf{A}=\begin{pmatrix} i\mathbf{k}\cdot\mathbf{J}_{\mathbf{k}}^{\alpha} \\ i\mathbf{k}\sum_{\beta\gamma}J_{\alpha\gamma}(k)\chi_{\gamma\beta}^{-1}(k)\delta\rho_{\mathbf{k}}^{\beta}\end{pmatrix}. \tag{III-4.19}$$

Then, the fluctuating force at $t = 0$ becomes

$$\mathbf{f}=\begin{pmatrix}0 \\ \Xi_{\mathbf{k}}^{\alpha}\end{pmatrix}, \tag{III-4.20}$$

where $\Xi_{\mathbf{k}}^{\alpha}$ is defined by

$$\Xi_{\mathbf{k}}^{\alpha}\equiv\dot{\mathbf{J}}_{\mathbf{k}}^{\alpha}-i\mathbf{k}\sum_{\beta\gamma}J_{\alpha\gamma}(k)\chi_{\gamma\beta}^{-1}(k)\delta\rho_{\mathbf{k}}^{\beta} \tag{III-4.21}$$

The fluctuating force is derived as

$$\mathbf{f}(t)\equiv e^{it(1-\mathcal{P})\mathcal{L}}\,\mathbf{f}=e^{it(1-\mathcal{P})\mathcal{L}}\begin{pmatrix}0 \\ \Xi_{\mathbf{k}}^{\alpha}\end{pmatrix} \tag{III-4.22}$$

The memory kernel (friction matrix): Using the definition in terms of the fluctuating force, we get the following expression for the memory kernel,

$$\mathbf{K}(t)\equiv\frac{(\mathbf{f},\mathbf{f}(t))}{(\mathbf{A},\mathbf{A})} = \begin{pmatrix}0 & 0 \\ 0 & N^{-1}\sum_{\gamma}J_{\alpha\gamma}^{-1}\left(\Xi_{\mathbf{k}}^{\gamma},\Xi_{\mathbf{k}}^{\beta}(t)\right)\end{pmatrix} \tag{III-4.23}$$

where $\Xi_{\mathbf{k}}^{\alpha}(t)\equiv\exp(it(1-\mathcal{P})\mathcal{L})\,\Xi_{\mathbf{k}}^{\alpha}$.

Generalized Langevin equations of molecular liquids

Substituting all the terms derived in Eqs. (III-4.18), (III-4.21), (III-4.22) into Eq. (III-3.7), the following equations are obtained.

$$\frac{d\delta\rho_{\mathbf{k}}^{\alpha}(t)}{dt} = i\mathbf{k}\cdot\mathbf{J}_{\mathbf{k}}^{\alpha}(t) \tag{III-4.24}$$

$$\frac{d\mathbf{J}_{\mathbf{k}}^{\alpha}(t)}{dt} = i\mathbf{k}\sum_{\beta\gamma} J_{\alpha\gamma}(k)\chi_{\gamma\beta}^{-1}(k)\delta\rho_{\mathbf{k}}^{\beta}(t) - \sum_{\beta\gamma}\int_0^t ds\mathbf{K}_{\mathbf{k}}^{\beta\gamma}(t-s)\cdot\mathbf{J}_{\mathbf{k}}^{\beta}(s) + \Xi_{\mathbf{k}}^{\alpha}(t) \tag{III-4.25}$$

where $\mathbf{K}_{\mathbf{k}}^{\beta\gamma}(s) = N^{-1}J_{\alpha\gamma}^{-1}(k)(\Xi_{\mathbf{k}}^{\gamma}, \Xi_{\mathbf{k}}^{\beta}(t))$. The first and second equations correspond, respectively, to the continuity equation and the equation of motion (or the momentum conservation) in the hydrodynamics. By taking the derivative of the first equation with respect to time t, and by substituting the second equation into $\dot{\mathbf{J}}_{k}^{\alpha}(t)$, one finds the GLE for the density field as follows.

$$\frac{d^2\delta\rho_{\mathbf{k}}^{\alpha}(t)}{dt^2} = i\mathbf{k}\cdot\left[i\mathbf{k}\sum_{\beta\gamma} J_{\alpha\gamma}(k)\chi_{\gamma\beta}^{-1}\delta\rho_{\mathbf{k}}^{\beta}(t) - \sum_{\beta\gamma}\int_0^t ds\mathbf{K}_{\mathbf{k}}^{\beta\gamma}(t-s)\cdot\mathbf{J}_{\mathbf{k}}^{\beta}(s) + \Xi_{\mathbf{k}}^{\alpha}(t)\right] \tag{III-4.26}$$

Multiplying $\delta\rho_{\mathbf{k}}(0)$, and taking the ensemble average of both sides, one gets,

$$\begin{aligned}\frac{d^2\left\langle\delta\rho_{\mathbf{k}}^{\alpha}(t)\delta\rho_{-\mathbf{k}}^{\beta}(0)\right\rangle}{dt^2} &= -\sum_{\gamma}\left\langle\omega_k^2\right\rangle_{\alpha\gamma}\left\langle\delta\rho_{\mathbf{k}}^{\gamma}(t)\delta\rho_{-\mathbf{k}}^{\beta}(0)\right\rangle \\ &\quad -\sum_{\gamma}\int_0^t ds\mathbf{K}_{\mathbf{k}}^{\alpha\gamma}(t-s)\cdot\left\langle i\mathbf{k}\cdot\mathbf{J}_{\mathbf{k}}^{\gamma}(t)\delta\rho_{-\mathbf{k}}^{\beta}(0)\right\rangle\end{aligned} \tag{III-4.27}$$

where $\langle\omega_k^2\rangle_{\alpha\beta}$ is defined by

$$\left\langle\omega_k^2\right\rangle_{\alpha\beta} = k^2\sum_{\gamma} J_{\alpha\gamma}(k)\left(\chi(k)^{-1}\right)_{\gamma\beta} \tag{III-4.28}$$

In the derivation, the orthogonality of $\delta\rho_k^{\beta}$ with the random force $\Xi_{\mathbf{k}}^{\alpha}(t)$, or $\langle\Xi_{\mathbf{k}}^{\alpha}(t)\delta\rho_k^{\beta}(0)^*\rangle = 0$, is taken into consideration. With the definition of the *intermediate scattering function*,

$$F_{\mathbf{k}}^{\alpha\beta}(t) \equiv N^{-1}\left\langle\delta\rho_{\mathbf{k}}^{\alpha}(t)\delta\rho_{\mathbf{k}}^{\beta}(0)^*\right\rangle, \tag{III-4.29}$$

Equation (III-4.27) can be rewritten as

$$\frac{d^2\mathbf{F}_{\mathbf{k}}(t)}{dt^2} = -\left\langle\omega_{\mathbf{k}}^2\right\rangle\mathbf{F}_{\mathbf{k}}(t) - \int_0^t ds\mathbf{K}_{\mathbf{k}}(t-s)\cdot\dot{\mathbf{F}}_{\mathbf{k}}(s) \tag{III-4.30}$$

As is understood from the definition given by Eqs. (III-4.3) and (III-4.6), $\delta\rho_k^{\alpha}(t)$ is the density fluctuation of atom α in the liquid, thereby $F_k^{\alpha\beta}(t)$ signifies the time dependence of the density pair correlation function between atoms α and β. Here, the function includes the cases $\alpha = \beta$ (self) and $\alpha \neq \beta$ (distinct). It is also noted that α and β in the distinct case belong either to the same molecule or to different molecules. The intermediate scattering function is the Fourier transform of the space-time correlation function, or the so called "van Hove function." In the following sections, we apply the theory to calculate the van Hove function of water in the diffusion limit.

III-5 Space-time Correlation Function (van Hove function) of Water

As an application of the theory developed in the previous section, let us solve Eq. (III-4.30) for water in the diffusion limit, where the friction loses the past memory (Markov process), and inertia term can be ignored, that is,

$$\tilde{\mathbf{K}}_{\mathbf{k}}(t-s) \approx \tilde{\mathbf{K}}_{\mathbf{k}}\delta(t-s) \tag{III-5.1}$$

and

$$\frac{d^2\tilde{\mathbf{F}}_{\mathbf{k}}(t)}{dt^2} = 0. \tag{III-5.2}$$

Then, the Eq. (III-4.30) is simplified considerably to read,

$$\frac{d}{dt}\tilde{\mathbf{F}}_{\mathbf{k}}(t) = -k^2\tilde{\mathbf{J}}(\mathbf{k})\tilde{\chi}(\mathbf{k})^{-1}\tilde{\mathbf{K}}_{\mathbf{k}}^{-1}\tilde{\mathbf{F}}_{\mathbf{k}}(t). \tag{III-5.3}$$

Defining the diffusion constant by $\tilde{\mathbf{D}}_{\mathbf{k}} \equiv \tilde{\mathbf{J}}(\mathbf{k})\tilde{\mathbf{K}}_{\mathbf{k}}^{-1}$, one gets a simple diffusion equation for a space-time correlation function of molecular liquids as,

$$\frac{d}{dt}\tilde{\mathbf{F}}_{\mathbf{k}}(t) = -k^2\tilde{\mathbf{D}}_{\mathbf{k}}\tilde{\chi}(\mathbf{k})^{-1}\tilde{\mathbf{F}}_{\mathbf{k}}(t), \tag{III-5.4}$$

In the equation, $\chi(\mathbf{k})$ is the Fourier transform of the site-site susceptibility defined by,

$$\tilde{\chi}_{\alpha\gamma}(\mathbf{k}) = \tilde{\omega}_{\alpha\gamma}(\mathbf{k}) + \rho^2\tilde{h}_{\alpha\gamma}(\mathbf{k}) \tag{III-5.5}$$

where $\tilde{\omega}_{\alpha\gamma}(\mathbf{k})$ and $\tilde{h}_{\alpha\gamma}(\mathbf{k})$ are, respectively, the site-site intramolecular and intermolecular correlation functions defined in Chapter II. The equation is referred to as "Site-Site Sumolchowski-Vlasov (SSSV)" equation (Hirata 1992).

van Hove function in the diffusion limit

Let's apply the SSSV equation to calculate the van Hove function of water. The van Hove function is defined as,

$$G_{\alpha\gamma}(\mathbf{r},t) = \frac{1}{\rho}\left\langle \delta\rho_\alpha(\mathbf{r},t)\delta\rho_\gamma(0,0)\right\rangle, \tag{III-5.6}$$

or its Fourier transform as,

$$\tilde{G}_{\alpha\gamma}(\mathbf{k},t) = \frac{1}{N}\left\langle \delta\tilde{\rho}_\alpha(\mathbf{k},t)\delta\tilde{\rho}_\gamma(0,0)\right\rangle, \tag{III-5.7}$$

where $\delta\rho_\alpha(\mathbf{r}, t)$ is the density fluctuation of atom α in the molecular liquid, defined by Eq. (III-4.3). Note that $\tilde{G}_{\alpha\gamma}(\mathbf{k}, t)$ is the same function as the intermediate scattering function $\tilde{\mathbf{F}}_{\mathbf{k}}(t)$ defined by Eq. (III-4.29), that is, $\tilde{\mathbf{G}}(\mathbf{k}, t) \equiv \tilde{\mathbf{F}}_{\mathbf{k}}(t)$. Therefore, the time evolution of the van Hove function in the diffusion limit is described by Eq. (III-5.4). We define the Laplace transform with respect to time by,

$$f(z) = \int_0^\infty f(t)e^{izt}dt \tag{III-5.8}$$

The Laplace transform of Eq. (III-5.4) reads,

$$\tilde{\mathbf{S}}(\mathbf{k},z)=\frac{\tilde{\mathbf{G}}(\mathbf{k},t=0)}{iz\mathbf{I}+k^2\tilde{\mathbf{D}}_\mathbf{k}\tilde{\chi}^{-1}(\mathbf{k})} \tag{III-5.9}$$

where **I** denotes the unit matrix. Equation (III-5.9) can be inversely transformed into the time domain to give

$$\tilde{G}(\mathbf{k},t)=\tilde{\Psi}(\mathbf{k},t)\tilde{G}(\mathbf{k},t=0) \tag{III-5.10}$$

In Eq. (III-5.10), $\tilde{\Psi}(\mathbf{k},t)$ is defined by,

$$\tilde{\Psi}(\mathbf{k},t)\equiv\frac{1}{2\pi i}\int_{-i\infty}^{i\infty}\left[z\mathbf{I}+\tilde{\mathbf{A}}(\mathbf{k})\right]^{-1}e^{zt}dz \tag{III-5.11}$$

where $\tilde{\mathbf{A}}(\mathbf{k})$ is defined as

$$\tilde{\mathbf{A}}(\mathbf{k})\equiv k^2\tilde{\mathbf{D}}_\mathbf{k}\tilde{\chi}^{-1}(\mathbf{k}) \tag{III-5.12}$$

The integral in Eq. (III-5.11) can be performed in the complex plane to lead for the (α,β)-element of the matrix Ψ,

$$\Psi_{\alpha\beta}=\sum_j^n\frac{(-1)^{\beta+\alpha}\,\Delta_{\beta\alpha}(z_j)\exp(z_jt)}{\Pi_{i(\neq j)}^n(z_j-z_i)} \tag{III-5.13}$$

where $\{z_i\}$ are the eigen values of the matrix $z\mathbf{I}+\tilde{\mathbf{A}}(\mathbf{k})$, and $\Delta_{\beta\alpha}$ denotes the minor of the (α,β)-element of the determinant $||z\mathbf{I}+\tilde{\mathbf{A}}||$. It is worthwhile to note that $\tilde{G}_{\alpha\gamma}(\mathbf{k},t)$ is a superposition of a number of exponentials, as many as the number of interaction sites, with different time constant. The different time constant originated from the matrix $\tilde{\mathbf{A}}(\mathbf{k})$ that reflects the spatial correlation of the density fluctuation of atoms, and that of the momentum. In the calculation, the diffusion constant matrix is approximated by $[\tilde{\mathbf{D}}_\mathbf{k}]_{\alpha\gamma}\sim D\delta_{\alpha\gamma}$, where $\delta_{\alpha\gamma}$ is the Cronekker delta-function, and D is the (translational) diffusion constant of water molecules, obtained from the molecular dynamics simulation.

Shown in Fig. III-5.1 is the space-time correlation function (solid lines, dashed lines, and dotted lines) calculated by the SSSV theory, compared with the results from a molecular dynamics simulation (○, +, ∗).

A. Oxygen-oxygen: The oxygen-oxygen correlation function includes contributions from the self part which is seen around $r=0$, and the distinct part which has a peak around $r=3.0$A. The distinct part is a manifestation of the intermolecular correlation. So, at $t=0$, it has essentially the same physical meaning with the pair correlation function for the O-O pair, exhibited in Fig. II-3.2 in Chapter II, except for the normalization factor. The SSSV results show a fair agreement with the experiment (MD) in general time dependency. The decay rate of the peak is somewhat faster in SSSV than in MD. It is worthwhile to note that there is a marked difference in initial decay of the self part between the two methods; the decay is slower in MD than in SSSV. It is likely that the inertia effect, which is neglected in SSSV, is responsible for the short time behavior (Eq. III-5.2). It is easy to verify that in a very short time, the time dependence of the mean square displacement of a molecule is proportional to t^2 rather than t. The behavior

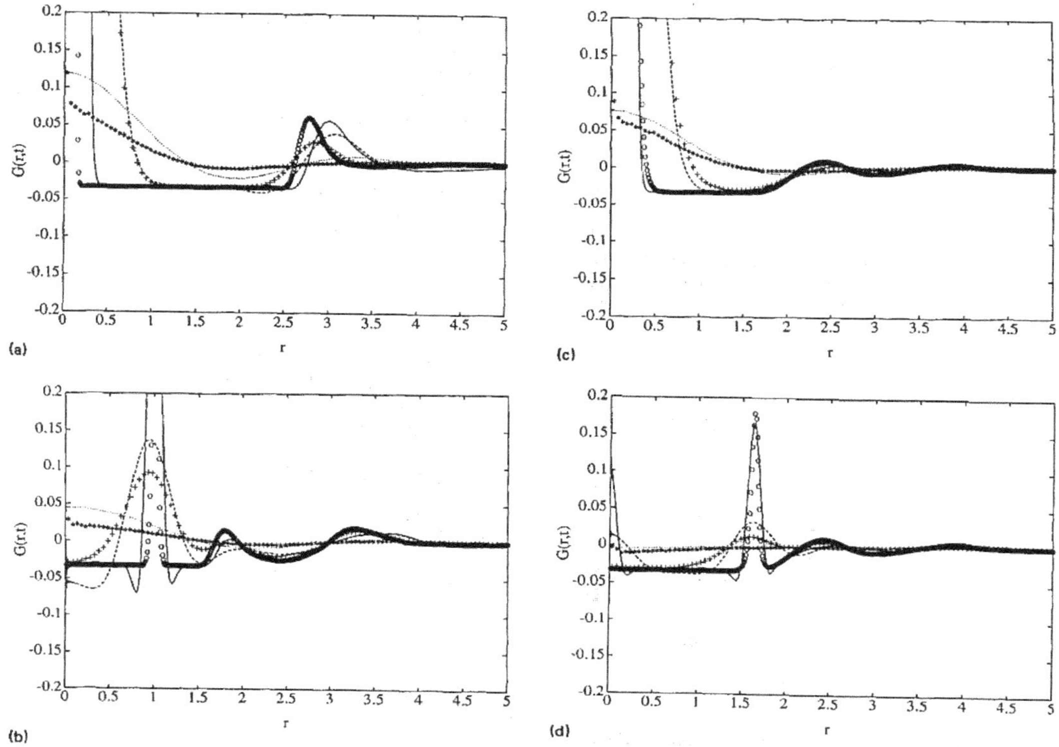

Fig. III-5.1: Space-time correlation function of water calculated by the SSSV theory: (a) oxygen-oxygen, (b) hydrogen-oxygen, (c) hydrogen-hydrogen (H-H), (d) hydrogen-hydrogen (H_1-H_2). Molecular dynamics results: ○, t = 0.01 ps; *, t = 0.1 ps, +, t = 1 ps. SSSV results: solid lines, t = 0.01 ps; dashed lines, t = 0.1 ps; dotted lines, t = 1.0 ps. Copyright (1992) American Institute of Physics.

is interpreted by the collision-free motion of a majority of molecules in the short time period. The difference in the peak positions concerning the distinct part between the two methods is essentially due to the problem inherited from the XRISM theory for the equilibrium pair correlation functions.

B. Hydrogen-oxygen: The O-H correlation function which is shown in Fig. (III-5.1b) features the strong peak around 1.0 A which corresponds to the intramolecular O-H bond, and the two small peaks around 1.8 A and between 3.0 and 4.0 A which correspond to the intermolecular O-H distance between nearest neighbor molecules. The first of those two peaks is assigned to O and H which are directly hydrogen-bonded, and the second to those which are not involved in the hydrogen bond. Regarding the peak corresponding to the intramolecular O-H distance, the broadening in the peak is caused by both the translational and rotational motions of the molecule. At 0.01 ps, the peak is sharper in MD than in SSSV. This may be due to the inertia effect mentioned previously. The effect is not so pronounced compared to the O-O correlation function. This is because a hydrogen atom is much lighter than an oxygen atom. Another interesting point to be made is the appearance of a peak around r = 0 at 1.0 ps, which is observed in MD and SSSV, although SSSV exhibits an exaggerated behavior. It is interesting because the intra- and inter-molecular O-H distances are finite, and the two atoms are never collapsed. The reason why $G(r, t)$ has the finite value is because the intramolecular correlation fades away at 1.0 ps, and the correlation becomes the one characteristic of the self-diffusion of a whole molecule.

C. Hydrogen-hydrogen: The hydrogen-hydrogen correlation in the site-site viewpoint can be classified into two different categories, one which includes the self part (H-H correlation) and the other which includes intramolecular distinct part ($H_1 - H_2$ correlation). The H-H correlation is shown in Fig. (III-5.1c). The curves have a self part which has the center around $\mathbf{r} = 0$, and the intermolecular distinct part around $\mathbf{r} = 2.5$ and 4.0 A. Almost quantitative agreement for the time dependence of the correlation function is observed between the SSSV and MD results. It may be because the contribution from the inertia term which is missing in SSSV is minor in the case of a hydrogen atom that has small mass. The $H_1 - H_2$ correlation function is shown in Fig. III-5.1(d). The peak centered around 1.63 A corresponds to the intra molecular H-H separation, while the peak around 2.5 and 4.0 A are assigned to the intermolecular H-H correlation. Again a qualitative agreement between SSSV and MD is obtained except for the region $r = 0$; the SSSV result has a distinct peak at r = 0 ar 0.01 ps, while no such peak is observed in the MD results. The behavior is obviously wrong in physics because two atoms in a rigid molecule may not occupy the same position in such a short time after they are found in the separate position. A possible cause of this behavior is the omission of inertia term. If the moment of inertia about an axis, say the axis bisecting the O-H-O angle, is zero, there may be a possibility of finding an H-atom at the same place within very short time due to rotation about the axis.

The theory just described requires the diffusion constant of solvent as an empirical parameter. Nonetheless, it provides a convenient tool to explore dynamics of molecular liquids. So, the theory has been applied to numerous problems that concern dynamics of molecular liquids, including the dynamics Stokes shift [Hirata et al. 1995, Nishiyama et al. 2000a, b, 2001, 2003], dynamics of electrolyte solutions [Iida and Sato 2012a], dielectric relaxation [Iida and Sato 2012b], and so on.

III-6 Collective Excitations in Water

In the previous section, a new picture to describe the dynamics of a molecule in liquids including water was presented, based on the interaction-site model. According to the picture, the dynamics of a molecule in solution is described by *translational* motions of *atoms* consisting the molecule, which are correlated with each other. But, how can the new picture of dynamics be reconciled with the conventional view of dynamics, or the rotational and translational motion of a molecule? The question is answered in this section [Ricci et al. 1989, Chong and Hirata 1998, 1999a, b].

For that purpose, let's go back to Eq. (III-4.30), describing the time evolution of density correlation functions. If one ignores the friction term, one finds,

$$\frac{d^2\mathbf{F}_{\mathbf{k}}(t)}{dt^2} = -\langle \omega_{\mathbf{k}}^2 \rangle \mathbf{F}_{\mathbf{k}}(t) \tag{III-6.1}$$

where $\langle \omega_k^2 \rangle$ is the collective frequency matrix defined by Eq. (III-4.28), or

$$\langle \omega_k^2 \rangle_{\alpha\beta} = k^2 \sum_{\gamma} J_{\alpha\gamma}(k) \left(\chi(k)^{-1} \right)_{\gamma\beta} \tag{III-6.2}$$

The Eq. (III-6.1) is *formally* identical to that describing a *harmonic oscillator*, if one identifies $\langle\omega_k^2\rangle$ as a *force constant*. The equation implies that *the restoring force exerted on density fluctuation is proportional to the density fluctuation*, and that the proportional constant (or a "force constant") is $\langle\omega_k^2\rangle$. It is as if the same law as "Hook's law" concerning the harmonic oscillator is governing the density fluctuation. Then, with an analogy to the normal mode analysis, "*normal modes*" of the density fluctuation of atoms, corresponding to "*eigen frequencies*," can be obtained by diagonalizing the matrix $\langle\omega_k^2\rangle$.

The diagonalization corresponds to a transformation of the dynamic variables from the density of atoms to their linear combination,

$$\rho_i(k) = \sum_\alpha x_{i\alpha}(k)\rho_\alpha(k) \tag{III-6.3}$$

where $\rho_i(k)$ denotes the i-th mode of the collective density fluctuation, $\rho_\alpha(k)$ denotes the density of atom α. In the case of water, α distinguishes an oxygen atom (O) and two hydrogen atoms (H_1, H_2). Since water has three atoms, we have three eigen modes. Obtained eigen values (frequency) and eigen functions (modes of atomic density fluctuation) of $\langle\omega_k^2\rangle$ for water are depicted against the wave number in Fig. (III-6.1). The theoretical results are compared with the MD simulation in Fig. (III-6.2).

Acoustic mode: One of the three eigen frequencies exhibits the following asymptotic behavior at the limit of $k \rightarrow 0$,

$$\omega_{acou}^2\left(k \rightarrow 0\right) = \frac{k_B T}{M\chi(0)}k^2 \tag{III-6.4}$$

where M denotes the mass of a molecule. The equation shows that the frequency (ω) at $k \rightarrow 0$ is proportional to the wave number $k = 1/\lambda$ (λ; wave length), that is,

$$\omega = ck \tag{III-6.5}$$

where c is defined as

$$c \equiv \sqrt{\frac{k_B T}{M\chi(0)}} \tag{III-6.6}$$

Here, $\chi(0)$ is related to the isothermal compressibility κ_T (Chapter II)

$$\rho k T_B \kappa_T = \chi(0) \tag{III-6.7}$$

From Eqs. (III-6.6) and (III-6.7), one gets

$$c = \sqrt{\frac{1}{\rho M \kappa_T}} \equiv \sqrt{\frac{1}{\rho_M \kappa_T}} \tag{III-6.8}$$

where ρ_M is the mass density defined from the number density as $\rho_M \equiv \rho \times M$. The expression suggests that the proportional constant c in Eq. (III-6.5) is essentially a *sound* velocity because it is the well-regarded hydrodynamic relation between the compressibility and the sound velocity. However, there is a subtle difference from the hydrodynamic expression. The compressibility included in the hydrodynamic expression is *adiabatic*, while the compressibility here is *isothermal*. Nonetheless, the essential physics involved in the two expressions is the same: the

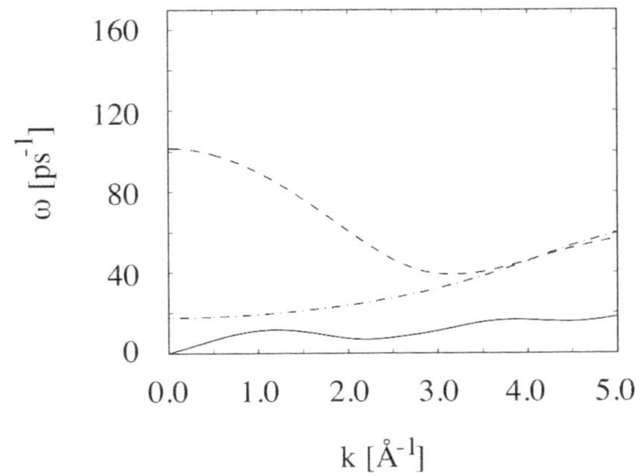

Fig. III-6.1: Dispersion relation of the collective excitation in water calculated by the GLE/RISM theory: solid line, acoustic mode; dashed line, optical mode I; dot-dashed line, optical mode II. Copyright (1999) American Institute of Physics.

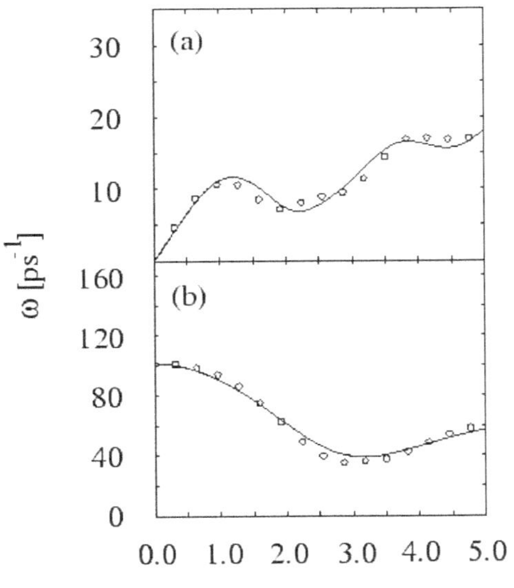

Fig. III-6.2: The dispersion relation calculated from the GLE/RISM theory, compared with the MD simulation: solid line, GLE/RISM; circles, MD (a) Acoustic mode. (b) Optical mode. Copyright (1999) American Institute of Physics.

sound velocity in a less compressible liquid is faster than that in a more compressible one. The mode satisfies the dispersion relation of the sound wave, $c = \lambda\omega$, thereby it is assigned to an acoustic mode. It is also understood that the mode is related to the *translational* motion of the molecule, since Eq. (III-6.4) includes the mass of a molecule, M.

Optical mode: On the other hand, the other two modes do not vanish in the $k \rightarrow 0$ limit as seen in Fig. (III-6.1). By an analogy to the solid-state physics, the two modes can be identified as "optical" modes. Those modes are concerned with the relative motions of atoms, whose characteristic frequencies do not vanish even in the $k \rightarrow 0$ limit. The limiting behavior of the two optical modes, referred to as "OM-I (optical mode I) and OM-II (optical mode II), are respectively given by

$$\omega^2_{OM-I}(k \to 0) = \frac{2k_B T}{3\chi^{(2)}(k=0)}(z_H - z_O)^2\left(\frac{1}{I_x} + \frac{1}{I_y}\right) \quad \text{(III-6.9)}$$

and

$$\omega^2_{OM-II}(k \to 0) = k_B T\left(\frac{1}{I_x} + \frac{1}{I_z}\right) \quad \text{(III-6.10)}$$

where I_x, I_y, and I_z are the moments of inertia around the principal x-,y-and z-axes, and z_O and z_H are the z-components of O and H atoms of a molecule in the body-fixed coordinate (see Fig. III-6.3). The optical modes are related to the rotational motion of molecules as can be inferred from the appearance of the moments of inertia in the above expressions. OM-I has the second moment of the density correlation function, $\chi^{(2)}(k = 0)$, which is closely related to the dielectric constant of liquids, as was clarified in the Section II-4 (see Eq. (II-5.23)). It indicates that OM-I has a *collective* character. On the other hand, OM-II is essentially a single-molecule mode, since the expression is not associated with any collective density fluctuation, but thoroughly determined by the single molecular properties such as the moment of inertia.

The contribution from each atom to the mode can be extracted in the following way. Diagonalizing the matrix $\langle\omega_k^2\rangle$ corresponds to changing the description of the system in terms of the density fluctuations $\rho_\alpha(k)$ to the one in terms of their linear combination, $x_O(k)\,\rho_O(k) + x_{H_1}(k)\,\rho_{H_1}(k) + x_{H_2}(k)\,\rho_{H_2}(k)$, where $x_O(k)$, $x_{H_1}(k)$, and $x_{H_2}(k)$ are the components of the eigen vector corresponding to the respective modes. Thus, by analyzing the sign and

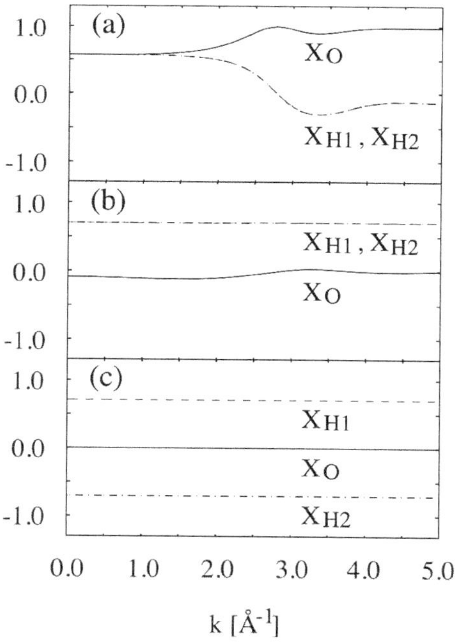

Fig. III-6.3: Contribution of atoms, O, H1 and H2 to (a) the acoustic mode, (b) OM-I mode. and (c) OM-2 mode. Copyright (1999) American Institute of Physics.

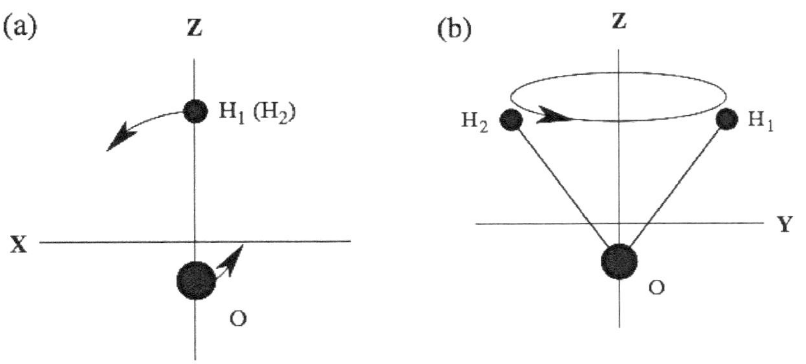

Fig. III-6.4: Two rotational modes of a water molecule. Copyright (1999) American Institute of Physics.

magnitude of $x_\alpha(k)$'s it is possible to obtain the information concerning how each atom contributes to the mode.

Plotted in Fig. III-6.3 is the contribution of each atom to the acoustic and two optical modes. It is seen from Fig. III-6.3(a) for the acoustic mode that $x_O(k) \approx x_{H_1}(k) = x_{H_2}(k)$ holds well in the small k region, which is consistent with the fact that the sound mode from the center-of mass translational motion of the molecules, that is, each atom in the molecule equally contributes to this mode. On the other hand, as can be seen from Fig. III-6.3(b), OM-I is governed by the lighter hydrogen atoms over the entire wavelength range because the rotational motion of a molecule is dominated by the motion of the lighter atoms that are located farther from the center of mass. Figure III-6.3(c) shows that OM-II is just related to the rotational motion in which two hydrogen atoms move in an out-of-phase fashion with oxygen atom fixed. More detailed analysis of the eigen vectors reveals that OM-I and OM-II are associated with the pitch and roll vibrational motions of molecules, respectively (Fig. III-6.4).

III-7 Site-site Mode Coupling Theory and its Applications

The generalized Langevin theory described so far in this chapter is based on an idea that few slow varying modes, or variables, are more or less decoupled in time from other modes which are varying fast. Motion of a Brownian particle, like a pollen particle in water, is a perfect example of such slow modes: relaxation of velocity and/or displacement of such a particle is very slow compared to change in the fluctuational or random force originated from the collisions of surrounding water molecules (see Section III-1). Such complete decoupling of the modes makes relaxation of the friction *memoryless*, or "Markovian." In the generalized Langevin formalism, we have selected the density field and its momentum as the slow variables, and projected all other variables onto the slow modes. However, the decoupling of modes in this case may not be so perfect as in the case of a Brownian particle because the density field and its conjugated momentum are also defined in the atomic resolution, as in the other variables. In such a case, the correlation time of *friction* concerning the density field may not be a delta-function in time, and it may have some *memory*. In this section, we briefly review an approximation to calculate the friction with memory. The theory is called "Mode Coupling Theory (MCT)" [Boon and Yip 1980, Hansen and McDonald 1986, Balucani and Zoppi 1994, Chong and Hirata 1998b, Hirata 2003].

In Section III-3, we have defined a random force by Eq. (III-3.9), that is a quantity in the space orthogonal to the density field. So, projection of the random force onto the density field should be zero by definition. However, square of the density field is not necessarily orthogonal to the random force. The mode coupling theory takes the projection of the random force onto the square density as the physical origin of the memory. Let us refer to the new dynamic variables as "secondary dynamic variables," whereas we call the dynamic variables in the preceding sections as "primary dynamic variables."

We define the secondary dynamic variables,

$$A_{\lambda\nu}(\mathbf{k},\mathbf{k}') \equiv \delta\rho_\lambda(\mathbf{k})\delta\rho_\nu(\mathbf{k}') \tag{III-7.1}$$

The projection of a variable X can be defined as

$$\mathcal{P}_2 X(k) = \frac{1}{2}\sum_k \sum_{\lambda,\mu,\lambda'\mu'} \frac{A_{\lambda\mu}\cdot\left(A_{\lambda'\mu'}, X(k)\right)}{(A_{\lambda\mu}, A_{\lambda'\mu'})} \tag{III-7.2}$$

Using the secondary projection operator, we approximate the projection of the random force onto the space complementary to the primary dynamic variables by

$$\exp[i(1-\mathcal{P})\mathcal{L}t] \approx \mathcal{P}_2 \exp(i\mathcal{L}t)\mathcal{P}_2 \tag{III-7.3}$$

The friction kernel in the mode coupling approximation can be written as

$$\mathbf{K}_{MCT}(k,t) \equiv (\mathcal{P}_2 R(\mathbf{k}), \exp(i\mathcal{L}t)\mathcal{P}_2 R(\mathbf{k}))\mathbf{J}^{-1}(k) \tag{III-7.4}$$

The denominator of Eq. (III-7.2), $(A_{\lambda\mu}, A_{\lambda'\mu'})$, contains the four-body density correlation functions. It is a usual approximation to decouple the correlation function into the pair correlation functions as

$$\begin{aligned}(A_{\lambda\mu}, A_{\lambda'\mu'}) &= \frac{1}{N}\left\langle \delta\rho_\lambda^*(\mathbf{k})\delta\rho_\mu^*(\mathbf{k}'-\mathbf{k})\delta\rho_{\lambda'}(\mathbf{k})\delta\rho_{\mu'}(\mathbf{k}'-\mathbf{k})\right\rangle \\ &\approx \frac{1}{N}\left\langle \delta\rho_\lambda^*(\mathbf{k})\delta\rho_{\lambda'}(\mathbf{k})\right\rangle\left\langle \delta\rho_\mu^*(\mathbf{k}'-\mathbf{k})\delta\rho_{\mu'}(\mathbf{k}'-\mathbf{k})\right\rangle\end{aligned} \tag{III-7.5}$$

Then, the memory kernel can be written in terms of the intermediate structure factors as

$$\begin{aligned}\left[K_{MCT}(k,t)\right]_{\alpha\beta} = \frac{\rho}{(2\pi)^3}\sum_{\lambda,\mu,\nu}\int d\mathbf{k}'\{&k_z'^2\left[\omega c(k')\right]_{\lambda\mu}\left[\omega c(k')\right]_{\beta\nu} F_{\mu\nu}(k',t)F_{\lambda\beta}(k-k',t) \\ &+ k_z'(k-k_z')\left[\omega c(k')\right]_{\lambda\mu}\left[\omega c(k-k')\right]_{\beta\nu} F_{\mu\beta}(k',t)F_{\lambda\nu}(k-k',t)J_{\alpha\lambda}(k)\end{aligned} \tag{III-7.6}$$

where $[\omega c(k)]_{\lambda\mu}$ is defined by

$$[\omega c(k)]_{\lambda\mu} \equiv \sum_\eta \omega_{\lambda\eta}(k)c_{\eta\mu}(k) \tag{III-7.7}$$

Dynamics of an ion in water

In the preceding section, we have seen how the two pictures of liquid dynamics, *translational* motion of interaction-sites (or atoms) and *translational and rotational* motions of a whole molecule, can be transformed from one to the other. In the following, the dynamics of an ion in water is analyzed from the view point of the interaction-site dynamics [Chong and Hirata 1998c, 1999c].

The dynamics of a small ion in water, such as alkali and halide ions, has attracted a lot of attention from chemists due to its practical interest in, say, the electrochemistry. The problem has also fascinated many theoretical physicists including Nobel Laureates such as M. Born and L. Onsager [Born 1920, Zwanzig 1963, Habbard and Onsager 1977, Wolynes 1978]. Why is the problem so attractive? It is because the dynamics of small ions exhibits a *non-Stokes-Einstein* behavior. It is well regarded that the friction exerted on a large ion from solvent follows the Stokes-Einstein law as

$$\varsigma_{Stokes} = 6\pi\eta\sigma$$

where η and σ denote the viscosity of solvent and the mass of solute. The law says that the friction increases linearly with the increasing size of solute. However, the friction on alkali-metal ions (Li^+, Na^+, K^+, Rb^+, Cs^+) and halide ions (F^-, Cl^-, Br^-, I^-) decreases with increasing ion size when the size is less, and then turns into increase, as illustrated by a cartoon in Fig. III-7.1.

The latter behavior can be explained in terms of the Stokes-Einstein law, while the behavior in the small ion-size is against the law. Two models have been proposed to explain the anti-Stokes-Einstein behavior of the ion friction (Fig. III-7.2). One is (a) "solventberg model" and the other (b) "dielectric friction model." The solventberg model that has been supported by many experimental chemists regards the Stokes radius as an effective radius of a solvated ion. If the strength of solvation is proportional to the electrostatic interaction between ion and solvent, the effective (or Stokes) radius of a solvated ion will increase with decreasing the size of bare ion. That explains the anti-Stokes-Einstein behavior in small ion size. On the other hand, the dielectric friction model, that was proposed and supported mainly by theoretical physicists including M. Born, sees the origin of the anti-Stokes-Einstein behavior in the rotational relaxation of solvent molecules around an ion. Let's imagine that an ion in polar solvent is in equilibrium. The ion would be in stable state by polarizing the surrounding solvent *isotropically.* Suppose that the ion is displaced abruptly due to a perturbation, such as an electrostatic field. The surrounding solvent may not follow the ion displacement instantaneously, to make the polarization anisotropic (see the left hand side of Fig. III-7.2(b)). The solvent polarization should relax (rotational or dielectric relaxation) to recover the equilibrium with new position of the ion, and to restore the isotropic polarization. Energy as well as momentum dissipation associated with the relaxation is claimed to be an origin of the friction.

Chong and Hirata have analyzed the friction exerted on an ion in water by means of the GLE/RISM theory, described above. In the theory, the friction is considered as a response of the collective excitation of solvent, described in the preceding section, to the perturbation or ion-

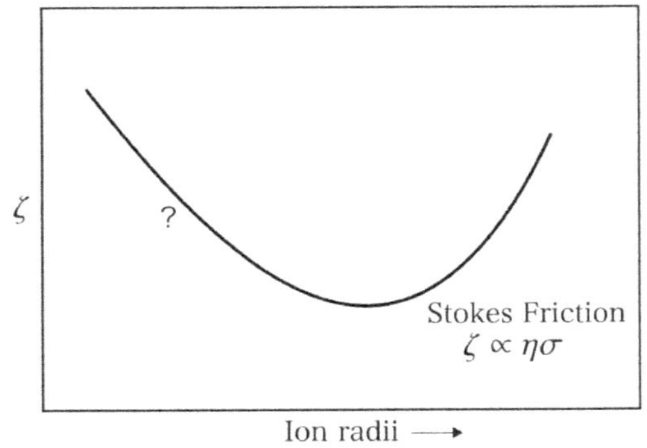

Fig. III-7.1: Illustration of the friction coefficient exerted on an ion in water.

(a) Solventberg model

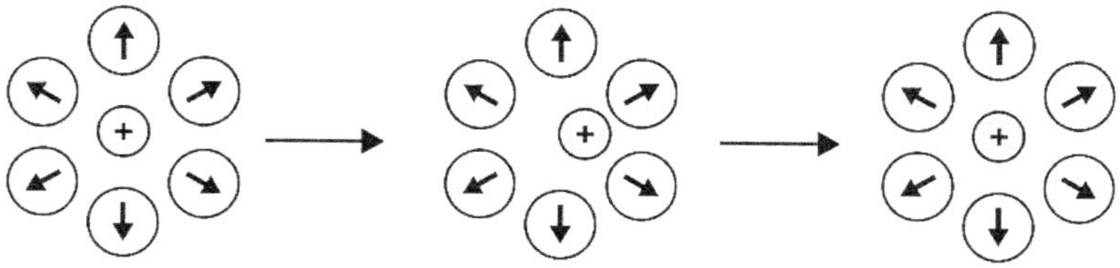

(b) Dielectric friction model

Fig. III-7.2: Two models for the friction exerted on an ion.

water interaction [Chong and Hirata 1998c, 1999c]. The ion-water interaction is decomposed into two contributions, the short-range interaction and the electrostatic interaction, while the response of the collective excitation is also decomposed into two contributions, the acoustic mode and the optical mode. Then, the friction coefficient is decomposed into the three contributions (the detail is skipped) as,

$$\varsigma = \varsigma_{NN} + \varsigma_{NZ} + \varsigma_{ZZ}$$

where ς_{NN} is the response of the acoustic mode of solvent to the short range interaction, that corresponds to the Stokes friction in the hydrodynamic model. ς_{ZZ} is the response of the optical mode to the electrostatic interaction, which corresponds to the dielectric friction in its physical meaning. Finally, ς_{NZ} is the cross term, namely the response of the acoustic mode to the electrostatic interaction and that of the optical mode to the short-range interaction. Depicted by circles in Fig. III-7.3 is the friction coefficient ς against the ionic radii. In the same figure, each component of ς is plotted by triangles (ς_{NN}), squares (ς_{ZZ}), and diamonds (ς_{NZ}). The overall friction coefficient (ς) exhibits an anti-Stokes-Einstein behavior in the small ion-size region, and follows the Stokes-Einstein law in the larger ion-size region, in accordance with the experimental observation. Then, how does each component of ς behave? The component from the optical mode ς_{ZZ} decreases monotonically with increasing the ion size both for cations and anions, indicating that the dielectric friction reduces as the electrostatic interaction decreases. The behavior proves that an origin of the anti-Stokes-Einstein behavior of ionic friction is the dielectric friction. On the other hand, ς_{NN} also shows anti-Stoke-Einstein behavior, except for Li^+, in the small ion-size region. This is due to reduction of the effective radii, represented by the direct correlation function, with increasing (bare) ion size. This indicates that the *solventberg* model also explains the anti-Stoke-Einstein behavior to some extent. However, the model breaks down for the Li^+. This may be because the effective ion radius is not proportional to the electric field for such an extremely small ion.

Chong and Hirata also calculated the velocity auto-correlation functions ($z(t) = \langle v(0)v(t)\rangle$) of ions in water, which is depicted in Fig. III-7.4. This is the first result for the velocity auto-

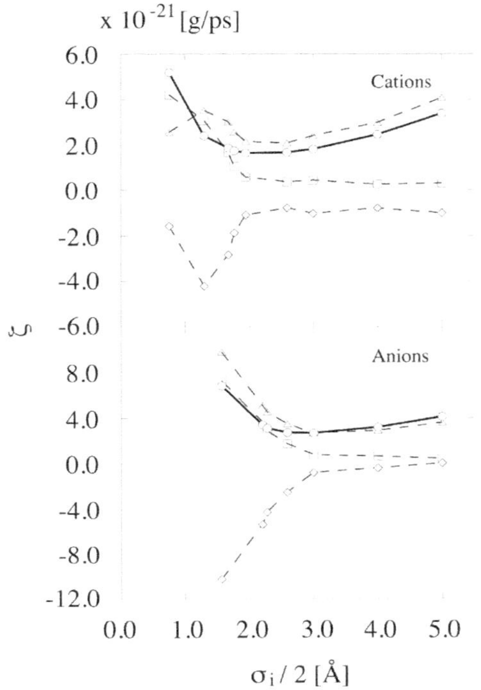

Fig. III-7.3: The friction exerted on small ions in water. Copyright (1999) American Institute of Physics.

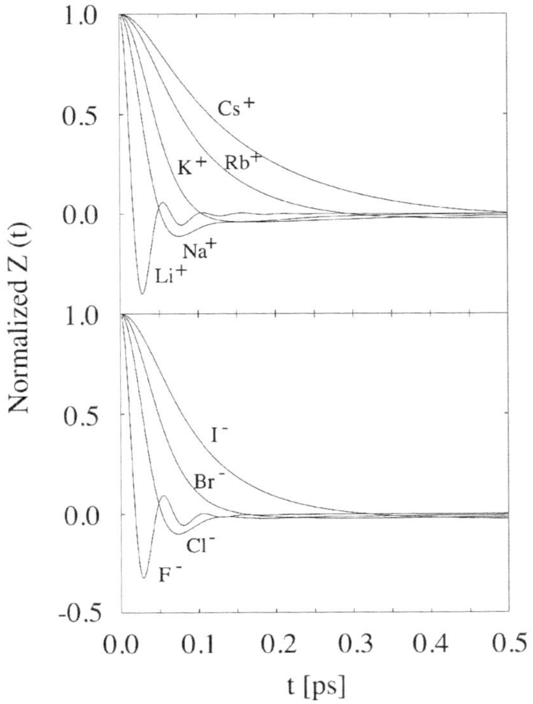

Fig. III-7.4: Velocity auto-correlation functions of ions in water. Copyright (1999) American Institute of Physics.

correlation function of ions in water, obtained entirely analytically by means of the statistical mechanics. The qualitative behavior looks very similar to the results from molecular dynamics simulation carried out by Koneshan et al. [Koneshan et al. 1998]. The most important feature seen in the figure is the oscillatory behavior observed in the auto-correlation functions for small ions, Li^+ and F^-. Those small ions are strongly coordinated by water molecules, and make a water-ion complex that has a considerably long lifetime compared with the other mono-valent ions. In other words, the water-ion complex as a unit makes diffusive motion in water for some period. The ion can oscillate within the complex for some period characterized by the relaxation time. The oscillatory behavior seen in $z(t)$ for the small ions reflects this physics.

III-8 Change of Equilibrium States Induced by a Perturbation

The generalized Langevin theory introduced in the previous section is a quite powerful tool to trace the time evolution of ensemble average of dynamic variables in an *equilibrium* system. However, the theory by itself cannot describe the change of equilibrium states, or the relaxation process. It is the linear response theory pioneered by R. Kubo in 1957 that can be applied to such a process [Kubo 1957]. The theory gives a beautiful expression for time evolution of dynamic variables of the system after getting a perturbation, in terms of the time correlation functions in an *equilibrium* system. Kubo's original version of the theory is developed in the general framework of the quantum mechanics. So, the derivation of equations is much more demanding than that required for applications to classical systems that is our main concern in this book. Such a theory applicable to classical many-body systems was developed by H.L. Friedman and his coworkers [Friedman 1964, 1985]. The theory was successfully applied to a variety of relaxation processes including the electric conductivity in solution, the dynamic Stokes shift, and so on. In the present section, we briefly review the classical version of the linear response theory. The theory is applied to the process of dynamic Stokes shift as an example.

III-8.1 Linear Response Theory

Let us express the probability distribution function of a representative point Γ in the phase space of an equilibrium state by

$$f_{eq}(\Gamma) = \frac{\exp\left(-\beta H^{(0)}\right)}{\int \exp\left(-\beta H^{(0)}\right) d\Gamma} \tag{III-8.1}$$

where $H^{(0)}$ denotes the Hamiltonian in equilibrium state. The equilibrium state will change if one applies a perturbation. If we apply a perturbation to the system, the Hamiltonian will be changed. We define the new Hamiltonian by

$$H(t) = H^{(0)} + H^{(1)} \tag{III-8.2}$$

where $H^{(1)}$ is the perturbation Hamiltonian defined by,

$$H^{(1)} = -AK(t) \tag{III-8.3}$$

The Liouville equation corresponding to the new Hamiltonian can be written as

$$\begin{aligned}\frac{\partial f(t)}{\partial t} &= -i\mathcal{L}f(t) \\ &= \left[H^{(0)} + H^{(1)}, f(t)\right] \\ &= \left[H^{(0)}, f(t)\right] + \left[H^{(1)}, f(t)\right] \\ &= -i\mathcal{L}^{(0)}f(t) - \left[A, f(t)\right] K(t)\end{aligned} \quad \text{(III-8.4)}$$

Now, let us denote the change of the distribution function due to the perturbation by $\delta f(t)$. Then,

$$f(t) = f^{(0)} + \delta f(t) \quad \text{(III-8.5)}$$

Substituting Eq. (III-8.5) into Eq. (III-8.4), and keeping the terms which are linear to the perturbation, one gets for the time evolution of the perturbed distribution function,

$$\frac{\partial \delta f(t)}{\partial t} = -i\mathcal{L}^{(0)}\delta f(t) - \left[A, f^{(0)}\right] K(t) \quad \text{(III-8.6)}$$

By integrating both sides of the equation with time, one finds,

$$\delta f(t) = -\int_{-\infty}^{t} \exp\left[-i(t-s)\mathcal{L}^{(0)}\right]\left[A, f^{(0)}\right] K(s)ds \quad \text{(III-8.7)}$$

The equation can be readily proved by differentiating both sides of the equation by t. Time evolution of a physical quantity $B(\Gamma)$ can be obtained by a non-equilibrium ensemble average of $\delta f(t)$ as

$$\begin{aligned}\langle \delta B(t)\rangle &= \int B(\Gamma)\delta f(t)d\Gamma \\ &= -\int d\Gamma B(\Gamma)\int_{-\infty}^{t} \exp\left[-i(t-s)\mathcal{L}^{(0)}\right]\left[A, f^{(0)}\right] \\ &= -\int_{-\infty}^{t} dsK(s)\int d\Gamma B(\Gamma)\exp\left[-i(t-s)\mathcal{L}^{(0)}\right]\left[A, f^{(0)}\right]\end{aligned} \quad \text{(III-8.8)}$$

where the Poisson bracket is defined by

$$\left[A, f^{(0)}\right] = \sum_{i=1}^{3N}\left[\frac{\partial A}{\partial q_i}\frac{\partial f^{(0)}}{\partial p_i} - \frac{\partial A}{\partial p_i}\frac{\partial f^{(0)}}{\partial q_i}\right] \quad \text{(III-8.9)}$$

where $f^{(0)}$ is the equilibrium distribution function,

$$f^{(0)} \sim C\exp\left(-\beta H^{(0)}\right) \quad \text{(III-8.10)}$$

By differentiating both sides of the equation with respect to the momentum p_i, one gets,

$$\frac{\partial f^{(0)}}{\partial p_i} = -\beta\frac{\partial H^{(0)}}{\partial p_i} f^{(0)} \quad \text{(III-8.11)}$$

Using Eq. (III-8.11), the right hand side of Eq. (III-8.9) can be written as

$$\begin{aligned}\left[A, f^{(0)}\right] &= -\beta\sum_{i=1}^{3N}\left[\frac{\partial A}{\partial q_i}\frac{\partial f^{(0)}}{\partial p_i} - \frac{\partial A}{\partial p_i}\frac{\partial f^{(0)}}{\partial q_i}\right] f^{(0)} \\ &= -\beta\left(i\mathcal{L}^{(0)}A\right)f^{(0)} \\ &= -\beta\dot{A}f^{(0)}\end{aligned} \tag{III-8.12}$$

Substituting Eq. (III-8.12) into the right hand side of Eq. (III-8.8),

$$\begin{aligned}\langle\delta B(t)\rangle &= \beta\int_{-\infty}^{t} dsK(s)\int d\Gamma B(\Gamma)\exp\left[-i(t-s)\mathcal{L}^{(0)}\right]\dot{A}f^{(0)}(\Gamma) \\ &= \beta\int_{-\infty}^{t} dsK(s)\left\langle B\dot{A}(t-s)\right\rangle \\ &= \beta\int_{-\infty}^{t} dsK(s)\Phi(t-s)\end{aligned} \tag{III-8.13}$$

where $\Phi(t)$ is defined by,

$$\Phi(t) \equiv \left\langle B\dot{A}(t)\right\rangle \tag{III-8.14}$$

III-8.2 Solvation Dynamics

Solvent dynamic response induced by abrupt change of the solute electronic structure, commonly referred to as "solvation dynamics," has been a target of intensive research in the field of femto-second molecular spectroscopy [Nomura and Kawaizumi 1994]. The most common way of probing the dynamics is to use the so-called "pump-probe" technique. In that technique, light with certain wavelength is first shined on a sample to *pump* solute molecules up to an electronic excited state. The process induces abrupt changes in the charged state of solute molecules. Since the new state of solute molecules is not equilibrium with configuration or polarization of surrounding solvent molecules, a relaxation dynamics of solvent molecules is induced to make solvent polarization to be equilibrated with the new state of solute. The process can be detected as the time resolved Stokes-shift of *probe* light with characteristic wave length. In that sense, the process is also called the "dynamics Stokes shift" in the community. The dynamic stokes shift is described by the dynamic response function defined by

$$S(t) = \frac{\nu(t)-\nu(\infty)}{\nu(0)-\nu(\infty)} \tag{III-8.15}$$

where $\nu(t)$ denotes characteristic wave length at time t of the spectra, or equivalently by

$$S(t) = \frac{E(t)-E(\infty)}{E(0)-E(\infty)} \tag{III-8.16}$$

Here, we introduce the theory of solvation dynamics developed by Friedman and his coworkers, which is based on the linear response theory described in the previous section, and the XRISM theory [Raineri et al. 1994, Friedman et al. 1995, Friedman 1985, Hirata and Rossky 1981].

Description of the model: We describe the electronic structure change of solute, as well as the dynamic response of solvent to the perturbation, in the regime of the classical mechanics. In that regime, the electronic-structure change of solute due to photo-excitation may be represented by the change in the partial charges of atoms. We also assume that the change of solute state due to the photo-excitation is abrupt, so that any dynamic process taking place inside solute is irrelevant to the solvation dynamics. Then, we may be able to write the Hamiltonian of the system as

$$H^D = H_{vv} + \hat{\Psi}_{uv}^D \tag{III-8.17}$$

where H_{vv} is the Hamiltonian of the bulk solvent molecules, and Ψ_{uv}^D means renormalized solute-solvent interaction. The superscript "D" labels the state of system before and after the photo-excitation, that is either "P" ($t < 0$; precursor state), or "S" ($t \to \infty$; successor state). With the definition of the renormalized Hamiltonian Eq. (III-8.17), the equilibrium distribution functions at $t < 0$ and $t \to \infty$ are expressed by,

$$f_{eq}^D(\Gamma) \propto \exp\left(-\beta H^D\right), \qquad D = P,\ S. \tag{III-8.18}$$

In the spirit of the linear response theory described in the previous section, we define the linearized distribution function by,

$$f_{eq}^D(\Gamma) \equiv f_{eq}^w(\Gamma)\left[1 - \beta\delta(\hat{\Psi}^D)\right], \qquad D = P, S \tag{III-8.19}$$

where $f_{eq}^w \propto \exp\left(-\beta H_w\right)$ is the equilibrium distribution function of bulk solvent. With Eq. (III-8.19), we define the average of a dynamic variable, $\hat{a}$, and its fluctuation by,

$$\left\langle \hat{a} \right\rangle \equiv \int d\Omega p(\Omega) \int d\Gamma f_{eq}^w(\Gamma)\hat{a} \tag{III-8.20a}$$

and

$$\delta\hat{a} \equiv \hat{a} - \left\langle \hat{a} \right\rangle. \tag{III-8.20b}$$

Therefore, $\delta(\hat{\Psi}^D)$ in Eq. (III-8.19) is given by $\delta(\hat{\Psi}^D) = \hat{\Psi}^D - \left\langle \hat{\Psi}^D \right\rangle$.

Time evolution after photoexcitation: The solvation dynamics following the photo-excitation is governed by the Liouville operator L^S associated with the Hamiltonian of solvent under renormalized field of the solute in the S-state. Thus, the time evolution of the solvent distribution function around the solute after the photo-excitation is governed by the following Liouville equation,

$$\left(\frac{\partial}{\partial t} + \mathcal{L}^S\right) f(\Gamma, t) = 0 \tag{III-8.21}$$

with the initial condition

$$f(\Gamma, 0) = f_{eq}^P(\Gamma) \tag{III-8.22}$$

in which $f_{eq}^P(\Gamma)$ is given by Eq. (III-8.18) with $D = P$.

Now, we separate the distribution function $f(\Gamma, t)$ and the Liouville operator into two parts, one that is independent of the solute-solvent interactions Ψ^D and the other that is linear in Ψ^D;

$$f(\Gamma, t) = f\ \ (\Gamma, t) + f\ \ (\Gamma, t) \tag{III-8.23a}$$

$$\mathcal{L}^S = \mathcal{L}_w + \mathcal{L}_\Psi^S \tag{III-8.23b}$$

where L_w is the Liouville operator of the homogeneous (bulk) solvent. Then, upto the first order in the solute-solvent interaction (linear approximation), one finds

$$\left(\frac{\partial}{\partial t}+\mathcal{L}_w\right)f^w(\Gamma,t)=0 \tag{III-8.24a}$$

$$\left(\frac{\partial}{\partial t}+\mathcal{L}_w\right)f^\Psi(\Gamma,t)=-\mathcal{L}_\Psi^S f^\Psi(\Gamma,t) \tag{III-8.24b}$$

which have the formal solution,

$$f^w(\Gamma,t)=\exp(-t\mathcal{L}_w)f^w(\Gamma,0) \tag{III-8.25a}$$

$$\begin{aligned} f^\Psi(\Gamma,t) &\equiv \exp(-t\mathcal{L}_w)y(t) \\ &= \exp(-t\mathcal{L}_w)y(0)-\int_0^t dt'\exp\left[-(t-t')\mathcal{L}_w\right]\mathcal{L}_\Psi^S f^w(\Gamma,t') \end{aligned} \tag{III-8.25b}$$

The first equality of Eq. (III-8.25b) defines the auxiliary function $y(t)$. The solutions can be readily verified by putting Eqs. (III-8.25a) and (III-8.25b) into Eqs. (III-8.24a) and (III-8.24b), respectively, and taking the derivative with respect to t.

The calculation is completed with the initial condition, (III-8.22), which separates according to the order in the renormalized interaction as,

$$f^w(\Gamma,0)=f_{eq}^w(\Gamma) \tag{III-8.26a}$$

$$f^\Psi(\Gamma,0)=-\beta f_{eq}^w(\Gamma)\delta\left(\Psi^P\right) \tag{III-8.26b}$$

When Eq. (III-8.26a) is substituted into Eq. (III-8.25a), we have

$$f^w(\Gamma,t)=\exp(-t\mathcal{L}_w)f_{eq}^w(\Gamma)=f_{eq}^w(\Gamma), \tag{III-8.27}$$

where $\mathcal{L}_w f_{eq}^w(\Gamma)=0$ is taken into account in deriving the second equality. Using this result, we can replace $\mathcal{L}_\Psi^S f^w(\Gamma,t')$ in the inhomogeneous term of Eq. (III-8.25b) by $\mathcal{L}_\Psi^S f^w(\Gamma)$, for which we can apply the relation,

$$\mathcal{L}_\Psi^S f_{eq}^w(\Gamma)=\beta f_{eq}^w(\Gamma)(\mathcal{L}_w\Psi^S) \tag{III-8.28}$$

Then, Eqs. (III-8.25b) and (III-8.26b) give

$$f^\Psi(\Gamma,t)=-\beta f_{eq}^w(\Gamma)\delta(\Psi^S)+\beta f_{eq}^w(\Gamma)\delta E(-t) \tag{III-8.29}$$

where

$$\hat{E}=H^S-H^P=\Psi^S-\Psi^P \tag{III-8.30}$$

is the renormalized energy gap between the precursor and successor states of the system, and $\delta\hat{E}(-t)$ is the result of applying the operator $\exp(-t\mathcal{L}_w)$ to the dynamic variable $\delta\hat{E}=\hat{E}-\langle\hat{E}\rangle$.

Substituting Eqs. (III-8.29) into Eq. (III-8.23a), one finds an expression for the time dependent distribution function as

$$f(\Gamma,t)=f_{eq}^S(\Gamma,t)+\beta f_{eq}^w(\Gamma)\delta\hat{E}(-t) \tag{III-8.31}$$

where $f^{S}_{eq}(\Gamma)$ is the equilibrium distribution function at the successor state ($D = S$), given by Eq. (III-8.19). In the derivation, Eq. (III-8.31), Eqs. (III-8.19) and (III-8.27) are taken into account.

It should be noted that the time evolution of an observable quantity can be calculated by taking an ensemble average of the corresponding dynamic variable using $f(\Gamma, t)$ in Eq. (III-8.31) as the probability distribution function. In the case of the solvation dynamics, we choose the renormalized energy gap defined by Eq. (III-8.30) as the dynamic variable. Then, the solvation time correlation function describing the dynamic Stokes shift corresponding to Eq. (III-8.16) can be defined with the renormalized energy gap (Eq. III-8.30) by,

$$S_E(t) \equiv \frac{\left\langle \hat{E}(t) \right\rangle - \left\langle \hat{E}(\infty) \right\rangle}{\left\langle \hat{E}(0) \right\rangle - \left\langle \hat{E}(\infty) \right\rangle} \tag{III-8.32}$$

Time correlation function: According to the linear response theory, the ensemble average of a dynamic variable at time t can be represented with its time correlation function as,

$$\left\langle \hat{G}(t) \right\rangle = \left\langle \hat{G} \right\rangle^{S} + \beta C_G(t) \tag{III-8.33}$$

where $C_G(t)$ is defined by

$$C_\varepsilon(t) = \left\langle \delta\hat{G}(t)\delta\hat{G}(0) \right\rangle = \int d\Gamma f^{w}_{eq}(\Gamma)\left[\exp(t\mathcal{L}_w)\delta\hat{G}\right]\delta\hat{G}(0) \tag{III-8.34}$$

Assuming that $C_G(t)$ vanishes at $t \to \infty$, it can be readily shown that

$$\left\langle \hat{G}(0) \right\rangle = \left\langle \hat{G} \right\rangle^{P}, \quad \left\langle \hat{G}(\infty) \right\rangle = \left\langle \hat{G} \right\rangle^{S}. \tag{III-8.35}$$

From Eqs. (III-8.32)~(III-8.35) with $\hat{G} = \hat{E}$, one finds the following expression for the time correlation function of solvation dynamics as,

$$S_E(t) = \frac{C_E(t)}{C_E(0)} = \frac{\left\langle \delta\hat{E}(t)\delta\hat{E}(0) \right\rangle}{\left\langle \delta\hat{E}(0)\delta\hat{E}(0) \right\rangle} \tag{III-8.36}$$

which corresponds to the experimental counterpart, Eq. (III-8.16). Here, it should be noted that the ensemble average $\langle \cdots \rangle$ included in the expression is taken with the probability distribution function concerning the homogeneous solvent without the perturbation from solute.

Renormalized potential by Interaction site model: So far, we have used the *renormalized potential* Ψ^D without providing a clear definition. Now, we identify what it is within the framework of the RISM theory. For that purpose, we first express Ψ^D in terms of the sum of the site-site renormalized interaction as

$$\Psi^D = \sum_{\lambda}\sum_{j}\int \hat{\rho}_{\lambda,j}(\mathbf{r})\phi^{D}_{\lambda j}(\mathbf{r})d\mathbf{r} \tag{III-8.37}$$

where $\phi^{D}_{\lambda j}(\mathbf{r})$ is the renormalized interaction between a solute site λ and a solvent site j, and $\hat{\rho}_{\lambda,j}(\mathbf{r})$ is the density field of the solvent site defined by

$$\hat{\rho}_{\lambda,j}(\mathbf{r}) = \sum_{a}\delta(\mathbf{r} - \mathbf{r}_{\lambda,aj}), \tag{III-8.38}$$

in which $\mathbf{r}_{\lambda,aj} \equiv \mathbf{r}_{aj} - \mathbf{r}_{\lambda}$ is the position of site j of a solvent molecule labeled with "a", relative to that of solute site λ. We also define the density fluctuation of solvent atom (site) j by

$$\delta\hat{\rho}_{\lambda,j}(\mathbf{r}) = \hat{\rho}_{\lambda,j}(\mathbf{r}) - \left\langle \hat{\rho}_{\lambda,j}(\mathbf{r}) \right\rangle, \tag{III-8.39}$$

where $\langle \rho_{\lambda,aj}(\mathbf{r}) \rangle = \rho$ is the average density of solvent. Accordingly, Ψ^D in Eq. (III-3.37) is expressed by

$$\Psi^D = \left\langle \Psi^D \right\rangle + \delta(\Psi^D) \tag{III-8.40}$$

where $\langle \Psi^D \rangle$ is defined as

$$\left\langle \Psi^D \right\rangle = \sum_{\lambda}\sum_{j} \rho \int \phi_{\lambda j}(\mathbf{r}) d\mathbf{r} \tag{III-8.41}$$

With the definition of the Fourier transform of $\phi_{\lambda j}(\mathbf{r})$,

$$\phi^D_{\lambda j}(\mathbf{k}) = \int \phi^D_{\lambda j}(\mathbf{r}) \exp\left(i\mathbf{k}\cdot\mathbf{r}\right) d\mathbf{r}, \tag{III-8.42}$$

Equation (III-8.30) can be written in the Fourier space as

$$\delta(\Psi^D) = \Psi^D - \left\langle \Psi^D \right\rangle = \frac{1}{(2\pi)^3} \int d\mathbf{k} \sum_{\lambda}\sum_{j} \delta\rho_{\lambda j}(\mathbf{k}) \phi^D_{\lambda j}(-\mathbf{k}) \tag{III-8.43}$$

where

$$\delta\hat{\rho}_{\lambda j}(\mathbf{k}) = \sum_a \exp\left[i\mathbf{k}\cdot\mathbf{r}_{\lambda,aj}\right] - (2\pi)^3 \rho\delta(\mathbf{k}) \tag{III-8.44}$$

or in a more convenient form,

$$\begin{aligned} \delta\hat{\rho}_{\lambda j}(\mathbf{k}) &= \exp\left[-i\mathbf{k}\cdot\mathbf{r}_{\lambda}\right]\left\{\left(\sum_a \exp\left[i\mathbf{k}\cdot\mathbf{r}_{aj}\right]\right) - (2\pi)^3 \rho\delta(\mathbf{k})\right\} \\ &= \exp\left[-i\mathbf{k}\cdot\mathbf{r}_{\lambda}\right]\delta\hat{\rho}_j(\mathbf{k}) \end{aligned} \tag{III-8.45}$$

The second equality in (III-8.45) defines the density field and its fluctuation in pure solvent.

Now we are ready to identify the renormalized potential $\phi_{\lambda j}(\mathbf{r})$ in terms of the interaction-site model of liquids. We first require the following relation to the density fluctuation of a solvent site j around a solute site λ in the D-state of solute.

$$\left\langle \delta\hat{\rho}_{\lambda,j}(\mathbf{k}) \right\rangle^D = \rho h^D_{\lambda j}(k) \tag{III-8.46}$$

where $h^D_{\lambda j}(k)$ is the density pair correlation function between solute site λ and solvent site j. The relation can be derived naturally by means of the "Percus trick" explained in Chapter II. The ensemble average in the left-hand-side of Eq. (III-8.46) can be evaluated with the distribution function defined by Eq. (III-8.19). After noting that $\langle \delta\hat{\rho}_{\lambda,j}(\mathbf{k}) \rangle = 0$, we have,

$$\begin{aligned} \left\langle \delta\hat{\rho}_{\lambda,j}(\mathbf{k}) \right\rangle^D &= \int d\Omega p(\Omega) \int d\Gamma f^D_{eq}(\Gamma) \delta\hat{\rho}_{\lambda,j}(\mathbf{k}) \\ &= \int d\Omega p(\Omega) \int d\Gamma f^W_{eq}(\Gamma)\left[1 - \beta\delta(\Psi^D)\right]\delta\hat{\rho}_{\lambda,j}(\mathbf{k}) \\ &= -\frac{\beta}{(2\pi)^3} \int d\mathbf{k}' \sum_{\lambda'}\sum_{l} \left\langle \delta\hat{\rho}_{\lambda,j}(\mathbf{k}) \delta\hat{\rho}_{\lambda' l}(-\mathbf{k}') \right\rangle \phi_{\lambda' l}(-\mathbf{k}') \end{aligned} \tag{III-8.47}$$

where in the third equality we have used Eq. (III-8.33). For the static average, using Eqs. (III-8.20a), (III-8.35), we evaluate

$$\begin{aligned}\left\langle \delta\hat{\rho}_{\lambda,j}(\mathbf{k})\delta\hat{\rho}_{\lambda' l}(-\mathbf{k}')\right\rangle &= \left\langle e^{-i\mathbf{k}\cdot\mathbf{r}_\lambda} e^{-i\mathbf{k}'\cdot\mathbf{r}_{\lambda'}}\right\rangle_\Omega \left\langle \delta\hat{\rho}_j(\mathbf{k})\delta\hat{\rho}_l(\mathbf{k}')\right\rangle \\ &= (2\pi)^3\delta(\mathbf{k}+\mathbf{k}')\chi_{jl}(k)\omega_{\lambda\lambda'}(k)\end{aligned} \tag{III-8.48}$$

In the second equality,

$$\chi_{jl}(k) \equiv \frac{1}{V}\left\langle \delta\hat{\rho}_j(\mathbf{k})\delta\hat{\rho}_l(\mathbf{k}')\right\rangle \tag{III-8.49}$$

is the static site-site structure factor of the homogeneous solvent, while

$$\omega_{\lambda\lambda'}(k) = \left\langle e^{-i\mathbf{k}\cdot\mathbf{r}_\lambda} e^{-i\mathbf{k}'\cdot\mathbf{r}_{\lambda'}}\right\rangle_\Omega = \int d\Omega p(\Omega) e^{-i\mathbf{k}\cdot(\mathbf{r}_\lambda - \mathbf{r}_{\lambda'})} \tag{III-8.50}$$

is the intramolecular site-site correlation function of solute. Combining Eqs. (III-8.47) and (III-8.48), one finds

$$\left\langle \delta\hat{\rho}_{\lambda,j}(\mathbf{k})\right\rangle^D = -\beta\sum_{\lambda'}\sum_l \omega_{\lambda\lambda'}(k)\chi_{jl}(k)\phi^D_{\lambda' l}(\mathbf{k}) \tag{III-8.51}$$

On the other hand, the right-hand side of Eq. (III-8.36) may be evaluated with the RISM equation for the solute-solvent site-site pair correlation functions $h^D_{\lambda j}(k)$. (Chapter II) For the solute at infinite dilution in the solvent, we have

$$\rho h^D_{\lambda j}(k) = \sum_{\lambda'}\sum_l \omega_{\lambda\lambda'}(k) c^D_{\lambda' l}(k)\chi_{lj}(k) \tag{III-8.52}$$

Combining Eqs. (III-8.46), (III-8.51), and (III-8.52), we identify the renormalized solute-solvent pair interaction $\phi^D_{\lambda j}(\mathbf{k})$ as proportional to the corresponding solute-solvent correlation function

$$-\beta\phi^D_{\lambda j}(k) = c^D_{\lambda j}(k) \tag{III-8.53}$$

with an equivalent relation in **r**-space.

Solvation Time correlation function: With the identification of the renormalized solute-solvent pair interactions, we define the energy gap $\delta\hat{E}$ to drive the solvation dynamics as,

$$\delta\hat{E} = \frac{1}{(2\pi)^3}\int d\mathbf{k}\sum_\lambda\sum_j \delta\hat{\rho}_{\lambda j}(\mathbf{k})\Delta\phi_{\lambda j}(-\mathbf{k}) \tag{III-8.54}$$

where

$$-\beta\Delta\phi_{\lambda j}(r) = \Delta c_{\lambda j}(r) \tag{III-8.55}$$

and

$$\Delta c_{\lambda j}(r) \equiv c^S_{\lambda j} - c^P_{\lambda j} \tag{III-8.56}$$

With Eq. (III-8.54), $C_E(t)$ in Eq. (III-8.56) is evaluated as

$$C_E(t) = \frac{1}{(2\pi)^6}\int d\mathbf{k}d\mathbf{k}'\left\langle \delta\hat{\rho}_{\lambda j}(\mathbf{k},t)\delta\hat{\rho}_{\lambda' l}(\mathbf{k}')\right\rangle \Delta\phi_{\lambda j}(-\mathbf{k})\Delta\phi_{\lambda' l}(-\mathbf{k}') \tag{III-8.57}$$

In the equation, $\langle \delta\hat{\rho}_{\lambda j}(\mathbf{k}, t)\delta\hat{\rho}_{\lambda' l}(\mathbf{k}')\rangle$ can be evaluated in the same manner as Eq. (III-8.48) to give,

$$\left\langle \delta\hat{\rho}_{\lambda j}(\mathbf{k},t)\delta\hat{\rho}_{\lambda' l}(\mathbf{k}')\right\rangle = (2\pi)^3\delta(\mathbf{k}+\mathbf{k}')F_{jl}(k,t)\omega_{\lambda\lambda'}(k) \qquad \text{(III-8.58)}$$

where

$$F_{jl}(k,t) = \frac{1}{V}\left\langle \delta\hat{\rho}_j(\mathbf{k},t)\delta\hat{\rho}_l(-\mathbf{k})\right\rangle \qquad \text{(III-8.59)}$$

is the dynamic structure factor of the homogeneous solvent (cf. Eq. (III-4.29)). The initial value $C_E(t)$ of the renormalized energy gap time correlation function is obtained from Eqs. (III-8.57) and (III-8.58) with $F_{jl}(k, t = 0) = \chi_{jl}(k)$. (cf. (III-8.49)) Thus, we find for the normalized solvation time correlation function,

$$S_E(t) = \frac{\int_0^\infty dkk^2\sum_{jl}F_{jl}(k,t)B_{jl}(k)}{\int_0^\infty dkk^2\sum_{jl}\chi_{jl}(k)B_{jl}(k)} \qquad \text{(III-8.60)}$$

where $B_{jl}(k)$ is defined as

$$B_{jl}(k) = \sum_{\lambda\lambda'}\omega_{\lambda\lambda'}(k)\Delta\phi_{\lambda j}(k)\Delta\phi_{\lambda' l}(k) \qquad \text{(III-8.61)}$$

According to Eqs. (III.8.60) and (III-8.61), the solvation dynamics is driven by the renormalized solute-solvent interaction gap $\Delta\phi_{\lambda j}(k)$ between the precursor and successor states of the system, and the dynamics follows the function $F_{jl}(k, t)$ of the homogeneous solvent without perturbation from the solute. It is a consequence of the linear response treatment of the dynamics.

III-8.3 Application of the Theory to Model Systems

Here, an example of the application of the theory just described to simple model systems is presented [Nishiyama et al. 2009]. The system examined concerns two types of solvent, water as a representative of hydrogen-bonded liquids, and acetonitrile as non-hydrogen bonded solvent. On the other hand, the solute molecule is modeled by a single sphere with or without a charge, *e*, at the center, where *e* is the electronic charge. The photo-excitation process of the solute is mimicked by abrupt change of the charged state from zero to *e*. Three different sizes of solute are examined; chloride ion, and small (S) and large (L) solutes. (Solutes "S" and "L" are hypothetical solutes examined by Maroncelli and Fleming by means of the molecular dynamics simulation.) All the solvent-solvent and solute-solvent interactions are represented by the interaction-site model, in which the site-site interactions are described by the (12-6-1) type potential function (see Eq. (II-3.21) in Chapter II).

Two methods are employed to evaluate $F_{jl}(k, t)$, which have been explained earlier in this chapter: the site-site Smoluchowski-Vlasov (SSSV) equation (Section III-5), and the site-site generalized Langevin equation with the mode coupling theory (SSGLE/MCT) (Section III-7). The results from SSGLE/MCT are also compared with the results from the molecular dynamics (MD) simulation obtained by Maroncelli and Fleming for the hypothetical solutes, "S" and "L", mentioned above. The results of calculation are depicted in Fig. (III-8.1). (Note that $S_S(t)$ is used in place of $S_E(t)$ in order make the notation consistent with the literature.) Details of the calculations are seen in the literature [Nishiyama et al. 2009].

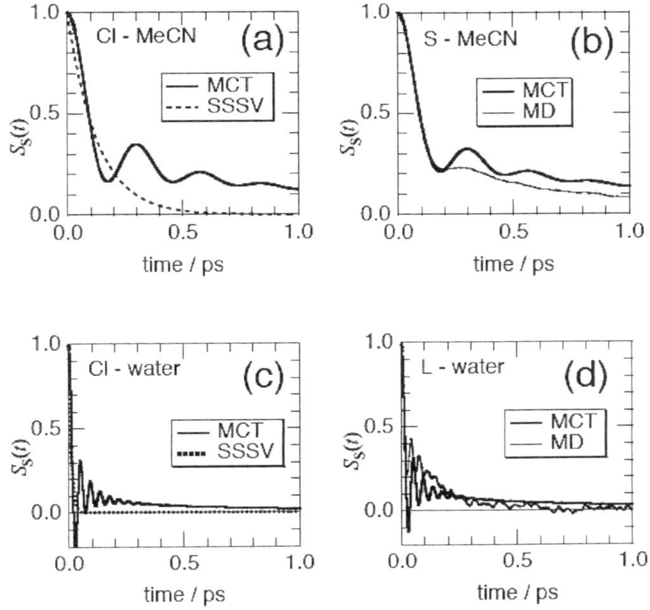

Fig. III-8.1: The solvation time correlation function calculated using Eq. (III-8.49). Copyright (2009) American Chemical Society.

***General feature of the dynamics*:** The general feature of the dynamics shared by all the methods and systems is the ultra-fast initial decay of the solvation energy. It may be attributed to the reorientation dynamics of solvent molecules along with the electrostatic field produced by the abrupt change of charge (0 → e) on the solute molecule. It is also apparent that the relaxation rate is quite different between the two solvents, acetonitrile (MeCN) and water: the relaxation rate of water is much faster than that of MeCN. The feature is captured by all the methods examined. Another important feature of the dynamics is the oscillatory behavior succeeding the initial decay, which is captured by SSGLE/MCT and MD, but not by SSSV. The reason why SSSV does not reproduce the behavior is simply because it assumes the diffusion limit. The oscillatory behavior apparently originated from the librational motion of solvent molecules around the solute. In that respect, the behavior has the physical origin that is common to the oscillation observed in the velocity auto-correlation function of a small ion in polar liquids, presented in Fig. (III-7.5) in the Section III-7.

In order to clarify the physics behind the initial decay and succeeding oscillation observed in $S_S(t)$, the quantity is decoupled into contributions from the acoustic mode, optical mode, and their coupling, following the method presented in the Section III-6, that is,

$$S_S(t) = C_{NN} S_{NN}(t) + C_{NZ} S_{NZ}(t) + C_{ZZ} S_{ZZ}(t), \tag{III-8.62}$$

where S_{NN}, S_{ZZ}, and S_{NZ} denote the contributions to the solvation time correlation function, respectively, from the acoustic mode, optical mode, and their coupling. The coefficients C_{NN}, C_{NZ}, and C_{ZZ} are the weight factor of each contribution. The results for the (0 → e) process of Cl^- in acetonitrile are depicted in Fig. (III-8.2). The results indicate that the short time relaxation of the time correlation function $S_S(t)$ is dominated by the optical mode $S_{ZZ}(t)$, that is, the rotational mode.

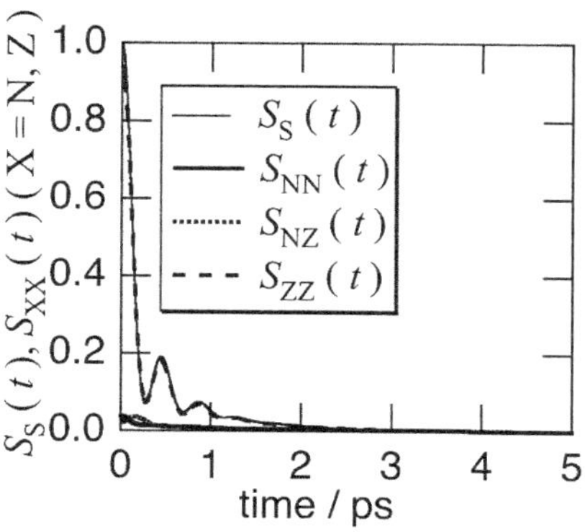

Fig. III-8.2: The solvation time correlation function decoupled into the acoustic and optical modes. Copyright (2009) American Chemical Society.

The Literature

Chandrasekhar gave a comprehensive description of the Brownian motion including Langevin theory and Fokker-Planck equation, which is summarized nicely by McQuarrie [Chandrasekhar 1943, McQuarrie 1976]. Friedman's book provides a good introduction to the irreversible statitistical mehcanics including the Liouville equation, generalized Langevin equation, and linear response theory in classical version [Friedman 1985]. For the mode coupling theory, one may find a good description in the book by Balucani and Zoppi [Balucani and Zoppi 1994].

CHAPTER IV

Theory of Biomolecular Solvation and Molecular Recognition

IV-1 Introduction

In living bodies, a variety of molecules are working in a concerted manner to maintain their life, and to pass on the genetic information from generation to generation. In that respect, life embodies two aspects, one as *matter* (or molecules) and the other as *information*.

Needless to say, protein, RNA, and DNA are among the main players of the *life theater*. Protein is a polymer generated from twenty different amino-acids through the condensation reaction that plays a variety of roles to maintain our life: catalyzing chemical reactions (enzyme); permeating molecules such as ions through membrane (molecular channels); conveying a molecule from an organ to the other such as hemoglobin, and so on (Table IV-1.1). Such a variety of functions played by protein originates from the specific molecular structure, which in turn is attributed to different number and sequence of amino-acids consisting the macromolecule (Fig. IV-1.1).

DNA is also a polymer made of four different nucleotides, adenine (A), guanine (G), cytosine (C), thymine (T), which stores genetic information of each species of biosystem that is inherited from generation to generation. The genetic information encoded in DNA is decoded in protein in such a way that a nucleotide triplet (or codon) corresponds to an amino-acid. DNA in a cell usually takes structure of helical double-strands, which are bound by hydrogen-bond between a pair of bases. A double strand wraps around protein called "histone" to make nucleosomes, which are further folded into the "chromosome." RNA is another polymer which is positioned between protein and DNA in the flow of the genetic information. RNA carries the sequence information corresponding to a protein, while DNA conveys the genetic information for all the proteins in a living body. The genetic information concerning a protein is transcripted from DNA onto RNA synthesized by a protein called *RNA synthetase*. The information is used as is or with some modification for synthesizing a protein. The two types of RNA concerned in the two steps are referred to as transfer RNA (*t*-RNA) and messenger RNA (*m*-RNA). The well-regarded hypothesis called "Central Dogma" concerns the flow of information in such a way as DNA ⇒ RNA ⇒ protein [Crick 1970].

Table IV-1.1: Amino acids.

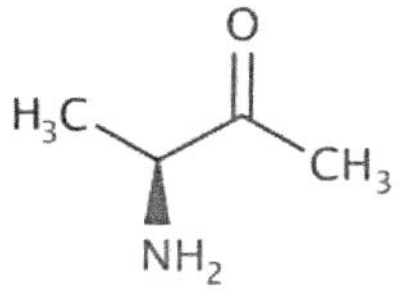

L-Alanine (Ala: A)

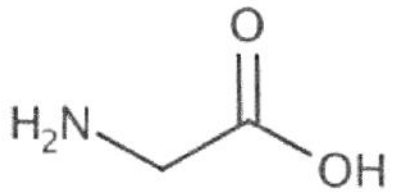

Glycine (Gly: G)

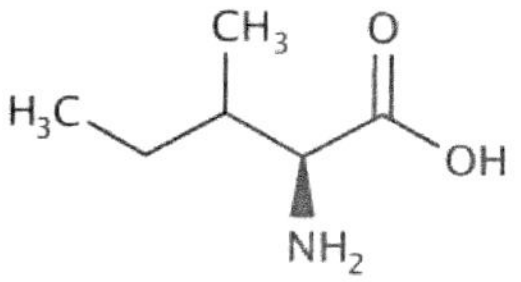

L-Isoleucine (Ile: I)

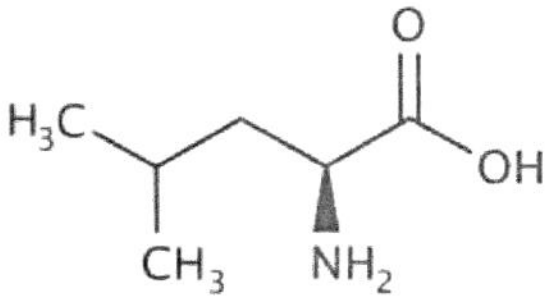

L-Leucine (Leu: L)

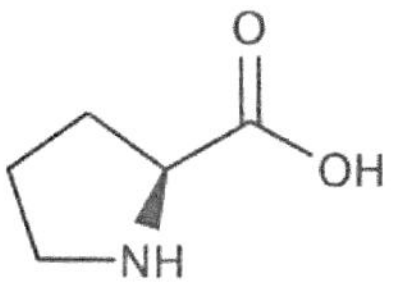

L-Proline (Pro: P)

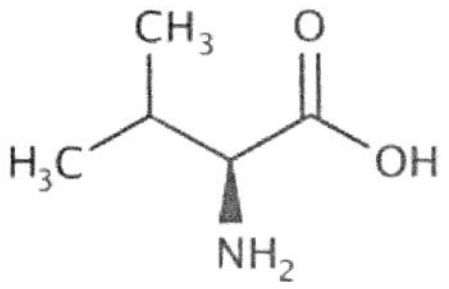

L-Valine (Val: V)

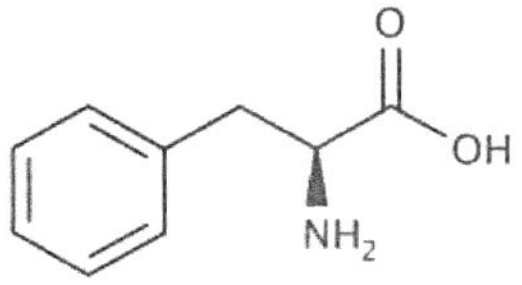

L-Phenylalanine (Phe: F)

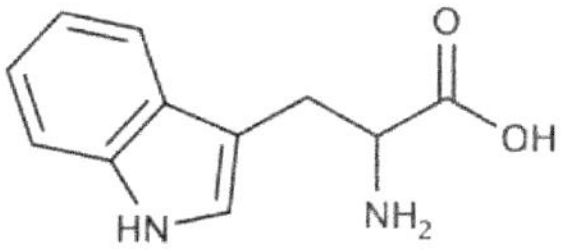

L-Tryptophan (Trp: W)

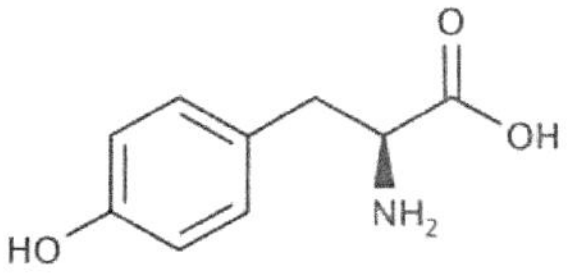

L-Tyrosine (Tyr: Y)

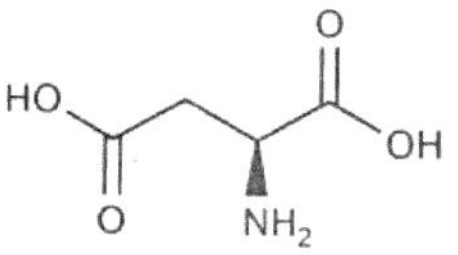

L-AsparaticAcid (Asp: D)

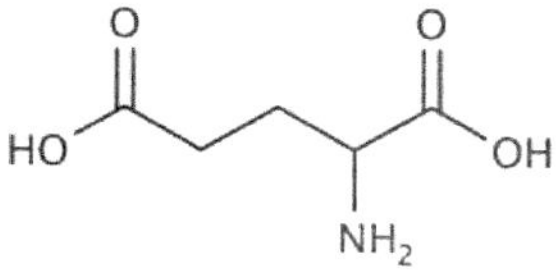

L-Glutamic Acid (Glu: E)

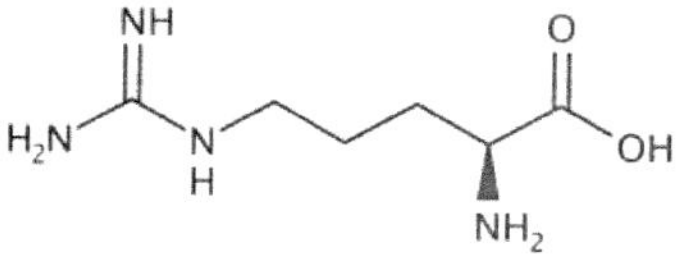

L-Arginine (Arg: R)

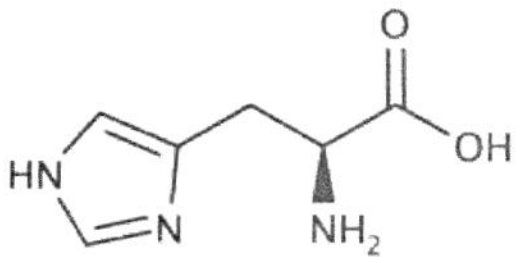

L-Histidine (His: H)

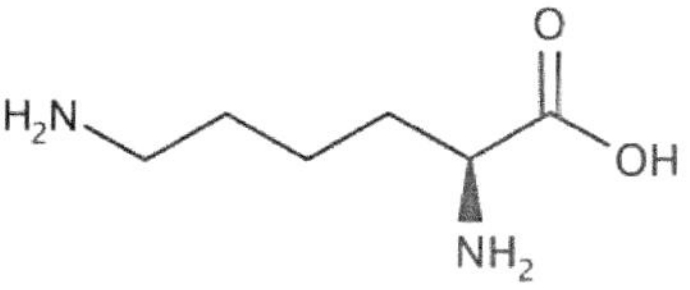

L-Lysine (Lys: K)

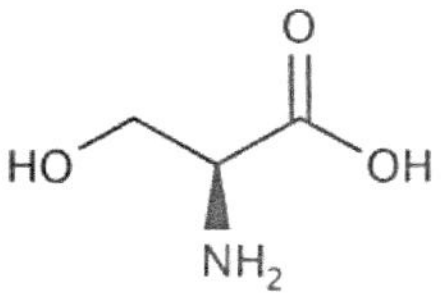

L-Serine (Ser: S)

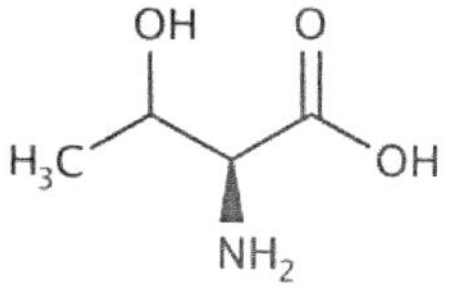

L-Threonine (Thr: T)

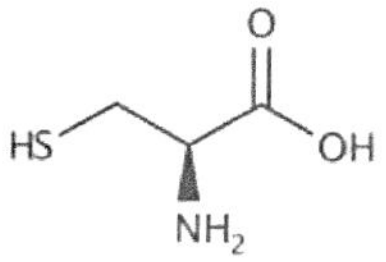

L-Cysteine (Cys: C)

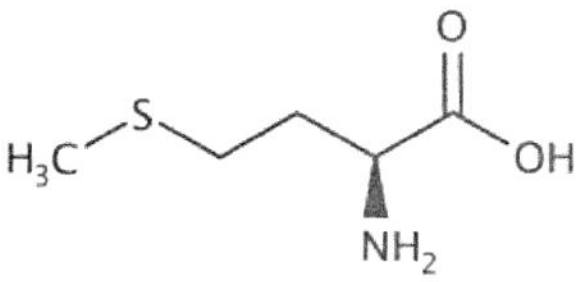

L-Methionine (Met: M)

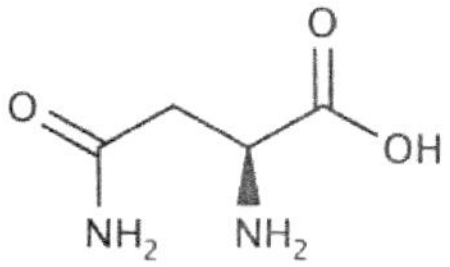

L-Asparagine (Arg: R)

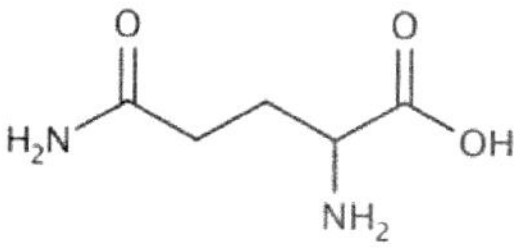

L-Glutamine (Gln: Q)

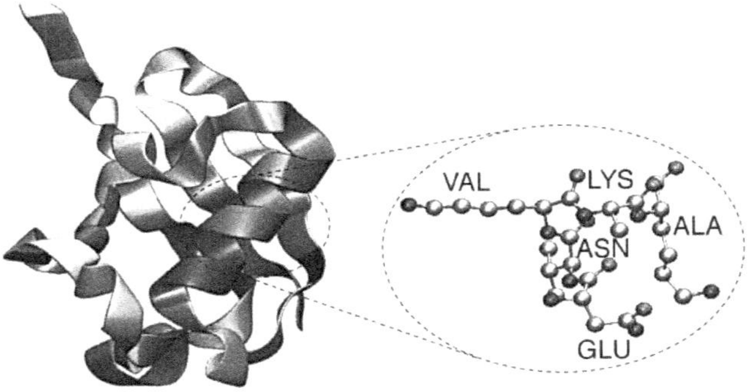

Fig. IV-1.1: Schematic view of protein structure.

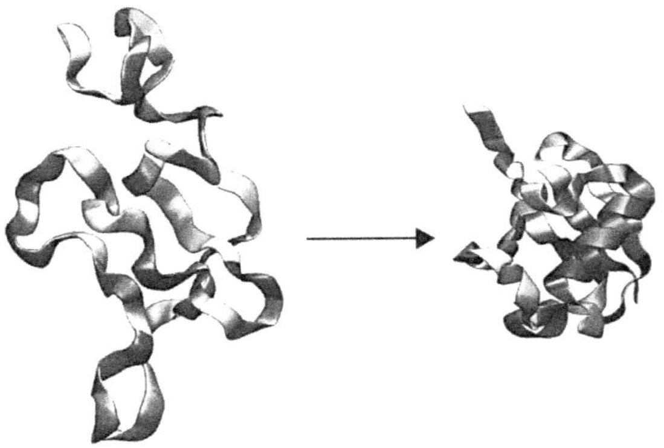

Fig. IV-1.2: Schematic view of protein folding.

If one focuses on the other aspect of life as *matter* or *molecules*, one will find two elementary physicochemical processes that are universal in all the bioactivities: the *self-organization* and the *molecular recognition processes* [Lehn 1990]. A typical example of the self-organization is the protein folding. Although protein is a polymer made up from amino acids through polymerization, it forms a specific conformation depending on its amino-acid sequence under a proper thermodynamic condition, unlike artificial polymers such as polyethylene [Anfinsen 1973]. The phenomenon is called "protein folding" (Figs. IV-1.2, IV-1.3).

Every protein has a specific structure to meet their specific rolls or functions in our body. It is just like constructions in human society: gas stations have a devise to supply gasoline to automobiles, restaurants have a kitchen to serve meal to their guests, and so on. Apparently, the process is not a spontaneous one if one just focuses on the structure of protein because the specific or native conformation has much less entropy compared to the non-native state having a random-coil structure. Why a protein can fold into a state of native conformation is because it is located in aqueous solutions, *not in vacuum*. We look into the problem in detail in the following chapter.

The other physico-chemical process universally seen in our body is the *molecular recognition* (MR). A process in which an enzyme binds substrate molecules in its reaction pocket is a typical example of the molecular recognition. A process in which an ion-channel accommodates an ion in its pore can also be regarded as a molecular recognition. Namely, the molecular recognition

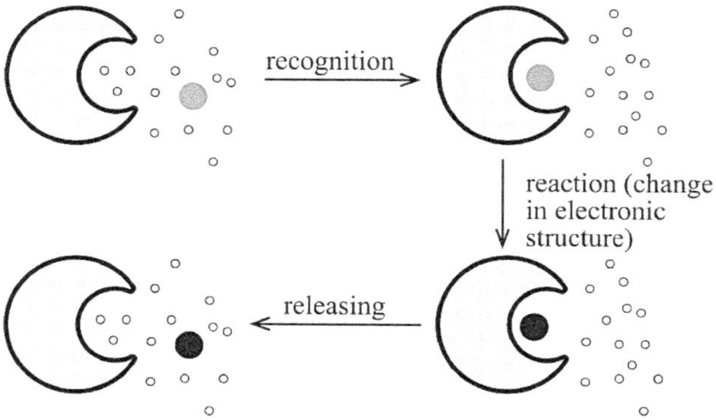

Fig. IV-1.3: A cartoon illustrating the enzymatic reaction. Gray crescent, enzyme; gray sphere, substrate before reaction; black sphere, substrate after reaction; small open spheres, water molecules. Note that water molecules are dehydrated from the reaction pocket of enzyme as the substrate comes in.

is an elementary process in life phenomena, through which a biomolecule should always pass, whenever the molecule performs its function. Let's think about an enzymatic reaction as an example of the molecular recognition (Fig. IV-1.3).

An enzymatic reaction is characterized essentially with the two equilibrium constants: one that concerns binding of substrate molecules at a reaction pocket, or the molecular recognition. The other equilibrium constant concerns a chemical reaction in the pocket, which is associated with a change in the electronic structure. The theory of reaction rate by Michaelis and Menten takes the two chemical equilibriums into consideration [Michaelis and Menten 1913]. Now, in general, the reaction pocket of an enzyme binds one or more water molecules and ions sometimes (see Fig. IV-1.3). Thereby, all or some of those water molecules and ions should be disposed from the reaction pocket to bulk solution, in order for a substrate molecule to be accommodated in the active site. On the other hand, a substrate molecule itself is more likely to be hydrated by water molecules. Whether or not a substrate molecule is desolvated upon the molecular recognition is determined by the entire free-energy change concerning the binding process, including the entropy change associated with the desolvation process. It indicates that the molecular recognition process is also governed by *water* [Yoshida et al. 2009].

It was emphasized so far that water molecules play essential roles in the two physicochemical processes governing life phenomena, "self-organization" and "molecular recognition." But, those are not only the roles water plays. In the enzymatic hydrolysis reaction, such as the hydrolysis of ATP, water molecules participate in the reaction as one of the reactants. Ions play crucial roles in many of chemical reactions in our body, but they can exist only in aqueous environment. A persistent belief that water does not play a crucial role for enzymatic reactions except for the "hydrolysis" is too naive. How many and in what configurations water molecules exist in a reaction pocket are essential factors to determine the reaction field of an enzyme. Any theory of chemical reactions disregarding the factors is senseless. The fact that approximately seventy percent of living body is occupied by water possesses a quite essential physicochemical significance.

Macromolecules in our body are distinguished from small molecules such as water or alcohol, treated in the previous chapter, by two aspects: large intramolecular degrees of freedom and extremely high spatial inhomogeneity at atomic level. Concerning the first aspect, let's estimate the conformational degrees of freedom for a protein molecule consisting of a hundred amino acids [Levinthal 1968]. If the protein molecule joined by peptide bonds can take three conformations at each residue, that creates ~ 3^{100} possible conformations. There is no problem if all the conformations are about the same. That is the case of a homopolymer consisting of the same unit monomers, or, segments. In such a case, the mean field type treatment, such as the random walk model described in the Chapter I of this book, works fine. Unfortunately, protein molecules in nature are not homopolymers in general that create a large variety in their structure. It in turn produces an inhomogeneous field in atomic scale that is different from protein to protein. In fact, such inhomogeneity is the most distinct feature of protein to perform their functions as an enzyme, ion channels, and so on. Therefore, a successful theory to treat the solvation of biopolymers should be able to account for the large degrees of freedom and the high inhomogeneity of a molecule at atomic scale. In the following section, we formulate a theory of biopolymer solvation called "3D-RISM/KH."

IV-2 Theory of Biomolecular Solvation (3D-RISM/KH Equation)

As was stated in the preceding section, the most important aspect which distinguishes the processes concerned with biomolecules, protein and DNA from ordinary chemical processes in solution is "inhomogeneity" of the field exerted on solvent molecules by the protein. For instance, the distribution of solvent, water and ions around a hydrophobic residue of protein is entirely different from that around a charged residue. The environment around an active site (or reaction pocket) may be even more different from that at the surface. Such inhomogeneity of the field at atomic level has been out-of-scope for the statistical mechanics for long time, and it is a target that has been bypassed by the conventional statistical mechanics including the ordinary or the one dimensional RISM theory [Chandler and Andersen 1972, Hirata and Rossky 1981, Hirata et al. 1982, 1983]. The situation was entirely revised by the 3D-RISM theory and its application to biomolecules. In particular, an application of the theory to protein made by Imai et al. has broken through a new horizon of biomolecular sciences [Imai et al. 2004, 2005b]. They could have probed water molecules bound in an active site of Lysozyme by means of the theory. The theory has been applied to a variety of the molecular recognition processes concerning functions of biomolecules, and has established its position as the theory of molecular recognition [Yoshida et al. 2009]. In the following, we briefly review the 3D-RISM/RISM theory [Beglov and Roux 1997, Kovalenko and Hirata 1998, Hirata 2003].

Derivation of 3D-RISM equation: As is briefly described in the Introduction, an "exact" equation referred to as the Ornstein-Zernike equation, which relates the pair correlation $h(1,2)$ with another correlation function called the direct correlation function $c(1,2)$, can be "derived" from the grand canonical partition function by means of the functional derivatives. Our theory to describe the molecular recognition starts from the Ornstein-Zernike equation generalized to a solution of polyatomic molecules, or the molecular Ornstein-Zernike (MOZ) equation,

$$\mathbf{h}^{uv}(1,2) = \mathbf{c}^{uv}(1,2) + \int \mathbf{c}^{uv}(1,2)\rho^{v}(3)\mathbf{h}^{vv}(3,2)d(3) \qquad \text{(IV-2.1)}$$

where $\mathbf{h}^{uv}(1,2)$ and $\mathbf{c}^{uv}(1,2)$ are the total and direct correlation functions, respectively, and the numbers in the parenthesis represent the coordinates of molecules in the liquid system, including both the position $\mathbf{R}$ and the orientation Ω. The boldface letters of the correlation functions indicate that they are matrices consisting of the elements labeled by the species in the solution. In the simple case of a binary mixture, the equation can be written down labeling the solute by "u" and solvent by "v" as follows. (It is straightforward to generalize the equations to the multi-component mixtures see Chapter II.)

The equations depend essentially on six coordinates in the Cartesian space, and it includes a six-fold integral. This integral is the one which prevents the theory from applications to polyatomic molecules. It is the interaction-site model and the Chandler-Andersen transformation, which is described in Chapter II. Here, we apply the transformation only to the coordinates of solvent, not to those of a biomolecule which is fixed in space (see Chapter II for the Chandler-Andersen transformation). First, we expand $\mathbf{h}^{vv}$ in the second term of the right hand side of Eq. (IV-2.1) in a series of ρ/Ω as

$$h^{vv}(1,2) = C\left[c^{vv}\middle|\rho/\Omega\right] \tag{IV-2.2}$$

where $C\left[c^{vv}\middle|\rho/\Omega\right]$ signifies the chain sum of c^{vv} (see Chapter II for the chain sum). Then, we apply the RISM approximation to $c^{vv}(1, 2)$,

$$c^{vv}(1,2) \sim \sum_{\alpha,\gamma} c^{vv}_{\alpha\gamma}(r) \tag{IV-2.3}$$

and the Chandler-Andersen transformation to the solvent coordinates.

$$h^{uv}_{\alpha}(\mathbf{r}) = \frac{1}{\Omega}\int \delta(\mathbf{R}_1)\delta\left(\mathbf{R}_2 + \mathbf{l}^{\alpha}_2 - \mathbf{r}\right)h^{uv}(1,2)d(2) \tag{IV-2.4}$$

The first two terms of the Chandler-Andersen transformation are explicitly given below for clarity. The first term reads,

$$I_1 = \int \delta(\mathbf{R}_1)\delta(\mathbf{R}_2 + \mathbf{l}^{\alpha}_2 - \mathbf{r})\sum_{\gamma} c^{uv}_{\gamma}\left(\mathbf{r}^{\gamma}_2 - \mathbf{R}_1\right)d\mathbf{R}_1 d\mathbf{R}_2 d\Omega_2 \tag{IV-2.5}$$

Defining the Fourier transform of $c^{uv}_{\gamma}(\mathbf{r})$ by

$$c^{uv}_{\gamma}(\mathbf{r}) = \int c^{uv}_{\gamma}(\mathbf{r})\exp\left[-i\mathbf{k}\cdot\mathbf{r}\right]d\mathbf{k} \tag{IV-2.6}$$

The first term can be transformed as follows

$$\begin{aligned}
I_1 &= \frac{\rho}{\Omega}\int \delta(\mathbf{R}_1)\delta(\mathbf{R}_2 + \mathbf{l}^{\alpha}_2 - \mathbf{r})\sum_{\gamma} c^{uv}_{\gamma}\left(\mathbf{r}^{\gamma}_2 - \mathbf{R}_1\right)d\mathbf{R}_1 d\mathbf{R}_2 d\Omega_2 \\
&= \sum_{\gamma}\frac{\rho}{\Omega}\iiiint \delta(\mathbf{R}_1)\delta(\mathbf{R}_2 + \mathbf{l}^{\alpha}_2 - \mathbf{r})\int c^{uv}_{\gamma}(\mathbf{k})\exp\left[-i\mathbf{k}\cdot\left(\mathbf{r}^{\gamma}_2 - \mathbf{R}_1\right)\right]d\mathbf{k}d\mathbf{R}_1 d\mathbf{R}_2 d\Omega_2 \\
&= \sum_{\gamma}\frac{\rho}{\Omega}\iiint \delta(\mathbf{R}_2 + \mathbf{l}^{\alpha}_2 - \mathbf{r})\int c^{uv}_{\gamma}(\mathbf{k})\exp\left[-i\mathbf{k}\cdot\mathbf{r}^{\gamma}_2\right]d\mathbf{k}d\mathbf{R}_2 d\Omega_2 \\
&= \sum_{\gamma}\frac{\rho}{\Omega}\iiint \delta(\mathbf{R}_2 + \mathbf{l}^{\alpha}_2 - \mathbf{r})\int c^{uv}_{\gamma}(\mathbf{k})\exp\left[-i\mathbf{k}\cdot\left(\mathbf{R}_2 + \mathbf{l}^{\gamma}_2\right)\right]d\mathbf{k}d\mathbf{R}_2 d\Omega_2 \\
&= \sum_{\gamma}\frac{\rho}{\Omega}\iint c^{uv}_{\gamma}(\mathbf{k})\exp\left[-i\mathbf{k}\cdot\left(\mathbf{r} - \mathbf{l}^{\alpha}_2 + \mathbf{l}^{\gamma}_2\right)\right]d\mathbf{k}d\Omega_2 \\
&= \sum_{\gamma}\rho\int d\mathbf{k}c^{uv}_{\gamma}(\mathbf{k})\frac{1}{\Omega}\int \exp\left[-i\mathbf{k}\cdot\left(\mathbf{l}^{\gamma}_2 - \mathbf{l}^{\alpha}_2\right)\right]d\Omega_2\exp[-i\mathbf{k}\cdot\mathbf{r}]d\mathbf{k} \\
&= \sum_{\gamma}\int d\mathbf{k}c^{uv}_{\gamma}(\mathbf{k})\rho\omega_{\gamma\alpha}(\mathbf{k})\exp[-i\mathbf{k}\cdot\mathbf{r}]d\mathbf{k} \\
&= \int d\mathbf{k}\sum_{\gamma}\left[\mathbf{c}^{uv}(\mathbf{k})\rho\mathbf{w}(\mathbf{k})\right]_{\gamma\alpha}\exp[-i\mathbf{k}\cdot\mathbf{r}]d\mathbf{k}
\end{aligned} \tag{IV-2.7}$$

The second term can be derived in the similar manner as,

$$
\begin{aligned}
I_2 &= \left(\frac{\rho}{\Omega}\right)^2 \iiiint\!\!\int \delta(\mathbf{R}_1)\delta(\mathbf{R}_2+\mathbf{l}_2^{\alpha}-\mathbf{r})\sum_{\gamma} c_{\gamma}^{uv}\left(\mathbf{R}_1-\mathbf{r}_3^{\gamma}\right)\sum_{\eta,\lambda} c_{\lambda\eta}^{vv}(\mathbf{r}_3^{\lambda}-\mathbf{r}_2^{\eta})d\mathbf{R}_1 d\mathbf{R}_2 d\Omega_2 d\mathbf{R}_3 d\Omega_3 \\
&= \left(\frac{\rho}{\Omega}\right)^2 \iiiint\!\!\int \delta(\mathbf{R}_1)\delta(\mathbf{R}_2+\mathbf{l}_2^{\alpha}-\mathbf{r})\sum_{\gamma}\int d\mathbf{k}c_{\gamma}^{uv}(\mathbf{k})\exp\left[i\mathbf{k}\cdot\left(\mathbf{r}_3^{\gamma}-\mathbf{R}_1\right)\right]\sum_{\eta,\lambda}\int d\mathbf{k}'c_{\lambda\eta}^{vv}(\mathbf{k}')\exp\left[-i\mathbf{k}'\cdot\left(\mathbf{r}_3^{\lambda}-\mathbf{r}_2^{\eta}\right)\right]d\mathbf{R}_1 d\mathbf{R}_2 d\Omega_2 d\mathbf{R}_3 d\Omega_3 \\
&= \sum_{\gamma}\sum_{\eta\lambda}\left(\frac{\rho}{\Omega}\right)^2 \iiiint \delta(\mathbf{R}_2+\mathbf{l}_2^{\alpha}-\mathbf{r})\int d\mathbf{k}c_{\gamma}^{uv}(\mathbf{k})\exp\left[i\mathbf{k}\cdot(\mathbf{R}_3+\mathbf{l}_3^{\gamma})\right]\int d\mathbf{k}'c_{\eta\lambda}^{vv}(\mathbf{k}')\exp\left[-i\mathbf{k}'\cdot\left((\mathbf{R}_3+\mathbf{l}_3^{\lambda}-\left\{\mathbf{R}_2+\mathbf{l}_2^{\eta}\right\}\right)\right]d\mathbf{R}_2 d\Omega_2 d\mathbf{R}_3 d\Omega_3 \\
&= \sum_{\gamma}\sum_{\eta\lambda}\left(\frac{\rho}{\Omega}\right)^2 \iiint \delta(\mathbf{R}_2+\mathbf{l}_2^{\alpha}-\mathbf{r})\int d\mathbf{k}c_{\gamma}^{uv}(\mathbf{k})\exp\left[i\mathbf{k}\cdot\mathbf{l}_3^{\gamma}\right]\int d\mathbf{k}'c_{\eta\lambda}^{vv}(\mathbf{k}')\exp\left[-i\mathbf{k}'\cdot\left(\mathbf{l}_3^{\eta}-\left\{\mathbf{R}_2+\mathbf{l}_2^{\lambda}\right\}\right)\right]\left\{\int\exp[i(\mathbf{k}-\mathbf{k}')\cdot\mathbf{R}_3]d\mathbf{R}_3\right\}d\mathbf{R}_2 d\Omega_2 d\Omega_3 \\
&= \sum_{\gamma}\sum_{\eta\lambda}\left(\frac{\rho}{\Omega}\right)^2 \iiint \delta(\mathbf{R}_2+\mathbf{l}_2^{\alpha}-\mathbf{r})\int d\mathbf{k}c_{\gamma}^{uv}(\mathbf{k})\exp\left[i\mathbf{k}\cdot\mathbf{l}_3^{\gamma}\right]\int d\mathbf{k}'c_{\eta\lambda}^{vv}(\mathbf{k}')\exp\left[-i\mathbf{k}'\cdot\left(\mathbf{l}_3^{\eta}-\left\{\mathbf{R}_2+\mathbf{l}_2^{\lambda}\right\}\right)\right]\delta((\mathbf{k}-\mathbf{k}')d\mathbf{R}_2 d\Omega_2 d\Omega_3 \\
&= \sum_{\gamma}\sum_{\eta\lambda}\left(\frac{\rho}{\Omega}\right)^2 \iiint \delta(\mathbf{R}_2+\mathbf{l}_2^{\alpha}-\mathbf{r})\int d\mathbf{k}c_{\gamma}^{uv}(\mathbf{k})\exp\left[i\mathbf{k}\cdot\mathbf{l}_3^{\gamma}\right]c_{\eta\lambda}^{vv}(\mathbf{k})\exp\left[-i\mathbf{k}\cdot\left(\mathbf{l}_3^{\eta}-\left\{\mathbf{R}_2+\mathbf{l}_2^{\lambda}\right\}\right)\right]d\mathbf{R}_2 d\Omega_2 d\Omega_3 \\
&= \sum_{\gamma}\sum_{\eta\lambda}\left(\frac{\rho}{\Omega}\right)^2 \iint \int d\mathbf{k}c_{\gamma}^{uv}(\mathbf{k})\exp\left[i\mathbf{k}\cdot\mathbf{l}_3^{\lambda}\right]c_{\eta\lambda}^{vv}(\mathbf{k})\exp\left[-i\mathbf{k}\cdot\left(\mathbf{l}_3^{\eta}-\left\{-\mathbf{l}_2^{\alpha}+\mathbf{r}+\mathbf{l}_2^{\lambda}\right\}\right)\right]d\Omega_2 d\Omega_3 \\
&= \sum_{\gamma}\sum_{\eta\lambda}\rho^2\int d\mathbf{k}c_{\gamma}^{uv}(\mathbf{k})c_{\eta\lambda}^{vv}(\mathbf{k})\left[\frac{1}{\Omega}\int\exp\left[i\mathbf{k}\cdot(\mathbf{l}_3^{\gamma}-\mathbf{l}_3^{\eta})\right]d\Omega_3\right]\left[\frac{1}{\Omega}\int\exp\left[i\mathbf{k}\cdot(\mathbf{l}_2^{\alpha}-\mathbf{l}_2^{\lambda})\right]d\Omega_2\right]\exp(-i\mathbf{k}\cdot\mathbf{r}) \\
&= \sum_{\gamma}\sum_{\eta\lambda}\rho^2\int d\mathbf{k}c_{\gamma}^{uv}(\mathbf{k})c_{\eta\lambda}^{vv}(\mathbf{k})\omega_{\gamma\eta}(\mathbf{k})\omega_{\alpha\lambda}(\mathbf{k})\exp(-i\mathbf{k}\cdot\mathbf{r}) \\
&= \sum_{\gamma}\sum_{\eta\lambda}\int d\mathbf{k}dc_{\gamma}^{uv}(\mathbf{k})\rho\omega_{\gamma\eta}(\mathbf{k})c_{\eta\lambda}^{vv}(\mathbf{k})\rho\omega_{\alpha\lambda}(\mathbf{k})\exp(-i\mathbf{k}\cdot\mathbf{r}) \\
&= \int d\mathbf{k}\sum_{\gamma}\left[\mathbf{c}^{uv}(\mathbf{k})\rho\mathbf{w}(\mathbf{k})\mathbf{c}^{vv}(\mathbf{k})\rho\mathbf{w}(\mathbf{k})\right]_{\gamma\alpha}\exp(-i\mathbf{k}\cdot\mathbf{r})
\end{aligned}
\tag{IV-2.8}
$$

Then, Eq. (IV-2.4) can be written as

$$
\begin{aligned}
h_{\alpha}(\mathbf{r}) &= \int d\mathbf{k}\sum_{\gamma}\left[\mathbf{c}^{uv}(\mathbf{k})\rho\mathbf{w}(\mathbf{k})\right]_{\gamma\alpha}\exp[-i\mathbf{k}\cdot\mathbf{r}] \\
&\quad + \int d\mathbf{k}\sum_{\gamma}\left[\mathbf{c}^{uv}(\mathbf{k})\rho\mathbf{w}(\mathbf{k})\mathbf{c}^{vv}(\mathbf{k})\rho\mathbf{w}(\mathbf{k})\right]_{\gamma\alpha}\exp(-i\mathbf{k}\cdot\mathbf{r})+\cdots \\
&= \int d\mathbf{k}\sum_{\gamma}\left[\mathbf{c}^{uv}(\mathbf{k})\left\{\rho\mathbf{w}(\mathbf{k})+\rho\mathbf{w}(\mathbf{k})\mathbf{c}^{vv}(\mathbf{k})\rho\mathbf{w}(\mathbf{k})+\rho\mathbf{w}(\mathbf{k})\mathbf{c}^{vv}(\mathbf{k})\rho\mathbf{w}(\mathbf{k})\right\}+\cdots\right]_{\gamma\alpha}\exp[-i\mathbf{k}\cdot\mathbf{r}] \\
&= \int d\mathbf{k}\sum_{\gamma}\left[\mathbf{c}^{uv}(\mathbf{k})\left\{\rho\mathbf{w}(\mathbf{k})+\rho\mathbf{h}^{vv}(\mathbf{k})\right\}\right]_{\gamma\alpha}\exp[-i\mathbf{k}\cdot\mathbf{r}]
\end{aligned}
\tag{IV-2.9}
$$

or it can be written in the real space as,

$$
h_{\alpha}^{vv}(\mathbf{r}) = \sum_{\gamma}\int \chi_{\alpha\gamma}^{vv}(|\mathbf{r}-\mathbf{r}'|)c_{\gamma}^{vu}(\mathbf{r}')d\mathbf{r}' \tag{IV-2.10}
$$

where $\rho^{v}\chi_{\alpha\gamma}^{vv}(r) = \omega_{\alpha\gamma}^{v}(r) + \rho^{v}h_{\alpha\gamma}^{vv}(r)$, and $\omega_{\alpha\gamma}^{v}(r)$ is the intramolecular correlation function of solvent defined by Eqs. (II-2.11) ~ (II-2.12) in Chapter II.

Physical meaning of the 3D-RISM equation

Purpose of the 3D-RISM/RISM theory is to find the atomic density or distribution of solvent around a solute or a protein. Let us consider the average density of solvent molecules at a position around a solute molecule (Fig. IV-2.1). When the position is far from the solute molecule so as to be regarded as the bulk, the density will be constant which is the same as in the pure solvent. On the other hand, when it is next to a solute, the density will be "perturbed" significantly by the field exerted by the solute molecule, and be different from that in the bulk,

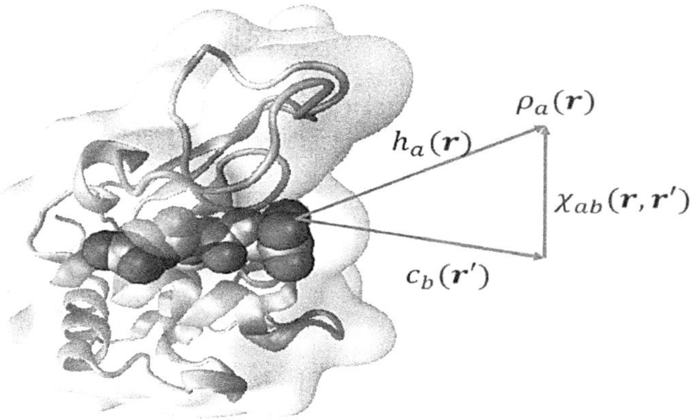

Fig. IV-2.1: Inhomogeneous density distribution of solvent around a protein.

depending on the strength of the perturbation. Physical meaning of the 3D-RISM equation can be readily understood in terms of a *non-linear* response theory as follows.

Let us denote the average density of solvent atom j at the bulk, the density nearby solute, and the density response to the perturbation as ρ_α, $\rho_\alpha(\mathbf{r})$, and $\Delta\rho_\alpha(\mathbf{r})$, respectively, where the subscript α specifies an atom of water. Then, statement above can be expressed as,

$$\rho_\alpha(\mathbf{r}) = \rho_\alpha + \Delta\rho_\alpha(\mathbf{r}) \tag{IV-2.11}$$

The density response to the perturbation can be expressed in terms of a non-linear response theory as

$$\Delta\rho_\alpha(\mathbf{r}) = \sum_\beta \int \rho_\alpha \chi_{\alpha\beta}(\mathbf{r},\mathbf{r}) c_\beta(\mathbf{r}) d\mathbf{r} \tag{IV-2.12}$$

where the $c_\beta(\mathbf{r})$ is a *non-linear* perturbation due to the solute molecule, and $\chi_{\alpha\beta}(\mathbf{r}, \mathbf{r}')$ is a response function. (The reason why we call it "non-linear" will be clarified in a moment.) The equation can be viewed as the 3D-RISM equation by identifying $c_\beta(\mathbf{r})$ as the direct correlation function and $\chi_{\alpha\beta}(\mathbf{r}, \mathbf{r}')$ as the correlation function of density fluctuation in bulk solvent, that is,

$$\rho^2 \chi_{\alpha\beta}(\mathbf{r},\mathbf{r}') = \left\langle \delta\rho_\alpha^{(0)}(\mathbf{r}) \delta\rho_\beta^{(0)}(\mathbf{r}') \right\rangle \tag{IV-2.13}$$

where $\delta\rho_\alpha^{(0)}(\mathbf{r})$ is the density fluctuation of atom a in the pure liquid defined by

$$\delta\rho_\alpha^{(0)}(\mathbf{r}) = \rho_\alpha^{(0)}(\mathbf{r}) - \rho_\alpha \tag{IV-2.14}$$

Several approximate equations have been devised for the direct correlation functions. For example, the hypernetted-chain approximation reads (Chapter II),

$$c_\alpha(\mathbf{r}) = \exp\left[-\beta u_\alpha(\mathbf{r}) + h_\alpha(\mathbf{r}) - c_\alpha(\mathbf{r})\right] - 1 - \left[h_\alpha(\mathbf{r}) - c_\alpha(\mathbf{r})\right] \tag{IV-2.15}$$

In the expression, $u_\alpha(\mathbf{r})$ is the direct interaction potential exerted on the solvent atom, α, from the solute molecule, and $h_\alpha(\mathbf{r})$ is the density fluctuation in solvent at position $\mathbf{r}$, normalized by the bulk density, namely,

$$h_\alpha(\mathbf{r}) = \Delta\rho_\alpha(\mathbf{r}) / \rho \tag{IV-2.16}$$

The three dimensional distribution function is defined from $h_\alpha(\mathbf{r})$ by

$$g_\alpha(\mathbf{r}) \equiv h_\alpha(\mathbf{r}) + 1 \qquad \text{(IV-2.17)}$$

It is not only the direct interaction $u_\alpha(\mathbf{r})$ with solute that perturbs the density of solvent at a certain position, but also that from solvent molecules at the other positions, whose density is also perturbed by the existence of the same solute. Such "indirect" perturbations are *renormalized* into the terms including $(h_\alpha(\mathbf{r}) - c_\alpha(\mathbf{r}))$. Such renormalization makes the perturbation highly "non-linear."

The procedure of solving the equations consists of two steps. We first solve the RISM equation Eq. (IV-2.2) for $h^{vv}_{\alpha\beta}(r)$ of solvent or a mixture of solvents in cases of solutions. Then, we solve the 3D-RISM equation, Eq. (IV-2.12), for $h_\alpha(\mathbf{r})$ of a protein-solvent (solution) system, plugging $h^{vv}_{\alpha\beta}(r)$ for the solvent into the equation, which has been calculated in the first step. Considering the definition ($g(\mathbf{r}) = h(\mathbf{r}) + 1$), $h(\mathbf{r})$ thus obtained is the three dimensional distribution of solvent molecules around a protein in terms of the interaction-site representation. So called solvation free energy can be obtained from the distribution function through the following equations corresponding, respectively, to the two closure relations described in Chapter II.

$$\Delta\mu_{\mathrm{HNC}} = \rho^v k_B T \sum_\gamma \int d\mathbf{r} \left[\frac{1}{2} h^{uv}_\gamma(\mathbf{r})^2 - c^{uv}_\gamma(\mathbf{r}) - \frac{1}{2} h^{uv}_\gamma(\mathbf{r}) c^{uv}_\gamma(\mathbf{r}) \right] \qquad \text{(IV-2.18)}$$

$$\Delta\mu_{\mathrm{KH}} = \rho^v k_B T \sum_\gamma \int d\mathbf{r} \left[\frac{1}{2} h^{uv}_\gamma(\mathbf{r})^2 \Theta(-h^{uv}_\gamma(\mathbf{r}) - c^{uv}_\gamma(\mathbf{r}) - \frac{1}{2} h^{uv}_\gamma(\mathbf{r}) c^{uv}_\gamma(\mathbf{r}) \right] \qquad \text{(IV-2.19)}$$

The equations have been used to evaluate the binding affinity of drug molecules to protein. Few examples of the application are provided later in the chapter.

The other thermodynamic quantities concerning solvation can be readily obtained from the standard thermodynamic derivative of the free energy except for the partial molar volume. The partial molar volume, which is a very important quantity to probe the response of the free energy (or stability) of protein to pressure, including so called "pressure denaturation," is not a "canonical" thermodynamic quantity for the (V,T) ensemble, since the volume is an independent thermodynamic variable of the ensemble. The partial molar volume of protein in its infinite dilution can be calculated from the Kirkwood-Buff equation generalized to the site-site representation of liquid and solutions [Kirkwood and Buff 1951, Imai et al. 2000] (see Chapter II). In the limit of the infinite dilution of protein, the 3D-RISM expression of the site-site Kirkwood-Buff equation reads,

$$\bar{V} = kT\chi_T \left[1 - \rho \sum_\gamma \int c_\gamma(\mathbf{r}) d\mathbf{r} \right], \qquad \text{(IV-2.20)}$$

where χ_T is the isothermal compressibility of pure solvent or solution, which is obtained from the site-site correlation functions of solutions. Imai et al. have applied the theory to calculate the partial molar volume of a variety of protein in order to demonstrate the robustness of the theory [Imai et al. 2004, 2005].

The partial molar volume of several proteins in water, which appear frequently in the literature of protein research, is plotted against the molecular weight in Fig. IV-2.2. By comparing the results with the experimental ones plotted in the same figure, one can readily

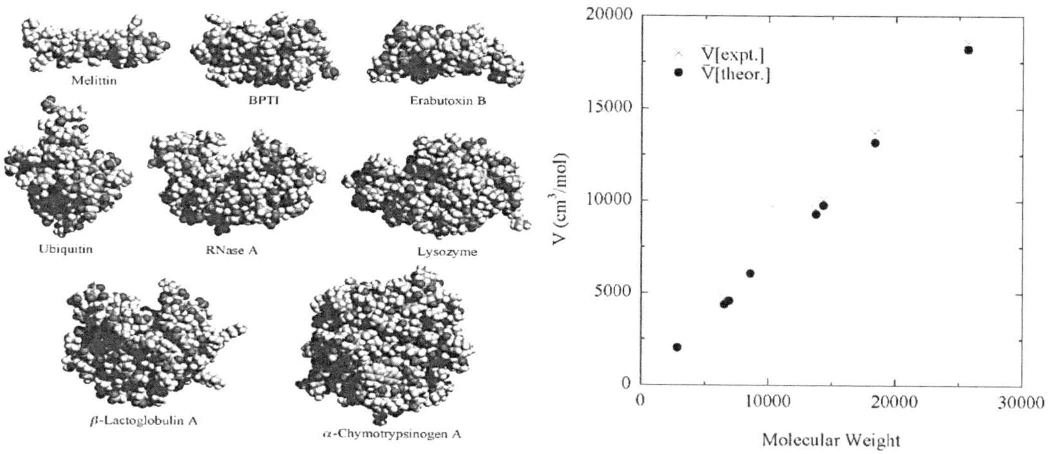

Fig. IV-2.2: The partial molar volume of proteins plotted against the molecular weight. The proteins calculated are illustrated on the left. The theoretical results (black circles) show quantitative agreement with the experimental ones (cross). Copyright (2005) American Chemical Society.

see that the theory is capable of reproducing the experimental results at a quantitative level. At first glance, the results seem to be reproduced by just simple consideration of protein geometry using a commercial software to calculate the *exclusion volume* of protein. However, it is never the case. The reason is because the partial molar volume is the "thermodynamic quantity," not the "geometrical volume" [Akasaka 2015]. The partial molar volume reflects all the solvent-solvent and solute-solvent interactions as well as all the configurations of water molecules in the system, while the geometrical volume accounts for just the simplified (hardcore type) repulsive interaction between solute and solvent. Other factors such as the attractive interactions between solute and solvent and the solvent reorganization are entirely neglected in the geometrical volume. The contributions from te solvent reorganization are of particular importance in the partial molar volume of protein because they are concerned with the so called volume of "cavity" in protein. As is well regarded, a protein has many internal cavities where water molecules can or cannot be accommodated [Akasaka 2015]. Let's carry out a simple *thought experiment* with respect to the partial molar volume of protein. The experiment is to dissolve a protein into water.

Upon the dissolution of protein in water, some of the cavities in the protein may be filled by water molecules, but others may not. If the cavity stays empty, then the empty space will contribute to increase the partial molar volume of the protein. On the other hand, if the space is filled by water molecules due to the reorganization of the solvent, it will contribute to the reduction of the entire volume of solution, and compensate the increase due to the cavity volume. This compensation is non-trivial: if a cavity can accommodate one water molecule, it gives rise to the reduction of the volume by 18 cm^3/mol. In this regard, unless a theory is able to describe the reorganization of water molecules induced by protein, it is useless to predict the partial molar volume. The nearly quantitative results shown in the figure demonstrate that the theory is properly accounting for all the solute-solvent and solvent-solvent interactions as well as solvent reorganization induced by protein, including the accommodation of some water molecules into the internal cavity.

IV-3 Application of the 3D-RISM Theory to Molecular Recognition

The molecular recognition (MR) process can be defined as a molecular process in which one or few guest molecules are bound in high probability at a particular site, a cleft or a cavity, of a host molecule in a particular orientation [Lehn 1990]. In this regard, MR is a molecular process determined by specific interactions between atoms in host and guest molecules. On the other hand, the process is a thermodynamic process as well, with which the chemical potential or the free energy of guest molecules in the recognition site and in the bulk solution are concerned. As an example, let's think about binding of a substrate molecule at some reaction pocket of a host protein (see Fig. IV-1.3). Usually, the reaction pocket is likely to be filled with one or few water molecules when there is no substrate. In order for a substrate molecule to come into the reaction pocket, one or some of the water molecules should be disposed from the pocket, while the substrate molecule itself should be partially or entirely dehydrated. The free energy changes associated with the processes are commonly called "dehydration penalty." When a guest molecule comes into a cleft or a cavity of a host molecule, it has to overcome a high entropy barrier because the space or the degree of freedom allowed to the guest molecule is so small compared to those in the bulk solution.

The molecular recognition can be probed in principle by any experimental methods practiced in the molecular science, including spectroscopy, magnetic resonance, diffraction, and scattering, that are capable of resolving small changes in the signal associated with the binding of a ligand by the host molecule: in the case of NMR, change in the chemical shift of a labeled atom in a amino-acid residue, or the *chemical-shift perturbation*. However, the task is not so easy in practice due to reasons that are different from method to method.

Probably, it is the X-ray diffraction that is currently the most popular method in probing a ligand bound by protein. In the most successful case, the method is able to determine not only the position of a ligand, but also its pose or orientation. However, the method has its own drawbacks that concern water molecules and small ions. The conventional procedure for determining positions (or distributions) of small legend including water and ions in a protein

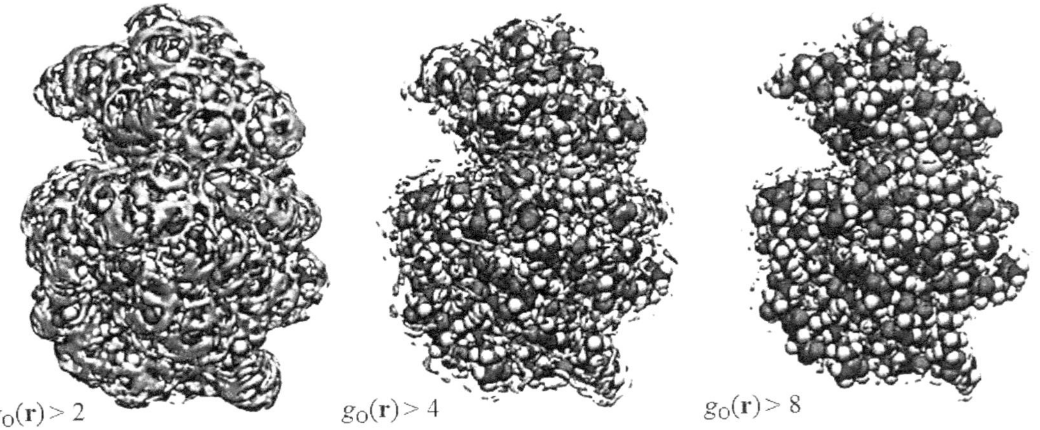

Fig. IV-3.1: Isosurface representation of spatial distribution of $g(\mathbf{r})$ of water around a lysozyme molecule. Copyright (2005) American Chemical Society.

Color version at the end of the book

from X-ray data is essentially trial and error, placing small ligands at some trial positions and determining where the overall electron density agrees with that observed by experiment. Sometimes, molecular simulation is employed to assist the trial-and-error procedure. However, simulation cannot help much in this case because the method, too, requires trial positions and orientations of small ligands as initial conditions. The number of configurations of the trial positions and orientations for a solution containing water and ions is so large that it is impossible to sample all possible configurations by simulation. It includes the number of water molecules and ions inside the active site and the number of ways of placing those molecules and ions, including position and orientation. These should be determined by the condition of the thermodynamic equilibrium, namely the equality of the chemical potentials of each species inside and outside the active site, which depends on the salt concentration. However, to determine the chemical potentials, we must place water molecules and ions at trial positions to perform any type of sampling, e.g., umbrella or thermodynamic perturbation. Thus, the procedure is self-contradictory. Therefore, the positions and orientations of water molecules and ions determined in this manner are never conclusive but are hypothetical, and must be verified by other means.

IV-3.1 Recognition of Water Molecules by Protein

Water molecules bound in protein play multiple roles for living systems. Some of the roles water plays in cells are as follows: (1) Water molecules control the stability of protein structure by their interactions with the biomolecule, especially due to the hydrogen-bonds: they sometimes work as a building block of a protein structure. The most striking example is a protein called "collagen" that is the major ingredient of our skin. In the protein, water molecules make bridges among the peptide chains to make the characteristic structure of triple helix. (2) Water molecules control enzymatic reactions by hydrating and dehydrating the active-cite or cavity of protein (see Fig. IV-1.3). The free energy change associated with the hydration or dehydration consists an important part of the Michaeris-Menten rate constant of the reaction. (3) A water molecule itself is a substrate in the enzymatic hydrolysis reaction.

The problem was first solved theoretically by Imai et al. by applying the 3D-RISM/KH method to a hen-eggwhite lysozyme [Imai et al. 2005, 2007]. They have carried out the 3D-RISM/KH calculation for the protein immersed in water and obtained the 3D-distribution function of oxygen and hydrogen of water molecules around and inside the protein. The native three-dimensional (3D) structure of the protein is taken from the protein data bank (PDB). The protein is known to have a cavity composed of the residues from Y53 to I58 and from A82 to S91, in which four water molecules have been determined by means of the X-ray diffraction measurement. In the 3D-RISM/KH calculation, those water molecules are not included explicitly.

In Fig. IV-3.1, depicted by green surfaces or spots are $g(\mathbf{r})$ of oxygen atoms of water molecules using isosurface representation, which is very similar to the electron density map obtained from the X-ray crystallography. The distribution $g(\mathbf{r})$ that is greater than a threshold value is drawn. The right, center, and left figures correspond, respectively, to the thresholds, $g(\mathbf{r}) \geq 2$, $g(\mathbf{r}) \geq 4$, and $g(\mathbf{r}) \geq 8$. Since $g(\mathbf{r}) = 1$ in the bulk, the left figure indicates that the probability of finding those water molecules at the surface is more than twice as large compared to the bulk water. The threshold value is slightly less than that in bulk, or $g(\mathbf{r}) \sim 3.0$, thereby the water molecules at the protein surface corresponding to the value are considered to be in diffusive motion, not those bound firmly to the protein. On the other hand, the water molecules depicted in the right figure has eight times higher probability to be found than those in the bulk. The water molecules are those bound firmly to some particular atoms of the protein due to, say, the hydrogen bonds, and they are quite rare as one can see from the figure. The results suggest that the X-ray and neutron diffraction communities have acquired a powerful theoretical-tool to analyze their data to locate the position and orientations of water molecules, since the theory also provides the distribution of hydrogen-atoms of water molecules.

The result depicted in Fig. IV-3.1 is what is expected before the calculation was carried out, although they were entirely new by themselves in the history of statistical mechanics. Entirely unexpected was some peaks of water distribution in a cavity "inside" the protein, which are surrounded by the residues from Y53 to I58 and from A82 to S91. The results are shown in Fig. IV-3.2.

The left picture in Fig. IV-3.2 shows the isosurfaces of $g(\mathbf{r}) > 8$ for water-oxygen (green) and hydrogen (pink) in the cavity. In the figure, only the surrounding residues are displayed, except for A82 and L83 which are located in the front side. There are four distinct peaks of water oxygen and seven distinct peaks of water hydrogen in the cavity. The spots colored by green and pink indicate water oxygen and hydrogen, respectively. From the isosurfaces plot, the most probable model of the hydration structure was constructed. It is shown in the center of Fig. IV-3.2, where the four water molecules are numbered in the order from the left. Water 1 (W1) is hydrogen-bonding to the main-chain oxygen of Y53 and the main-chain nitrogen of L56. Water 2 (W2) forms hydrogen bonds to the main-chain nitrogen of I56 and the main-chain oxygen of L83, which is not drawn in the figure. Water 3 and 4 (W3, W4) also form hydrogen bonds with protein sites, the former to the main-chain oxygen of S85 and the latter to the main-chain oxygen of A82 (not displayed) and of D87. There is also a hydrogen bond network among Water 2, 3, and 4. The peak of the hydrogen between Water 3 and 4 does not appear in the figure because it is slightly less than 8, which means the hydrogen bond is weaker or looser than the other hydrogen-bonding interactions. Although the hydroxyl group of S91 is located in the center of the four water molecules, it makes only weak interactions with them.

It is interesting to compare the hydration structure obtained by the 3D-RISM/KH theory with crystallographic water sites of X-ray structure [Wilson et al. 1992]. The crystallographic water molecules in the cavity are depicted in the right of Fig. IV-3.2, showing four water sites in the cavity, much as the 3D-RISM/KH theory has detected. Moreover, the water distributions obtained from the theory and experiment are almost identical to each other. Thus, it was verified

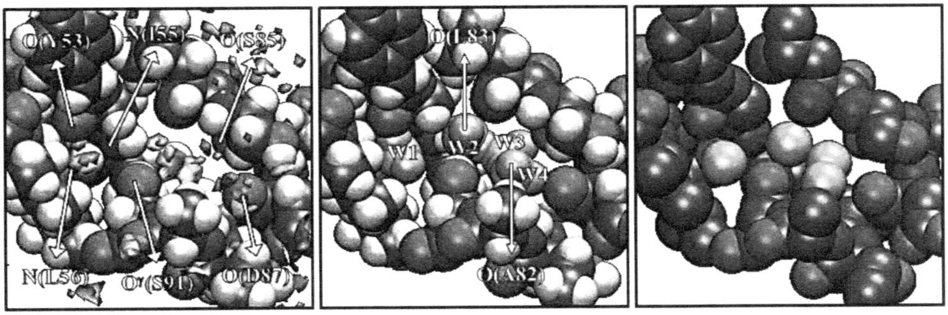

Fig. IV-3.2: The spatial distribution ($g(\mathbf{r})$) of water inside the active site of lysozyme. Green surface, oxygen; purple surface, hydrogen. Copyright (2005) American Chemical Society.

Color version at the end of the book

unambiguously that the 3D-RISM/KH theory can predict the *recognition* of water molecules by a cavity of protein.

It should be noted that one peak of the 3D-distribution function does not necessarily correspond to one molecule. If a water molecule transfers back and forth between two sites in the equilibrium state, two peaks correspondingly appear in the 3D-distribution function. In fact, the number of the water molecules within the cavity calculated from the 3D-distribution function is 3.6. It is less than the number of the water-binding sites, and includes decimal fractions. To explain that, a molecular dynamics (MD) simulation was carried out using the same parameters and under the same thermodynamic conditions as the 3D-RISM calculation [Imai et al. 2007a]. Only one exception was that the four crystallographic water molecules in the cavity as well as the other crystallographic water molecules were initially put at their own sites in the MD simulation. The result of MD simulation also shows the hydration number is less than 4, that is, 3.5. From the MD trajectory, it is found that two inner water molecules, Water 1 and 2, stay at their own sites during all the simulation time, and make only small fluctuation around the sites. On the other hand, two outer water molecules, Water 3 and 4, sometimes enter and leave the sites, and by chance exchange with other water molecules from the bulk phase. As a result, the number of water molecules at the outer sites is 1.5 on average. Thus, the 3D-RISM/KH theory can provide a reasonable hydration number including decimal fractions through statistical-mechanical relations, even though the theory takes no explicit account of the dynamics of molecules.

IV-3.2 Noble-gas Bound by Protein

The methodology described in the previous section can be applied to the molecular recognition process with a slight modification, and provides a powerful theoretical-tool to realize the ligand binding by protein [Imai et al. 2007b]. The modification to be made is just to change the solvent from the pure water to an aqueous solution containing ligand molecules. In this section, we present the results for noble gases which are the simplest model of non-polar ligands.

Figure IV-3.3 shows the 3D distribution functions of xenon and water (oxygen and hydrogen) around lysozyme, calculated by the 3D-RISM theory for lysozyme in water-xenon mixture at the concentration of 0.001 M. The molecular surface of the protein is painted blue. The regions where $g(\mathbf{r}) > 8$ are painted with different colors for different species: yellow, xenon; red, water oxygen; white, water hydrogen. Of course, the surface painted blue is covered by water molecules weakly bound to the protein, which are not shown. A number of well-defined peaks, yellow and red spots, are found for xenon and water oxygen at the surface of the protein, which are separated from each other. The result demonstrates the great capability of the 3D-RISM theory to predict the "preferential binding" of ligands. The distributions of ligand and water are simultaneously found in this result, which means that the peak of either the ligand or water is found at each site depending on the ratio of their affinities to the site. Actually, Fig. IV-3.3 indicates that there are water- and xenon-preferred sites on the protein surface. Similar results are obtained for the other gases and the other concentrations.

It is interesting to compare the distribution of xenon obtained by the 3D-RISM theory with the xenon sites in the X-ray structure [Pange et al. 1998], even though their conditions are different: the former is aqueous solution under atmospheric pressure, while the latter is crystal under xenon gas pressure of 12 bar. There are two binding sites of xenon in lysozyme: one corresponds to the binding pocket of native ligands, which is referred to as the substrate binding site, and the other is located in a cavity inside the protein, which is referred to as the internal site. Figure IV-3.3 compares the theoretical result of the 3D distribution of xenon with the X-ray xenon site at the substrate binding site. The location of a high and sharp peak found by the theory is in complete agreement with the X-ray xenon site depicted by orange color. Shown in Fig. IV-3.3 are the result of the internal site. The xenon peak found there is actually a minor one; nevertheless, the location is again consistent with the X-ray site. It is interesting to note that the peaks of water are shifted off from the xenon binding site.

Shown in Fig. IV-3.4 is the size dependence of the coordination number of noble gases at the two binding sites, which is calculated at the concentration of 0.001 M. At the substrate

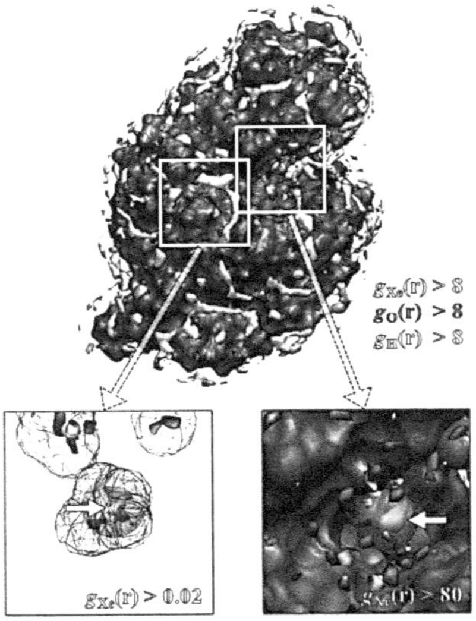

Fig. IV-3.3 The spatial distribution ($g(\mathbf{r})$) of Xe and water molecules in Lysozyme: red surface, water oxygen; white surface, water hydrogen; yellow surface, xenon. The X-ray results are painted by orange surface. Copyright (2007) American Chemical Society.

Color version at the end of the book

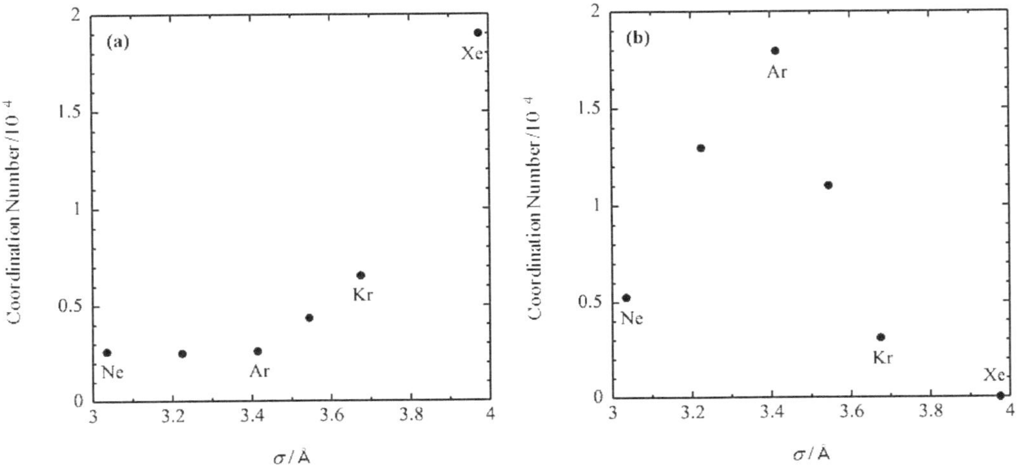

Fig. IV-3.4: The size dependence of noble gasses bound at (a) the substrate binding site, and (b) the internal site. Copyright (2007) American Chemical Society.

binding site, the coordination number becomes exponentially larger as the size of gas increases (Fig. IV-3.4a). At the internal site, the coordination number becomes larger with increasing the gas size up to s ~ 3.4 Å, while it decreases in the region where $\sigma \geq 3.4$ Å (Fig. IV-3.4b). As a result, argon has the largest binding affinity to the internal site. These results demonstrate that the 3D-RISM theory has the ability to describe size selectivity in the binding, or molecular recognition. Although there are no corresponding experimental data, the present results serve as a representative test case.

IV-3.3 Selective Ion-binding by Protein

Ion binding is essential for a variety of physiological processes. The binding of calcium ions by some protein triggers the process to induce the muscle contraction and enzymatic reactions [Herzberg 1985, Ikura 1992]. The initial process of the information transmission through the ion channel is the ion-binding by channel protein [Hille 2001]. The ion-binding plays an essential role, sometimes to the folding process of a protein by inducing the secondary structure [Ikura et al. 1992]. Such processes are characterized by the highly selective ion recognition by proteins. It is of great importance, therefore, for the life science to clarify the origin of the ion selectivity in molecular detail.

Presented in this section are the theoretical results for the ion binding by human lysozyme obtained by Yoshida et al. through basically the same procedure as that described in the preceding section, but changing the solution from noble gas to ionic solutions [Yoshida et al. 2006, 2007]. Yoshida et al. first prepared the correlation functions for the bulk solutions by solving Eq. (11), then they plugged those functions into the 3D-RISM/KH equation to obtain the 3D-distribution of ions along with water molecules. A special attention, however, should have been paid to the treatment of the bulk solution as the reference state because the ion-ion interactions in the solutions are the Coulomb interaction, and their contribution to the "dehydration penalty" should not be disregarded even in low concentration. In order to make sure that the free energy due to ion-ion interaction is reasonably accounted, they have calculated the excess chemical potential, or the mean activity coefficient, of ions in solutions. The results are shown in Fig. IV-3.5. The results in general show fair agreement with the experimental results. Especially, the theory discriminates the divalent ion from the monovalent ions quite well. Apparently, the concentration dependence of the two monovalent ions is not resolved well. This may be due

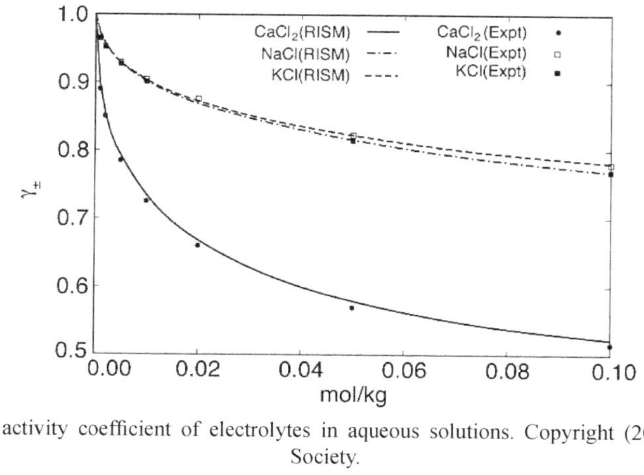

Fig. IV-3.5: The mean activity coefficient of electrolytes in aqueous solutions. Copyright (2007) American Chemical Society.

to the potential parameters for the ions. However, it will not have serious influence on the results for the ion recognition by protein because the process is determined primarily by the free energy difference of the same ion inside protein and bulk solutions.

The 3D-RISM calculation was carried out for aqueous solutions of three different electrolytes, $CaCl_2$, NaCl, and KCl, and for four different mutants of the protein: wild type, Q86D, A92D, Q86D/A92D that have been studied experimentally by Kuroki and Yutani [Kuroki and Yutani 1998]. Thermodynamic conditions of solutions, potential parameters, and the structural data of proteins are listed in Table IV-3.1.

Table IV-3.1: Thermodynamic conditions, potential parameter, PDB files. Copyright (2007) American Chemical Society.

Temperature	298 K	
Concentration	0.01 M	
Dielectric constant	78.4973	
Potential parameter		
Water		SPC[49,a]
Cl^-, K^+, Na^+, Ca^{2+}		OPLS[50,51]
Protein		Amber99[52]
Protein structure		
Wild type		1LZ1[53]
Q86D-lysozyme		1I1Z[48]
A92D-lysozyme		1I20[48]
holo-Q86D/A92D-lysozyme		3LHM[54]

[a] Lennard-Jones parameters of water hydrogen were added with $\sigma = 0.4$ Å and $\varepsilon = 192.5$ J/mol.

In Fig. IV-3.6, shown are the spatial distributions function $g(\mathbf{r})$ of water molecules and the cations inside and around the cleft under concern, which consists of amino acid residues from Q86 to A92. The area where $g(\mathbf{r})$ is greater than five is painted with a color for each species: oxygen of water, red; Na^+ ion, yellow; Ca^{2+} ion, orange; K^+ ion, purple. For the wild type of protein in the aqueous solutions of all the electrolytes studied, $CaCl_2$, NaCl and KCl, there are no distributions greater than five ($g(\mathbf{r}) > 5$) observed for the ions inside the cleft, as is seen in the figure in upper left. The Q86D mutant exhibits essentially the same behavior with that of the wild type, but with the water distribution changed slightly (upper-center figure). (There is a trace of yellow spot that indicates a slight possibility of finding a Na^+ ion in the middle of the binding site, but it would be too small to make a significant contribution to the distribution.) Instead, the distribution corresponding to water oxygen is observed as is shown by the red color in the figure. The distribution covers faithfully the region where the crystallographic water molecules have been detected, which are shown with the spheres colored gray. There is a small difference between the theory and the experiment, which is the crystallographic water bound to the backbone of Asp-91. The theory does not reproduce the water molecule by unidentified reasons. Except for the difference, the observation is consistent with the experimental finding, especially, that the protein with the wild type sequence binds neither Na^+ nor Ca^{2+}.

The A92D mutant in the NaCl solution shows conspicuous distribution of a Na^+ ion bound in the recognition site, which is in accord with the experiment (upper-right figure). The Na^+ ion is apparently bound to the carbonyl oxygen-atoms of Asp-92, and is distributed around the moieties. There is water distribution observed in the active site, but the shape of the distribution is entirely changed from that in the wild type. The distribution indicates that the Na^+ ion bound in the active site is not naked, but is accompanied by hydrating water molecules. The mutant does not show any indication of binding K^+ ion. (The results are not shown.) This suggests

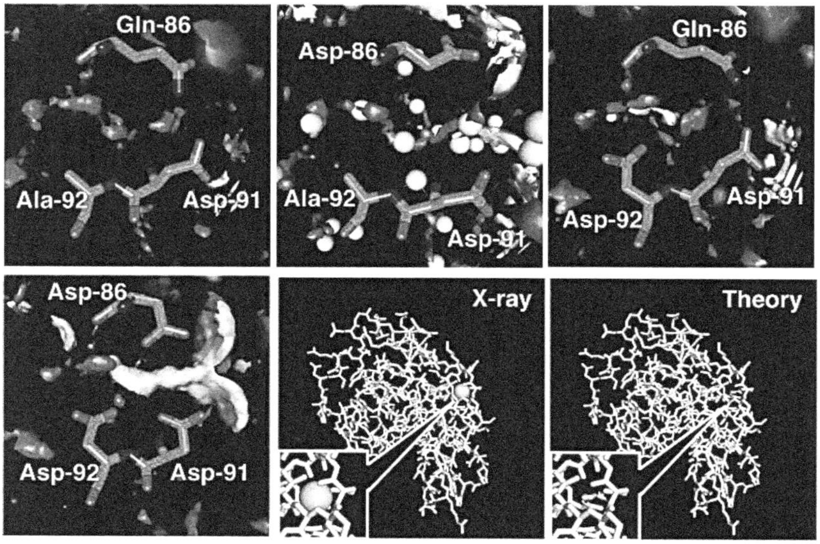

Fig. IV-3.6: The spatial distribution functions ($g(\mathbf{r})$) of ions in lysozyme. Copyright (2007) American Chemical Society.

Color version at the end of the book

that the A92D mutant discriminates a Na^+ ion from a K^+ ion. The finding demonstrates the capability of the 3D-RISM theory to realize the ion selectivity by protein.

In the lower panels, shown are the distributions of Ca^{2+} ions and of water oxygen in the ion binding site of the holo-Q86D/A92D mutant. The mutant is known experimentally as a calcium binding protein. The protein, in fact, exhibits a strong calcium binding activity as can be inspected in the figure. The calcium ion is recognized by the carboxyl groups of the three Asp residues, and is distributed around the oxygen atoms. The water distribution at the center of the triangle made by the three carbonyl oxygen atoms is reduced dramatically, which indicates that the Ca^{2+} ion is coordinated by the oxygen atoms directly, not with water molecules in between. The Ca^{2+} ion, however, is not entirely naked because persistent water distribution is observed at least at two positions where original water molecules were located in the wild type of the protein.

IV-4 Molecular Channels

There is a class of proteins called molecular channels in our body, which control the permeation of small molecules including water and ions through membranes. The channels play crucial roles in the signal transduction among cells, which of course is indispensable in the physiological process. The present section is devoted to applications of 3D-RISM/KH to selected topics concerning molecular channels including water, ion, and proton channels.

IV-4.1 Aquaporin (Water Channels)

The transportation of water across cell membrane is a fundamental process to maintain life. The homotetramic water-channels, aquaporins (AQPs), facilitate the rapid transport of water across cell membranes in response to osmotic gradients [Preston et al. 1992, Zeidel et al. 1992]. AQPs are believed to play an important role in many physiological processes, and their mutations are responsible for human disease ranging from diabetes insipidus to congenital cataract formation [King et al. 2004]. A wide range of members of AQP family such as AQP0-AQP9 and AQPZ has been found in many cell types from human to bacteria [Borgnia et al. 1999]. Recently, the tetrameric structure of AQP was revealed from the X-ray crystallography for few members of aquaporin family: AQPZ, SoPIP2;1 [Jiang et al. 2006, Törnroth-Horsefield et al. 2006]. However, the distinct close and open states in the tetramer are found only in AQPZ.

In spite the fact that intensive efforts have been devoted to clarify the molecular mechanism of the aquaporin functions, there are many questions that remain unanswered. Among the questions, the following are most sharply focused on the experimental as well as theoretical studies.

(1) Gating mechanism of AQPs. (2) Why the proton is excluded from conduction through the channel? (3) Are ions permeated through the channel? (4) What is the possibility of small molecules such as carbon dioxides and ammonia to be permeated through the channel?

Some of the question are answered in the following based on the 3D-RISM/KH theory [Phongphanphanee et al. 2007, 2008a].

***Gating mechanism of AQPs*:** The water permeability of AQPs is regulated by various mechanisms which are distinct in each member of the AQP family [Johansson et al. 2000, Nemeth-Cahalan et al. 2004, Zelenina et al. 2003, Gunnarson et al. 2005]. As is the case in other proteins, the function of aquaporin, or the conduction mechanism of water, is considered to be closely related to its conformation and structural fluctuation. So, in order to be able to clarify the conduction mechanism of aquaporin, it is essential to have the entire conformation of a tetramer and the water distribution in each monomer. Here we use AQPZ, which is an aquaporin found in *E. coli*, for the purpose of clarifying the gating mechanism of AQP. The X-ray structure of tetrameric AQPZ with the resolution 3.2 Å gives a distinct conformation of the side chain R189 [Jiang et al. 2006]. According to the results, the amide group of R189 in each monomer rotates either outward or inward from the channel surface. The experiment found that one among four monomers has R189 rotated outward, while the other three have R189 rotated inward. It is quite natural to relate the rotation of the residue to the gating mechanism of the channel: R189 rotated inward will shut the water flow, and that rotated outward will let water go through. So, the experimentalists named the residue rotated inward as "closed" structure, and that rotated outward as "open" structure. In order to verify the postulate of gating mechanism, the distribution of water inside the channel should be clarified. The X-ray crystallography has shown positions of several water molecules inside the channel. However, the resolution of the measurement is not quite enough to make solid conclusion.

The distribution functions of water around and inside an AQPZ tetramer was calculated using the 3D-RISM/KH method [Phongphanphanee et al. 2007]. The results are depicted in Fig. IV-4.1 by blue surface, in which the picture labeled "a" corresponds to the open structure, and those labeled "b", "c" and "d" to the closed structures. The position of water molecules determined by the X-ray crystallography is also shown in the figures by red spheres. The water molecules distribute continuously along the channel in the open conformation, whereas the distribution is interrupted by the large gap at the location of the side chain R189. The results indicate that water cannot pass through the AQPZ channel in its close conformations. The density profile of water in each conformation calculated by the theory is consistent with

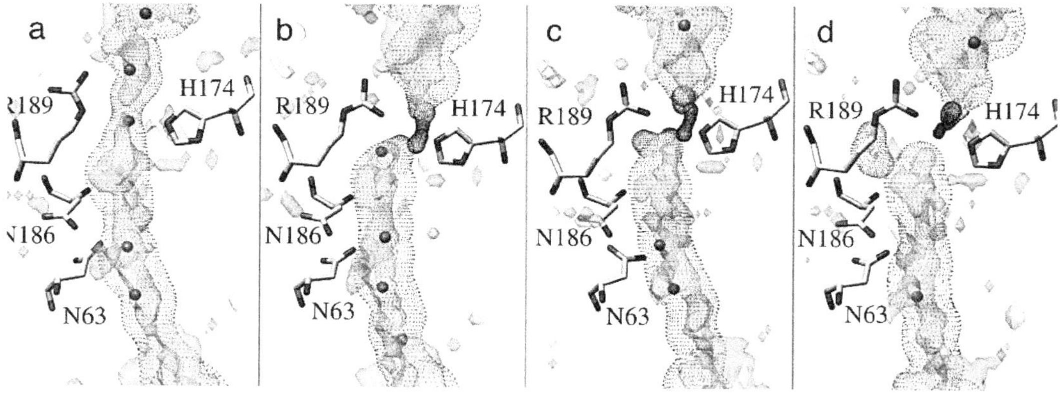

Fig. IV-4.1: Distribution of water inside AQPZ. Copyright (2007) Elsevier B.V.

Color version at the end of the book

the positions determined by the X-ray crystallography. However, the spatial resolution of the experiment is obviously far lower than the theoretical prediction that we cannot make quantitative comparison. The results confirm the experimental postulate for the role of R189 in the gating mechanism of AQPZ.

***Possibility of Ion permeation through AQPs*:** Due to their prevailing population and diversity in our body, the possibility of the aquaporin family to be participating in the trans-membrane conduction of small ligands such as ions, carbon-dioxide, and ammonia has been a focus of many studies. However, the results are still controversial both in experiments and simulations, probably due to the same physical reason, "low concentration." In the physiological conditions, the concentration of ions, or sodium chloride, is typically about thousand times lower than that of water, while that of CO_2 and NH_3 is just trace. Fortunately, for ions, people can employ the technique well established in the study of ion channel, or the "patch clump" method, to measure the rate of ion current quite accurately. However, even in that case, they may not be able to identify the molecular mechanism of the ion-conduction because the concentration of ions inside channel is so low that the distribution of ions cannot be probed by the current experimental method such as the X-ray crystallography. In the case of uncharged ligands, the situation is even worse. All they can do is to compare the concentration of the ligands under concern by, for example, measuring the change of pH, before and after the conduction experiments. Since the concentration of ligands in solution is already very low, it requires extremely high sensitivity to measure the change in the concentration. Of course, determining the distribution of those ligands inside the channels is just beyond the scope for the current experimental techniques. The problem of "low concentration" imposes a similar difficulty on the molecular simulations. As can be readily understood, the conduction mechanism of small ligands involves the "molecular recognition" process at the entrance and exit of the channel. The molecular recognition process is a rare event that involves crossing of a free energy barrier associated with the dehydration process. When the concentration of ligands is low, the process becomes an "extremely" rare event because the chance of getting ligands at the channel entrance is extremely small.

The 3D-RISM method does not suffer from any of those difficulties as demonstrated in the preceding sections because the sampling of the configuration space of water and ligands is made entirely analytically due to the statistical mechanics. In the following, the studies by Phongphanphanee et al. concerning the ligand permeation through the two members of aquaporins, AQP1 and GlpF, are presented [Phongphanphanee et al. 2008, 2010]. Depicted in Fig. IV-4.2 are the contour maps of the electrostatic potential due to atoms of the channel proteins along their axes. In the case of AQP1, the region with positive electrostatic potential is extended throughout the channel; the potential is especially high in the middle region as is indicated by the red color. On the other hand, the positive potential is distributed only near the Ar/R and NPA regions in the case of GlpF. The difference in the electrostatic potential and that in the diameters of channel pore between the two channels produces a marked variety in the distribution of the ions.

In Fig. IV-4.3, the 3D-distribution of ions, H_3O^+, Na^+, and Cl^-, are plotted along the channel axis. The one-dimensional profile of the potential of mean force (PMF) of the solution components including water is also depicted in the same figure. As can be seen from the PMF, water inside the channels is a little more stable than that outside (or bulk), and there is no appreciable barrier to prevent the molecules from diffusing through the channels for both AQP1 and GlpF. The finding is consistent with the observations that water is extremely permeable through those channels, and with the X-ray measurements which indicate the existence of more or less stable water molecules inside the channel pores. The finding, however, is in contrast with the simulation results which show the positive PMF throughout the channel pore [De Groot and Grubmuller 2001].

The 3D-distributions and PMFs of the ions inside the channels exhibit much more variety than those of water, which are explained as follows.

Aquaporin1 (AQP1): In AQP1, the cations are excluded from a large area, extending more than 10 Å along the axis, which spans amino-acid residues from R197 (or Ar/R) to the NPA region. Although these gap areas exclude the hydronium ion, water is still distributed along these areas. It is also understood from the figure that the distribution of the hydronium ion is determined essentially by the electrostatic potential inside the channel: the hydronium ion is excluded from the channel primarily by the positive electrostatic repulsion. The gap of the distribution of sodium ions is greater than those of hydronium ions. It is probably because the size of sodium ions is slightly larger than that of hydronium ions. So, the electrostatic and steric effects seem to be working in concerted manner to exclude sodium ions from the channel. The large gap in the

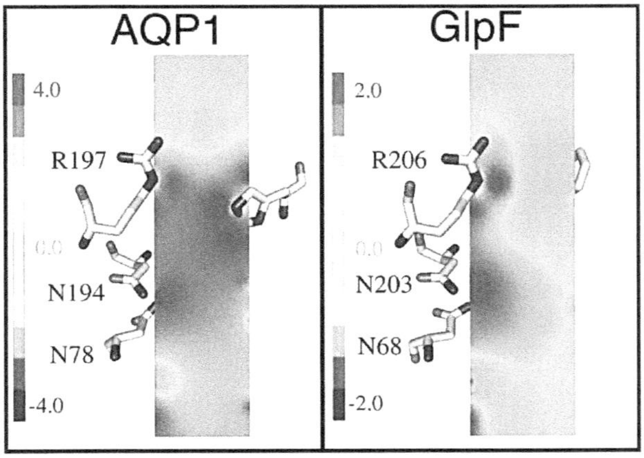

Fig. IV-4.2: Contour maps of electrostatic potential inside aquaporins. Copyright (2008) Elsevier B.V.

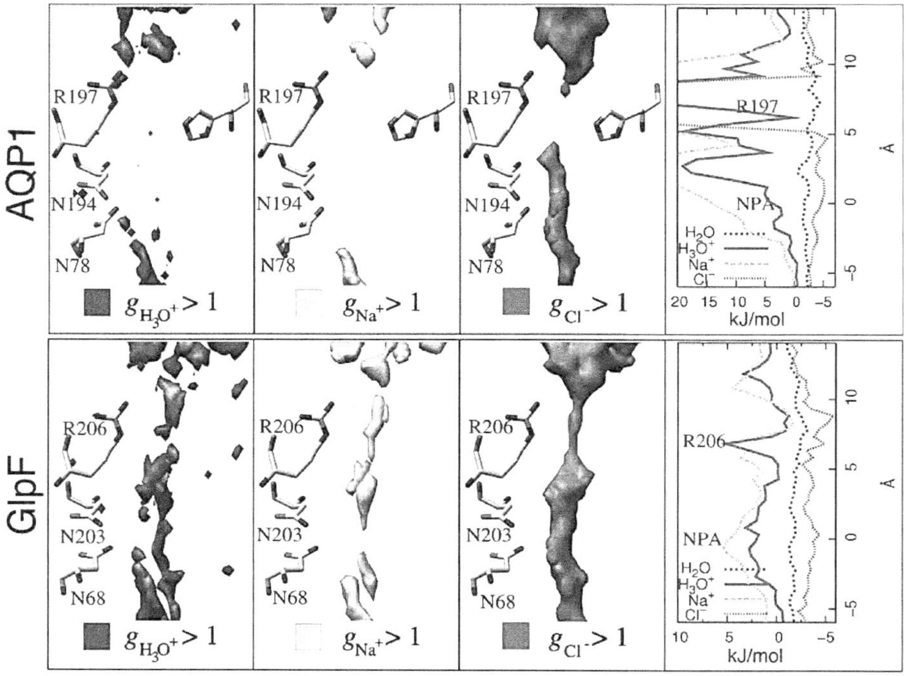

Fig. IV-4.3: Spatial distribution ($g(\mathbf{r})$) and the potential of mean force of ions inside aquaporins. Copyright (2008) Elsevier B.V.

chloride distribution around the R197 residue should have a different interpretation from the case of cations because the electrostatic potential and the charge of the ion have opposite signs. In this case, the steric effect due to the bulkiness of ion, approximately 4.4 Å, dominates over the electrostatic attraction to exclude the ion from the SF region of the channel. The results are consistent with the experimental findings which indicate that none of those ions are permeable through the AQP1 channel [Preston et al. 1992].

Glyceroporin (*GlpF*): The distributions of the two cations, sodium and hydronium ions, in GlpF have a similar shape: both cations have three peaks in the PMF located around NPA, Ar/R and the area surrounded by T137, G199 and I44. While the distribution is almost nothing at the three regions in the case of sodium ion, it vanishes only at the Ar/R region in the case of hydronium ion. The distribution function of the chloride ion also has a minimum at Ar/R region. The minimum in the case of chloride ions, though, is not as low as those in the cations. The finding suggests that there is a slight possibility that the anion can permeate through the channel. However, to our best knowledge, such experimental result, which suggests the possible chloride permeation through the channel, has not been reported. The results are suggestive of another mechanism working for chloride ions to be blocked from the permeation, which is a "trapping" mechanism. By taking a look at the PMF of Cl^- in GlpF, one may notice that it has a long-lasting down-hill slope from the intracellular exit to the SF region, which turns uphill after passing through the deep valley around the SF region. The depth of the valley relative to the bulk is about –6 kJ/mol, about twice the thermal energy at the ambient condition, which is deep enough to trap a Cl^- ion inside the channel when there is no electrostatic field. That may be the reason why the ion is not permeable through the channel in the physiological condition. The explanation, however, is just qualitative. The depth of the valley, –6kJ/mol, may change if one takes the membrane around the channel into account, which we have disregarded in this study.

A special attention should be paid to the transport of a hydronium ion in the GlpF channel, since it may give some hint for the question why a proton is excluded from permeation through the aquaporin channels. In the case of AQP1, the mechanism may not work at all since the hydronium ion is excluded from the region with large gap. A proton has no chance to transport through this gap either "riding on" a water molecule as a hydronium ion or by the Grotthuss mechanism. On the contrary, the gap of hydronium ion distribution in GlpF is limited in the small area around R206. A proton may have a chance to move across this area by the Grotthuss mechanism, although it may not be easy for a hydronium ion to pass through the gap. Water molecules which have substantial distribution around this region can make a "bridge" for a proton to transport by the Grotthuss mechanism. In this respect, the possibility of proton transfer around R206 should not be excluded. Then, another question is raised. Why and how a proton is banned by the channel from permeation? An answer to the question was reported in the JACS communication by Phonephanphanee et al. [Phonephanphanee et al. 2008]. For the Grotthuss mechanism to be effective, water molecules should take proper orientation so that they can make hydrogen-bond among themselves. The oxygen of a water molecule at the NPA motif is coordinated by two amide groups (NH_2), N203 and N68 of GlpF, N194 and N78 of AQP1. The bipolar coordination creates a high potential barrier for the reorientation of the water molecule, and prevents it from making the proper hydrogen-bond chain for the Grotthuss mechanism to be effective.

The physics of the proton exclusion in AQP1 and GlpF stated above are illustrated schematically in Fig. IV-4.4. Although there are two mechanisms to prevent a proton from permeating through GlpF, the electrostatic repulsion and the bipolar coordination of orientation of water, the mechanisms may not completely eliminate the possibility of proton conduction through the channel. The high barrier for the reorientation of a water molecule at the NPA region prevents a proton from transfer via the Grotthuss mechanism, but the hydronium ion may diffuse across this region. On the other hand, the small gap for the hydronium ion around R206 due to the electrostatic repulsion and/or the steric hindrance does not completely

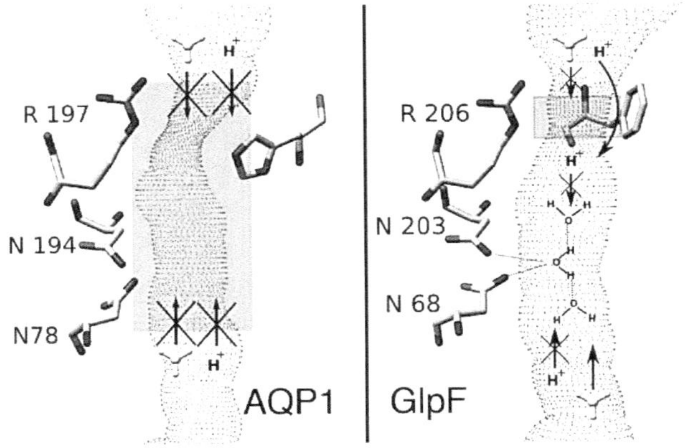

Fig. IV-4.4: Schematic illustration of the mechanisms of proton exclusion in AQP1 and GlpF. The transparent pale red surface represents the gap region in the distribution of hydronium ions. The molecules colored depict the hydronium ions. Copyright (2008) Elsevier B.V.

Color version at the end of the book

eliminate the possibility of a proton to transfer across the gap via the Grotthuss mechanism. The interpretation of the results is basically in accordance with the picture proposed by G. Voth, and is in harmony with the experiments which indicate the negligibly small conductivity of a proton through GlpF [Ilan et al. 2004, Chen et al. 2007].

IV-4.2 Ion Channels

Ion channels are proteins that control ion permeation through a membrane [Hille 2001]. The protein is akin to the pipeline, a device for transporting fluid such as water and petrol from one place to the other. The molecular machine operates, in general, two functions, "gating" and selecting ion species, which correspond to "valve" and "filter," respectively, in the usual machinery of a pipeline. The gating controls the *timing* of opening and closing the gate of entrance and exit of the channel in a manner harmonic to up- and down-streams of a signal transduction. Failure of the gating will cause a malfunction of the signal-transduction. The mechanism of gating is different from channel to channel, electromotive force, ion concentration, pressure, and so on. We will look at an example of the gating mechanism in the next section, where a proton channel in virus is investigated.

The other function of ion channels, or "selective filter," is to screen an ion species from the other kind of ions. For example, a potassium channel has a region that screens K^+ from other kind of ions such as Na^+ and Li^+. Failure of screening will cause illness or death of a living body. In the following, the mechanism of filtering an ion from the others is investigated, taking a potassium channel called KcsA as an example.

The potassium channel permeates K^+ efficiently, while not only larger Cs^+ but smaller Na^+ is also not allowed to pass through the selectivity filter. The mechanism underlying the ion selectivity has been studied extensively for more than several decades, and still remains elusive. To understand the underlying molecular mechanism of ion selectivity of channel proteins, various hypotheses have been proposed. Among the underlying mechanisms, the snug-fit hypothesis has been a leading concept, in which the channel accepts ions only if ions fit the size of the vacant space in the channel pore. This hypothesis was consolidated when the crystal structure of the KcsA potassium channel was revealed.

The first crystallized channel is the KcsA potassium channel of bacterial origin [Doyle 1998]. The KcsA gene encodes 160-residue protein, and the functional channel is formed as homo-tetrameric assembly. The transmembrane pore domain is the most important part for ion channels. The ion permeating pore exists along the symmetric axis of the channel. The radius of the pore is not uniform, and the pore near the extracellular space is very narrow (1.5 Å), serving as the selectivity filter. The pore becomes wider after 12 Å length of the selectivity filter, named the central cavity. The wide entryway connects to the intracellular bulk solution.

The selective filter constitutes 5 residues, TVGYG, from each subunit, but the inner surface of the SF is lined by the backbone carbonyls. The structure of carbonyls in SF mimics the environment of an ion in bulk water. For example, in the snug fit model, the size of pore and the size of an ion have been considered to be the principal of selectivity. From the X-ray structure, the distance from K^+ to carbonyl oxygen is near the distance from the ion in bulk to the surrounding water of first hydration shell resulting in the structure of SF compensating the dehydration of K [Doyle 1998]. However, in the case of Na^+, the channel is constrained from collapsing and bringing the carbonyl oxygen close to the ion. Therefore, the pore is fit for K^+, but not fit for Na^+. In addition, over coordination model is another hypothesis which considers the coordination number of an ion. In bulk liquid, K^+ is coordinated by number of water around 7–9, but it is just 4–7, in the case of Na^+ [Morais-Cabral et al. 2001, Ando et al. 2005]. At the binding sites of K^+, the ion is coordinated by 8 carbonyl oxygen which correspond to environment in the bulk.

Here, the 3D-RISM/KH method is applied to answer the questions as to how ions are stabilized in the open-filter structure; what distribution do they take? How does water contribute to the ion bindings? Examining these issues will lead to understand how highly permeable K^+ and poorly permeable Na^+ and Li^+ interact in the open-filter structure to express the selectivity. Calculated first is the distinct configuration of cations in the selective filter based on the 3D-RISM/KH theory. It will be shown that small ions such as Na^+ and Li^+ are bound in the open-filter structure of SF region for KcsA with high affinity. Then the free energy profile of K^+ and Na^+ in SF is calculated to consider the mechanism of the ion-selectivity based on the 3D distribution functions [Phonghamphanee et al. 2014a, b].

***Distribution function of water and ions in the channel*:** Shown in Fig. IV-4.5 is the contour of the spatial distribution function (SDF: $g(\mathbf{r})$) of H_2O, K^+, Na^+, and Li^+ in SF and the central cavity. The distribution of water inside SF is similar to that of K^+, and the binding sites of water and K^+ in SF coincide well with those from the X-ray crystallography [Zhou et al. 2001]. Outside SF, confined water is surrounded by E71 and D80, the peak positions in DF are in agreement with the crystallographic results.

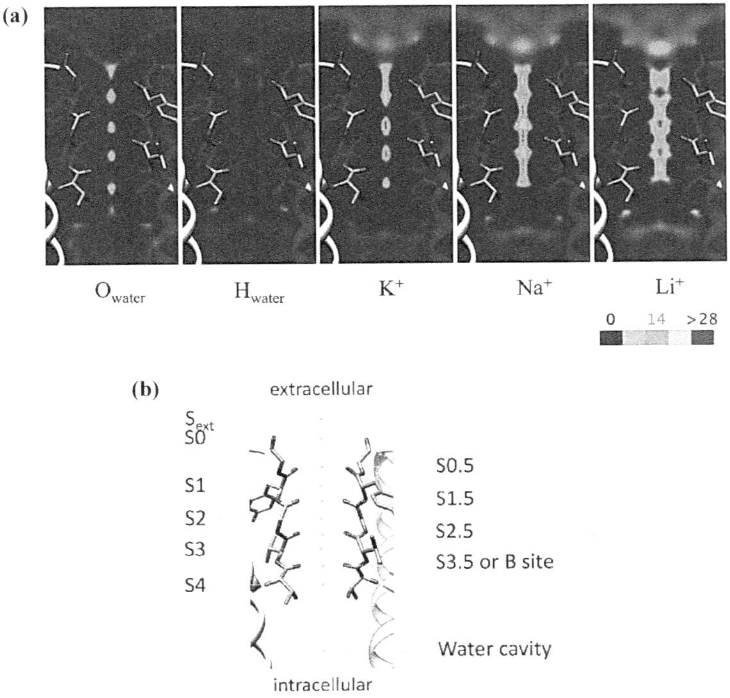

Fig. IV-4.5: (a) Contour of the spatial distribution function ($g(\mathbf{r})$) of O and H of water, K^+, Na^+, and Li^+ in the selective filter (SF) region. Red color denotes the region of high probability. (b) Illustrative view of the selective filter (SF) region. S0, S1, etc., specify the position of ion-binding sites. Copyright (2014) IUPAC & De Gruyter.

Color version at the end of the book

The peaks of SDFs of K^+ reveal that the binding sites of the K^+ ion are located at S1, S2, S3, and S4, (Fig. IV-4.5b), which are at the center of the octuplex made by eight carbonyl or threonine hydroxyl oxygen-atoms (the cage site). On the other hand, the SDF for Li^+ is nearly a reverse image of that for K^+. The binding sites of Li^+ are located in between the binding sites of K^+. Li^+ is located at the center of a plane formed by four carbonyl oxygen-atoms (quadruplex). The sites are referred to as "plane-sites," and they are marked as S0.5, S1.5, S2.5 and S3.5 (Fig. IV-4.5b). In contrast to the high density on the plane, the density of Li^+ was emptied at the center of the cage, and SDF between two planes is connected but off-centered. In case that Li^+ would move through SF, it may trace winding routes along SF. It should be noted that there is a high barrier at the extracellular entrance between S0.5 and S_{ext}. The SDF for Na^+ is similar to that of Li^+ but significant density is observed even at the cage positions, leading to the elongated continuous density throughout SF. The barrier at the external entrance is lower than that for Li^+. In any case, the results from 3D-RISM/KH unequivocally show that not only K^+ but also Li^+ and Na^+ have quite large binding-affinity to SF of KcsA, although the position and pattern of the distribution is different from each other.

Explicit ions at the binding sites: In order to clarify how K^+, Na^+ and Li^+ are bound at their respective binding-sites, Phongphanphanee et al. calculated SDF of water molecules with an ion (K^+, Na^+ or Li^+) placed explicitly at the binding-site: S1, S2, S3 and S4 for K^+, and S0.5, S1.5, S2.5 and 3.5 for Li^+ and Na^+. In the case of K^+, the highest peak in SDF of water-oxygen appears at the cage-sites adjacent to K^+ occupying the cage. For example, when S2 is occupied by K^+, the distribution peak of water-oxygen appears at the center of S1, S3, and S4 (Fig. IV-4.6). Corresponding distributions of hydrogen atoms are seen at the positions where they can make hydrogen-bonds with the carbonyl oxygen atoms belonging to the adjacent cage-sites.

In the case of Li^+, two water molecules appear at the close vicinity of Li^+ ion, and a complex of Li^+ and two water molecules fit into the two cages. For Li^+ placed at S2.5, the oxygen distributions are nearly centered at S2 and S3. Those water molecules form hydrogen-bonds with carbonyl oxygen-atoms in S1.5 and S3.5 sites. In next further cages (S1 and S4) from the Li^+ bound at the plane site, water occupies the center similar to those in adjacent cages for K^+.

For Na^+, water distribution similar to that for Li^+ was observed. Among the plane sites, we found a specific distribution of water for Na^+ bound at S2.5. In contrast to the other sites, water molecules were not bound to Na^+ and almost no density of water oxygen was observed in the cages (S2 and S3).

Free energy profile of ion permeation: Phongphanphanee et al. have examined the binding of small ions in the open conformation of the SF region of the KcsA channel based on the 3D-RISM/KH theory. The structure of SF serves distinct binding sites for both highly permeable K^+ and least permeable Li^+, where K^+ is stabilized in the cage sites and Li^+ is bound in the alternative plane sites exclusively, while poorly permeable Na^+ binds to both sites (Fig. IV-4.7a). They refer to this feature as the ambivalent snug-fit sites presented by SF, remarking the intriguing structural characteristics. While the cage is optimized for K^+ binding in accordance with the snug-fit hypothesis, coexistence of the plane site suggests that SF is not exclusively K^+ selective, but would possibly become selective even for Na^+ and Li^+. The paradoxical ambivalent snug-fit sites claim that the snug-fit hypothesis unique to K^+ was refuted at least for the equilibrium binding selectivity.

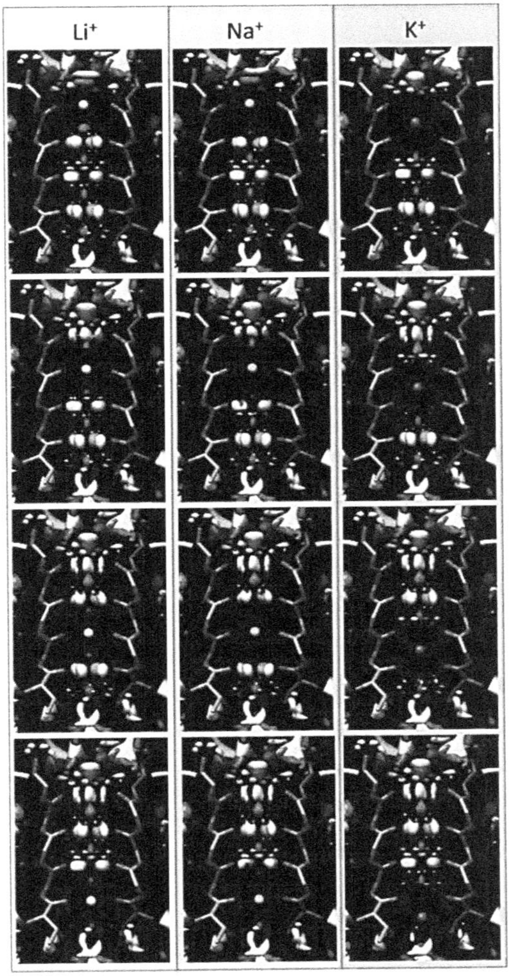

Fig. IV-4.6: Spatial distribution function ($g(\mathbf{r})$) of O and H of water around an ion fixed at the binding sites: $g_O > 3$ (red), $g_H > 2$ (light blue). Li^+, Na^+, and K^+ are colored, respectively, by yellow, green, and purple. The ions are fixed at S0.5, 1.5, 2.5 and 3.5 for Li^+ and Na^+, at S1, S2, S3, and S4 for K^+. Copyright (2014) IUPAC & De Gruyter.

Color version at the end of the book

Then, what is the underlying mechanism of the selectivity? In order to touch this question, a result for the free energy profile of Na^+ and K^+ along the channel axis of SF is shown. The free energy profile ΔG_{ion}(bulk→channel) to transfer an ion from bulk water to a position along the channel axis can be calculated by the relation

$$\Delta G_{ion}(\text{bulk}\rightarrow\text{channel}) = \Delta\mu_{sol}(\text{KcsA-ion}) - \Delta\mu_{sol}(\text{KcsA}) - \Delta\mu_{sol}(\text{ion}) + (E^1 - E^0),$$

where $\Delta\mu_{sol}$(KcsA-ion) is the solvation free energy of the complex of KcsA and an ion, $\Delta\mu_{sol}$(KcsA) is the solvation free energy of KcsA, $\Delta\mu_{sol}$(ion) is the solvation free energy of an

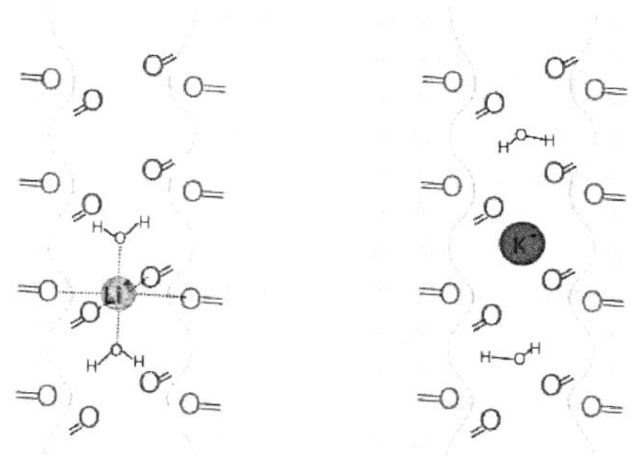

(a)

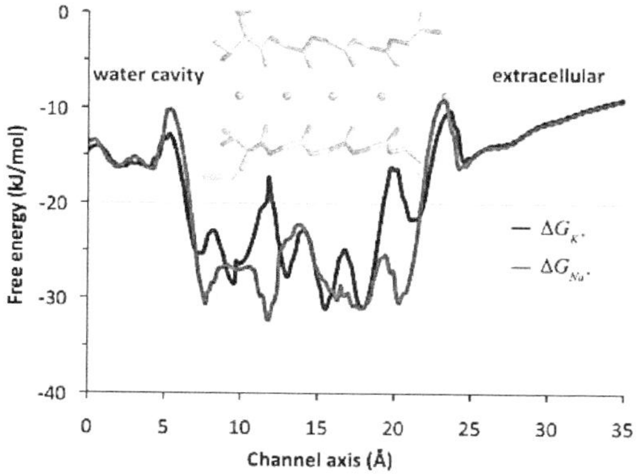

(b)

Fig. IV-4.7 (a) Illustrative picture of the binding of ion in SF. Red colored O denotes the backbone carbonyl. (b) Free energy profile of K^+ (blue) and Na^+ (red) ions along the selective filter. Copyright (2014) IUPAC & De Gruyter.

Color version at the end of the book

ion, and $(E^1 - E^0)$ is the binding energy in vacuum that takes account of conformation change of protein. The result was obtained from 3D-RISM/KH for the KcsA structure with and without an explicit ion in SF, by optimizing the structure of amino-acid residues at SF with a fixed other part of the channel. The potential profile, thus, represents single-ion permeation rather than multi-ion permeation.

As can be seen in Fig. IV-4.7b, K^+ and Na^+ have quite different free energy profiles along the channel axis. K^+ has a profile which is periodic with the wavelength characteristic of the distance between the center of two cage sites. The barrier between the wells apparently reflects the moderate core-repulsion of the ion with the carbonyl groups, which can be overcome by structural fluctuations. On the other hand, the profile of Na^+ does not show such a clear periodicity, which is consistent with SDF of the ion. The most distinct difference between the two profiles is seen at both ends of the SF region. For example, at the extracellular mouth of SF, the barrier height to be overcome for K^+ to exit from inside to outside SF is ~ 10 kcal/mol, while it is ~ 20 kcal/mol for Na^+. The reduced barrier height in the profile of K^+ is due to the existence of the intermediate local minima, which play a role of steps along the route of *rock climbing*. The profiles at the mouth on the other side have similar characteristics. It can be speculated that the difference in the barrier heights discriminates the two ions upon exiting from SF. (The model has some similarity to that proposed earlier by Kiss et al. based on an electrophysiological experiment as long as K^+ is concerned.) The barrier height for entering into the SF region is also in favor of K^+ by about 2 to 3 kcal/mol. To summarize, Na^+ is more difficult to get into the SF region compared to K^+, but it is even harder to get out, if the ion is once bound in the region. A similar but even more significant effect may take place in the case of Li^+, although the analysis concerning Li^+ is not performed yet.

From the results of ion distribution and the preliminary result for the free energy profile of K^+ and Na^+ inside SF, one may draw the following picture concerning the ion permeation through SF in KcsA.

Small alkali-ions, K^+, Na^+, and Li^+, get into the SF region of open state KcsA channel from the mouth of either side of the region, driven by relatively weak membrane potential, although Na^+ and Li^+ require higher potential compared to K^+. Once they are accommodated into the channel, the ions can stay in the region comfortably with their respective binding modes: K^+ distributes periodically in the cage sites, Li^+ stays also periodically in the planar sites, while Na^+ distributes more diffusively compared to the other two ions. A potassium ion K^+ requires the free energy which is about a half of that required for Na^+, to get out of SF through the mouth on either side of the region depending on the direction of the force. It may be speculated that the difference is the physical origin of the ion-selectivity of KcsA channel.

The picture concerning the ion selectivity can also explain why coexisting Na^+ depresses the K^+ conductivity through the channel, and why the conductivity recovers with higher electrostatic field. When Na^+ are coexisting with K^+ in solution, both ions may get into the channel, although the barrier height for Na^+ is higher than that for K^+. Once Na^+ gets into the SF region, it is so hard to get out of the channel, and it will block the permeation route of K^+. This may be the reason why the K^+ conductivity is depressed by coexisting Na^+. However, the blocking Na^+ can be "punched through" by applying greater electrostatic field. This may be the reason why the K^+ conductivity is recovered with a higher membrane electrostatic field. The picture is consistent with that proposed by Nimigean and coworkers [Thompson et al. 2009, Nimigean and Allen 2011].

IV-4.3 Proton Channel (M2)

The M2 protein channel, which is found on viral membrane of the virus, has received a great deal of attention as a target of drug development due to its important role of proton transport and viral replication [Lamb et al. 1994, Helenius 1992]. It is known that an Amantadine drug family is effective against influenza A by blocking M2 channel, which disturbs viral proton conduction and consequently causes inhibition of viral replication [Pinto et al. 1992]. However, the underlying mechanism of proton blockade activity by Amantadine is not clarified yet. In addition, many new strains of influenza virus are resistant to Amantadine. Therefore, understanding the mechanism of proton transportation through the virus membrane is one of the central issues for drug design.

Several electrophysiological results have shown that the M2 channel is highly selective for proton, and that its gating is controlled by pH [Lear 2003, Mould et al. 2000a, b, Chizhmakov et al. 1996, Vijayvergiya et al. 2004]. A number of researchers has explored the relationship between functions and molecular structures of the M2 channel, and has indicated important roles of His37 in gating mechanism. Pinto et al. 1992 have demonstrated that a mutant of M2 channel with the His residue substituted by Gly, Ser, or Thr loses the proton selectivity, and that the selectivity is restored upon addition of imidazole; this has suggested that the imidazole group plays an important role in proton selectivity [Venkataraman 2005]. Concerning the pH-controlled gating mechanism, Hu et al. have made important suggestions based on their experiments that the pKa associated with four histidines at the gating region is different from each other, and that the open forms are dominated by triply and quadruply protonated histidine [Hu et al. 2006]. These experimental results suggest that His 37 acts as a pH sensor switch to open and close the gate and as a selective filter to allow the permeation of proton, but not other cations. The experiments also indicate that Trp 41 has an important role in the gating. Some mutants of the protein, in which the Trp41 is replaced by a residue with smaller size, Ala, Cys or Phe, have higher proton conductivity compared with its wild type [Okada et al. 2001]. The results from UV Raman spectra show the interaction between protonated imidazole of His37 and the indole of Trp41, or cation-pi interaction [Okada et al. 2001, Takeuchi et al. 2003]. These investigations suggest that the indole of Trp41 has a role to occlude the channel pore. This residue behaves as a "door" to turn open or close the pore, which is controlled by protonated His due to the cation-pi interaction.

To investigate the mechanism of proton transfer in the M2 channel, Phongphanphanee et al. considered the distributions of water and hydronium ion inside the channel as a function of pH which regulates the protonated state (PS) of histidine tetrad from the non-PS (0H) to the quadruple PS (4H) in the decreasing order of pH [Phongphanphanee et al. 2010]. For each state, they randomly picked up the coordinates of M2 protein from a trajectory of the MD simulation which has been carried out. The spatial distribution function (SDF; $g(\mathbf{r})$) of water and hydronium ion with $g > 1$ in the five different protonated states of M2 channel is depicted in Fig. IV-4.8. The figure indicates that the accessibility of water (cyan in Fig. IV-4.8) to the channel pore increases with the protonated state of the channel in the order 0H < 1H < 2H < 3H < 4H. This result can be explained readily in terms of the pore diameter which is widened due to the electrostatic repulsion among the protonated histidines His37, and the bulky indole ring of Trp41 which can turn to block or open the gate. The results suggest that there are two distinct states in the channel conformation, or "open" and "closed" forms. The 0H, 1H and 2H forms are considered as closed forms, since water distributions are not observed at the selective filter regions of His37 (yellow stick in Fig. IV-4.8) and at the gating region with the Trp41 residues (orange sphere and stick in Fig. IV-4.8). On the other hand, the 3H and 4H forms with a continuous water distribution along the pore are identified as open forms (Fig. IV-4.8). In addition, the narrowest SDF of water in the channel, also the narrowest of pore, is seen at the Trp41 region.

The PMFs of water corresponding to the distribution function are shown in Fig. IV-4.9, in which high barriers are found only in the 0H, 1H, and 2H states. It is obvious that in the closed form, water cannot overcome the high barrier made up of the steric hindrance between the channel atoms and water molecules. On the other hand, PMF of water in 3H and 4H is negative along the entire channel pore, which indicates that water molecules in the channel are more stable than those in the bulk, and that water is permeated through the channel. The results are

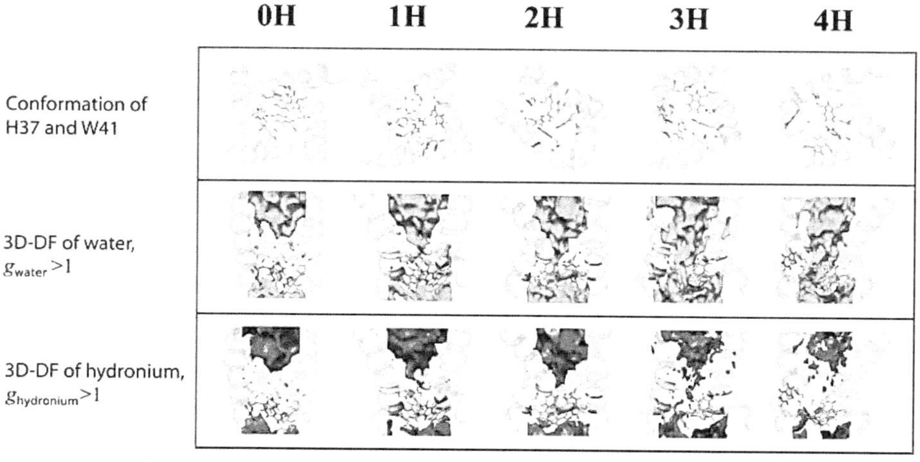

Fig. IV-4.8: Structure of Trp41gating (orange) of different protonated histidine tetrad, and SDF ($g(\mathbf{r})$) of water (cyan) in the channel and hydronium ion (red), with $g > 1$. Copyright (2010) American Chemical Society.

Color version at the end of the book

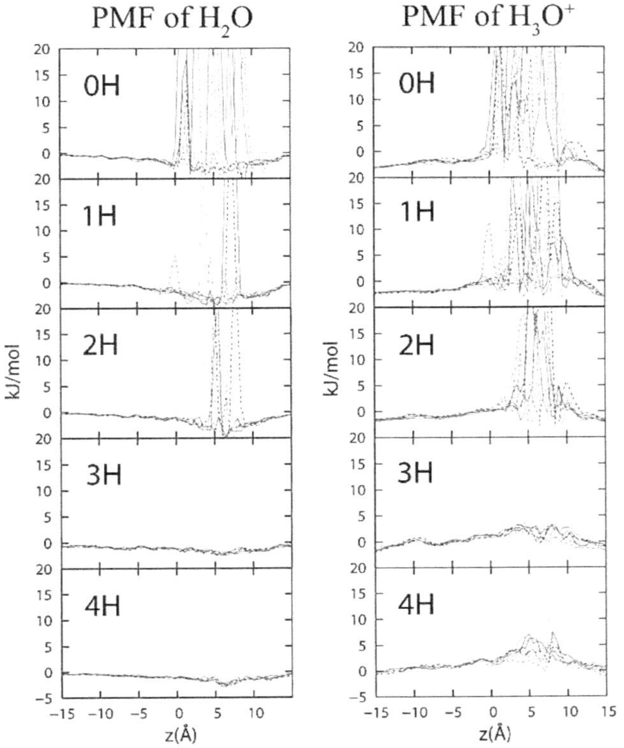

Fig. IV-4.9: PMFs of water and hydronium ion in five protonation states, each line represents each conformation in the state. Copyright (2010) American Chemical Society.

in harmony with the earlier theoretical studies carried out by different methods [Intharathep et al. 2008, Kas and Arkin 2005, Chen et al. 2007].

In the three states of closed gate, or 0H, 1H, 2H, hydronium ions exhibit a behavior similar to water, but with lower distribution, and higher barrier in PMFs compared to those of water (Fig. IV-4.9). It indicates that a hydronium ion, or a proton, cannot be distributed in the channel and is prevented from transporting across the channel. The results are consistent with those for the closed conformations of the M2 channel at high pH values reported in many experimental and theoretical studies [Lear 2003, Mould et al. 2000a, b, Chizhmakov et al. 1996, Vijayvergiya et al. 2004, Chen et al. 2007].

As was already seen, water continuously distributes throughout the 3H and 4H channels (Fig. IV-4.8) indicating the dramatic structural change of the channel from closed to open conformations. In contrast, the distributions of hydronium ion in the 3H and 4H states do not look much different from those in 0H to 2H. However, it is just apparent, if one takes look at Fig. IV-4.8, where PMF of the hydronium ion in 3H and 4H exhibits entirely different behavior than that of 0H to 2H. Protons in the 0H to 2H forms have extremely high barriers due to the same cause in the case of water, and no chance to exist in the gating region of the channel, while the barrier heights in the 3H and 4H forms are just 2–3 and 5–7 kJ/mol, respectively, which are comparative to the thermal energy. There is another interesting observation in the figure. The barrier height for protons is higher in 4H than in 3H, which is against the heuristic argument based on the pore size around the gating region.

All those observations are suggestive of two competing factors working on the proton distribution as the protonated state of the channel is increased from 0H to 4H. One of those is the channel opening due to the increased repulsion among the protonated His residues, which tends to enhance the distribution of proton as well as water in the channel. The other factor is the electrostatic repulsion between protons and the protonated His residues, which will reduce the proton distribution with increasing number of the protonated His. The two effects are balanced at the 3H state to make the distribution of proton in the channel maximum. These results are consistent with those by Voth and co-authors [Chen et al. 2007].

IV-5 Escaping Pathway of CO from Myoglobin

Myoglobin (Mb) is a globular protein which has important biological functions, like oxygen storage [Wittenberg and Wittenberg 2003]. Due to its biochemical function, many researchers have made intensive efforts to identify the escaping pathway of the ligand, experimentally and theoretically [Perutz and Mattews 1966, Huang and Boxer 1994, Sakakura et al. 2001, Nishihara et al. 2004]. Thirty years ago, Perutz et al. proposed that the entry and exit pathway of ligand into the active site of Mbs involves rotation of the distal histidine to form a short and direct channel between the heme pocket and solvent [Perutz and Mattews 1966]. Recent experimental studies by the time-resolved x-ray crystallography have shown consistent results with the earlier studies [Schotte et al. 2004]. Unfortunately, their reports were limited to the conditions under low temperature and crystal structures. Several groups have argued that the detaching scheme could be different in the physiological conditions due to the conformational fluctuation of the protein. Lately, they have reached the consensus that the histidine gate may not be the primary pathway for ligand movement into and out of myoglobin [Huang and Boxer 1994, Scott et al. 2001, Schotte et al. 2004].

It has been well-recognized that there are several intermediate states separated by activation barriers along the escaping pathway, which are referred to as "Xe-sites" [Ewing and Maestas 1970, Tilton et al. 1984]. The experimental results indicate that CO spends some time at the Xe sites before escaping to solvent at room temperature. In the mixed Xe and CO solutions, the difference in affinity between Xe and CO to each Xe trapping site makes the CO escaping pathway different depending on Xe concentration. It is believed that the dissociated CO escapes to the solvent through the Xe1 trapping site predominantly under the Xe-free condition. On the other hand, CO escapes through the Xe4 site in a Xe-rich solution.

Recently, Terazima and his coworkers have proposed a new method to observe the time resolved thermodynamics in solutions based on the transient-grating (TG) spectroscopy [Terazima 2000, 2016]. The authors applied the method to the photo-dissociation process of carbon mono-oxide (CO) from Mb in solutions to identify not only the escaping pathway but also the thermodynamics at each step along the pathway. The study has revealed that the ligand migrates into internal cavities of Mb upon the photo-dissociation, and that the population of the two Xe sites is associated with protein relaxation that occurs after the photolysis [Sakakura et al. 2001, Nishihara et al. 2004].

Terazima and co-workers have measured the partial molar volume (PMV) change of the system along the pathway of CO escaping process under Xe solution based on the TG method. They hypothesized that the intermediate of the pathway is through Xe4 site, since the experiments were carried out under the Xe-rich condition.

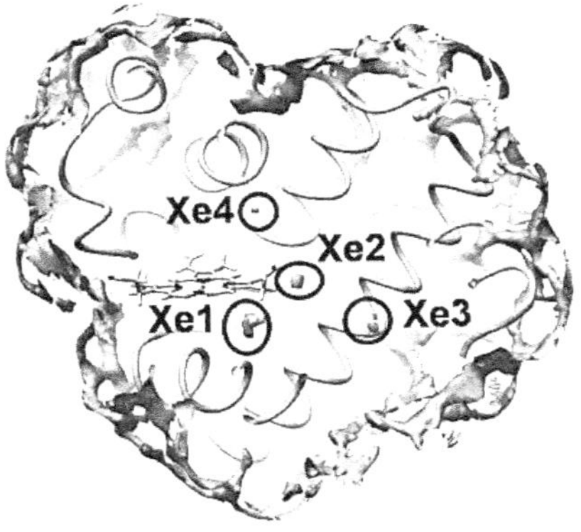

Fig. IV-5.1: The spatial distribution function ($g(\mathbf{r})$) of Xe inside and around Mb. The surface of $g(r) \geq 3$ is depicted. Circles in the figure denote Xe trapping cavities. Copyright (2009) American Chemical Society.

There were some theoretical attempts based on the molecular dynamics simulation to verify the results of TG and the hypothesis made by Terazima and coworkers. However, it was not so successful, since it is not so easy to evaluate the PMV change of the CO-Mb complex by the method.

Kiyota et al. applied the 3D-RISM/KH theory and the Site-Site Kirkwood-Buff theory described in Chapter II to identify the CO escaping pathway of Mb and to find the PMV change of the Mb-CO complex [Kiyota et al. 2009, Imai et al. 2000]. As mentioned above, the ligand dissociating process of Mb occurs from heme to solvent through some specific cavities. The CO escaping pathway from Mb along the pathway can be identified by examining the spatial distribution functions (SDF) of the ligand. To examine the distribution of a Xe ligand in the Xe trapping sites, the relaxed structure of Mb in solution was optimized with the 3D-RISM theory. In the calculation, a Mb molecule was immersed in a mixture of water and Xe. Therefore, the probability of existence of Xe atoms is discussed in terms of SPD of solvent Xe around solute Mb. The geometry search was started from 1MBC [Tilton et al. 1984] structure taken from PDB. The SPD of Xe evaluated by 3D-RISM is shown in Fig. IV-5.1. At a glance, four conspicuous peaks of the Xe distribution can be found. These positions correspond to the Xe trapping cavities which are reported by the experiments [Sakakura et al. 2001, Nishihara et al. 2004].

Shown in Fig. IV-5.2 are the coordination numbers (CNs) of CO and Xe in each Xe site of the RISM optimized structure, which are calculated from the radial distribution function at each Xe cavity. Since the structure was optimized in the Xe-water mixture, these results are regarded as the ligand affinity in *Xe-rich condition*. As seen in the figure, CNs of both ligands show similar behavior, namely Xe1 site has the largest affinity while Xe4 site has the smallest. Note that only in Xe1 site, the affinity of Xe is greater than those of CO unlike others. This result indicates that CO is excluded from Xe1 site in the *Xe-rich condition* due to the preferential binding of Xe. On the other hand, CO is preferentially bound to the Xe4 site in the *Xe-rich condition*. These results are consistent with the experimental observations indicating that the Xe4 site is the intermediate of the escaping pathway of CO in the *Xe-rich condition* [Nishihara et al.].

As is stated above, Terazima and his coworkers have determined experimentally the PMV change from an intermediate state to the next along the escaping pathway. Kiyota et al.

calculated the PMV along the pathway theoretically based on the 3D-RISM/KH theory, and it became the first attempt to reproduce the time-resolved solvation-thermodynamics. In order to evaluate the PMV change, they have calculated the PMV of each intermediate structure of Mb along the escaping pathway of CO. Here, "intermediate structure" means the structure in which a CO molecule is trapped in each specific Xe site of Mb. Therefore, the CO molecule is regarded as a part of solute species. The position of the trapped CO molecule is determined by the peak positions of SDF. Depicted in Fig. IV-5.3(a) are the SDF of CO in the Xe1 to Xe4 sites in the optimized Mb, calculated by the 3D-RISM/KH method. The results show conspicuous distributions of carbon and oxygen atoms of CO in the sites. Not only the molecular position but also their orientation can be deduced from the peak positions. From the peak positions of the distribution functions, one can reproduce the 3D-structure of Mb-CO complexes in which a CO molecule is bound in one of the Xe-sites, which is depicted in Fig. IV-5.3(b) taking the Xe1-site as an example. It was found that the orientation of a CO molecule in the Xe1 site shown in Fig. IV-5.3(b) is consistent with the configuration of polar groups of amino-acid residues surrounding the molecule, such as the hydroxyl groups. The structure of the Mb-CO complex with CO in the Xe4 site was determined in a similar manner.

The PMVs of the CO attached (MbCO), intermediate (Mb:CO) and CO dissociated (Mb+CO) states are shown in Table IV-5.1. The PMV of Mb:CO(Xe1) is smaller than that of Mb:CO(Xe4). Since the size of cavity of Xe1 is large enough, it can accommodate water

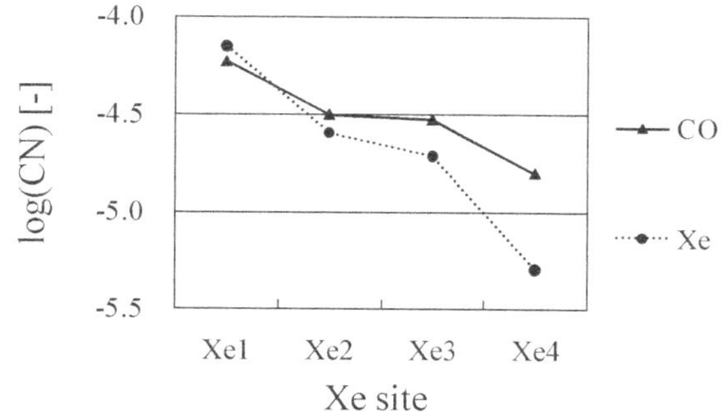

Fig. IV-5.2: Coordination number of CO and Xe molecules in each Xe site, which are calculated from the radial distribution function at each Xe cavity. Copyright (2009) American Chemical Society.

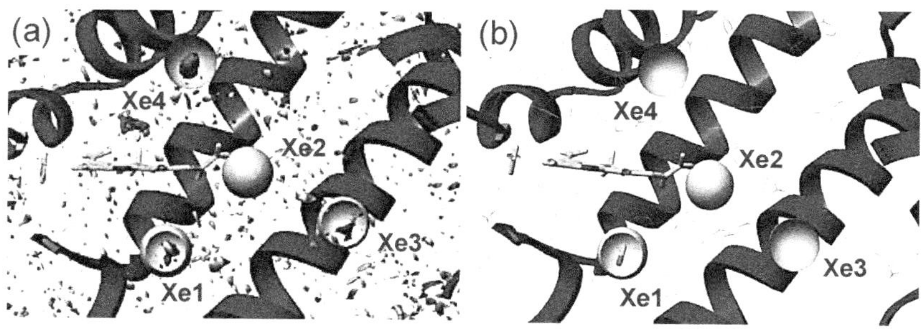

Fig. IV-5.3: (a) CO molecules in Xe1, Xe3, and Xe4 sites which are in the same plane. (b) Intermediate model for which a CO molecule exists in Xe1 ate reproduced by CO distribution. Copyright (2009) American Chemical Society.

Color version at the end of the book

even after trapping CO to be filled up. In contrast, since the size of Xe4 cavity is larger than CO molecule but is too small to accommodate water simultaneously, the extra void is created in the cavity. This extra void causes the increases in PMV of Mb:CO(Xe4). The changes of PMV through Xe4 site show excellent agreement with those by the experiment [Nishihara et al. 2004].

Although both ligand species, CO and Xe, show similar tendency, the dependence of affinity on each Xe trapping site is different. This difference indicates that the CO prefers Xe4 to Xe1 sites in the Xe-rich condition. The PMV changes estimated from the Site-Site Kirkwood Buff theory indicate that the CO escaping pathways through Xe4 are dominant. This result supports the conjecture made by Terazima and his coworkers.

Table IV-5.1: PMVs of each model and their changes. ΔV_1 represents partial molar volume change from MbCO to Mb:CO (Xe4), and ΔV_2 represents change from Mb:CO (Xe4) to Mb+CO. ΔV_{total} represents change from MbCO to Mb+CO. Copyright (2009) American Chemical Society.

Models			PMV[cm^3/mol]
MbCO			9029.0
Mb:CO(Xe1)			9030.4
Mb:CO(Xe4)			9032.8
Mb+CO			9019.6
	Xe1	Xe4	(exptl.[5])
V_1 =	1.4	3.8	(3 ± 1)
V_2 =	–10.8	–13.2	(–12.6 ± 1.0)
V_{total} =	–9.4	–9.4	(–10.7 ± 0.5)

IV-6 Identifying Molecular Tunnels: Ammonia Pathway in Purine Biosynthesis

Enzymatic reactions are undoubtedly the most important function proteins perform in our body. In a sense, our life is characterized by a temporal and spatial network of chemical reactions that are catalyzed by a variety of enzymes. Since the life requires so many different chemical reactions, the diversity of enzyme is enormous. However, there are two physicochemical features that are common in all the enzymatic reactions, as is described in the beginning of this chapter. One of those is the chemical reaction that is associated by change in the electronic structure. That feature is shared by all chemical reactions, whatever the condition is, solution, solid surface, and so on. Another feature that distinguishes the enzymatic reaction from other reactions is the *molecular recognition*. All the molecules concerning the reaction, including substrates, should be bound or *recognized* at the active site inside the enzyme in order for the reaction to be completed. The present and following sections are devoted to two topics concerning the enzymatic reaction: purine synthetase and the restriction enzyme. Purine is a nucleotide that plays a variety of roles in maintaining our life. The molecule is a building block of DNA and RNA. It is also a part of ATP which, of course, is an important molecule to supply energy for many activities performed by molecules in our body, such as chemical reactions and signaling molecules in regulatory pathways.

Formylglycinamide ribonucleotide amidotransferase (PurL) is an enzyme that participates in the purine synthesis. PurL has two domains, each of which catalyses a different chemical

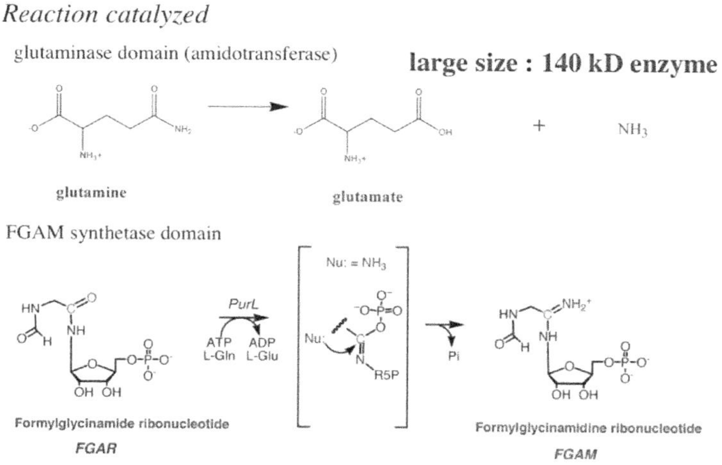

Scheme 1: Conversion of FGAR to FGAM. Copyright (2014) American Chemical Society.

reaction in the purine synthesis [Massiere and Badet-Denisot 1998, Raushel et al. 1999]. The glutaminase domain called "amidotransferase" produces ammonia by decomposing glutamine. The ammonia molecule is used to synthesize the formylglycinamidine ribonucleotide (FGAM) from the formylglycinamide ribonucleotide (FGAR) in the other domain. The two domains are connected by an ammonia channel. The ammonia molecule produced in the first domain should be protected from "hydrolysis" in order to be used as a substrate in the second catalytic reaction. So, the ammonia molecule should not spill out of the protein to bulk solution before it is used as a substrate. This is the reason why the enzyme has a tunnel or channel to transport ammonia molecules from the first to the second catalytic domain [Miles et al. 1999, Huang et al. 2001, Milani et al. 2003, Spivey and Ovádi 1999, Mouilleron and Golinelli-Pimpaneau 2007].

Although existence of such an ammonia tunnel has been known empirically, the position inside the protein has not been clarified yet [Anand et al. 2004]. There are two candidates for the pathway, which have been identified by the X-ray crystallography, one of which should be for the ammonia, while the other can be that of water. Unfortunately, it is impossible by the experimental method itself to clarify which is the ammonia tunnel, since the method can just probe the electronic distribution which does not have any signature concerning molecular species, ammonia or water. The question has been solved using the 3D-RISM/KH method combined with the molecular dynamics (MD) simulation [Tanwar et al. 2015]. The MD simulation was used in order to account for the structural fluctuation of the protein.

The 3D-RISM/KH method was applied to the aqueous solution of ammonia including infinite dilution of a wild-type PurL to calculate the distribution of NH_3 molecules around and inside the protein. The structure of protein required for the 3D-RISM/KH calculation was taken from the 100 snapshots of a 5 ns MD trajectory (each 50 psec). The snapshots of the distribution functions from 3D-RISM/KH were analyzed to identify the NH_3 pathway by a stochastic analysis referred to as *RismPath* [Sindhikara and Hirata 2013]. Shown by green spheres in Fig. IV-6.1 is the NH3 pathway of the PurL identified by the 3D-RISM/KH method, which roughly overlaps with Path 2 found by the crystallography. On the other hand, no NH_3 pathway is seen in the region of Path 1.

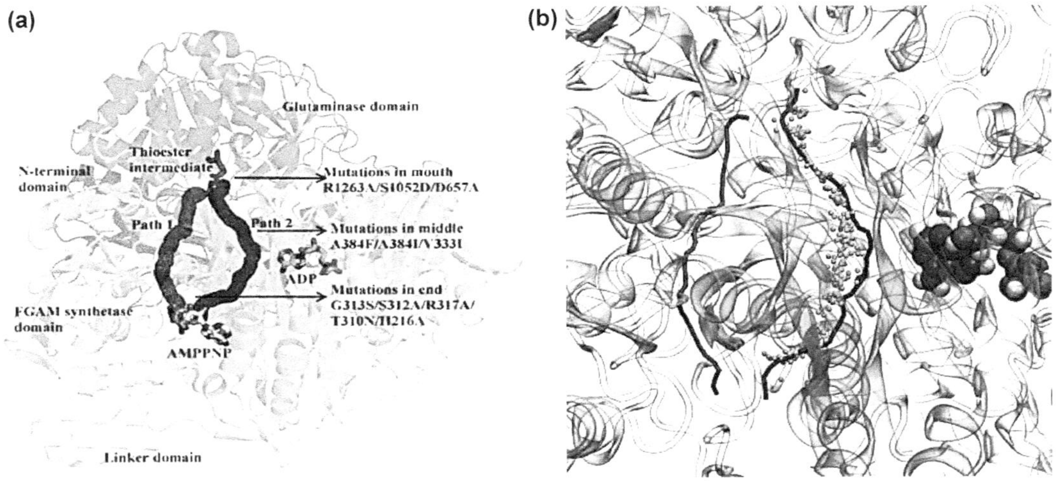

Fig. IV-6.1: Ammonia channel in StPurL: (a) Two predicted pathways for ammonia channel in StPurL shown in red and blue color mesh for path 1 and path 2, respectively. StPurL structure is depicted in cartoon with N-terminal domain in blue, linker domain in yellow, glutaminase domain in red, and FGAM synthetase domain in green color. Ligands are shown in stick representation. (b) Results of *RismPath* calculations on PurL WT trajectory. Crystallography purported Path 1 (red) and Path 2 (blue) are shown as tubes inside PurL (Cartoon, translucent). Copyright (2014) American Chemical Society.

Color version at the end of the book

Although the RISM/KH method could have identified the ammonia pathway, it is not a trivial task to verify the results empirically. (In fact, the X-ray crystallography could not identify which one of Path 1 and Path 2 is the HN_3 tunnel.) A common strategy in such a case is to apply the technique of the mutation or amino-acid substitution to the theoretical prediction and experiments, and to compare the results for the quantities that probe the effect of mutation.

In the theoretical prediction, the twelve amino acids along Path 2 were substituted, and the structures of mutants were determined *in silico* using the MD simulation. The NH_3 tunnel was identified by the same method with the case in the wild type. In Table IV-6.1, listed are the percentage of pathway found by 3D-RISM/KH and *RismPath* for the mutants created *in silico*. The first column in the table lists the mutants investigated, and the second column tabulates the number of samples out of 100 MD snapshots, for which the NH_3 pathways were identified. Shown in the rest of the columns is the number of samples for which the pathway was identified in each candidate tunnel. As is the case in the wild type, no pathway was identified in Path 1. It sounds trivial because the amino acids along Path 1 were not substituted. However, it is not trivial in fact, since the amino-acid substitution in general changes not only the structure along Path 2, but also that of the other part of the protein. The NH_3 distribution along Path 2 suffered from significant change in either direction, increasing or decreasing: increasing side, S312W, S312F, V333F; decreasing side, S312V, S312I, S312L, A384F, A384I, A384L, A384V, V333I, V333L. The results were more or less expected before the analysis was carried out. Entirely unexpected was the appearance of Path 3 due to the mutations: S312V, S312I, S312F, A384F, A384I, A384L, A384V. Although the numbers are small, it suggests strongly there is an alternative pathway for NH_3 as is indicated in Fig. IV-6.2, which might have played the role of the NH3 pathway in the evolution history of the protein, or may play a *backup* pathway in case the protein suffers from a significant change.

Table IV-6.1: Apparent paths determined by RismPath calculations on PurL WT and mutants. Copyright (2014) American Chemical Society.

	% path success	Path1	Path2	Path3
WT	23	0	23	0
S312W	55	0	55	0
S312V	12	0	10	2
S312I	13	0	10	3
S312L	11	0	11	0
S312F	90	0	88	2
A384F	8	0	7	1
A384I	9	0	6	3
A384L	16	0	15	1
A384V	17	0	16	1
V333F	28	0	28	0
V333I	12	0	12	0
V333L	13	0	13	0

Fig. IV-6.2: NH_3 pathways in the PurL mutants, identified by 3D-RISM/KH: green, wild type; brown, mutants. Copyright (2014) American Chemical Society.

Color version at the end of the book

Experimental studies corresponding to the theoretical study could have been carried out by measuring the enzymatic activities for only a limited number of the mutants, S312F and V331I that represent those with increased and decreased NH_3 distributions, respectively, along the pathway (Fig. IV-6.3). Unfortunately, the experiments for the other mutants could not have been carried out due to some reasons, for example, the structure of mutant is unstable in the cases of A384F and A384I. In the case of S312F, the enzymatic activity was increased by 3 ~ 4 percent, in accordance with the increased NH3 transport activity shown in Table IV-6.2. On the other hand, the activity was decreased by about 10 percent in the case of V331I. The result is also in accordance with the theoretical prediction because in that case the predicted ammonia-transportability was decreased to about a half.

In any case, the results demonstrate the ability of the 3D-RISM/KH method to analyse the enzymatic reactions.

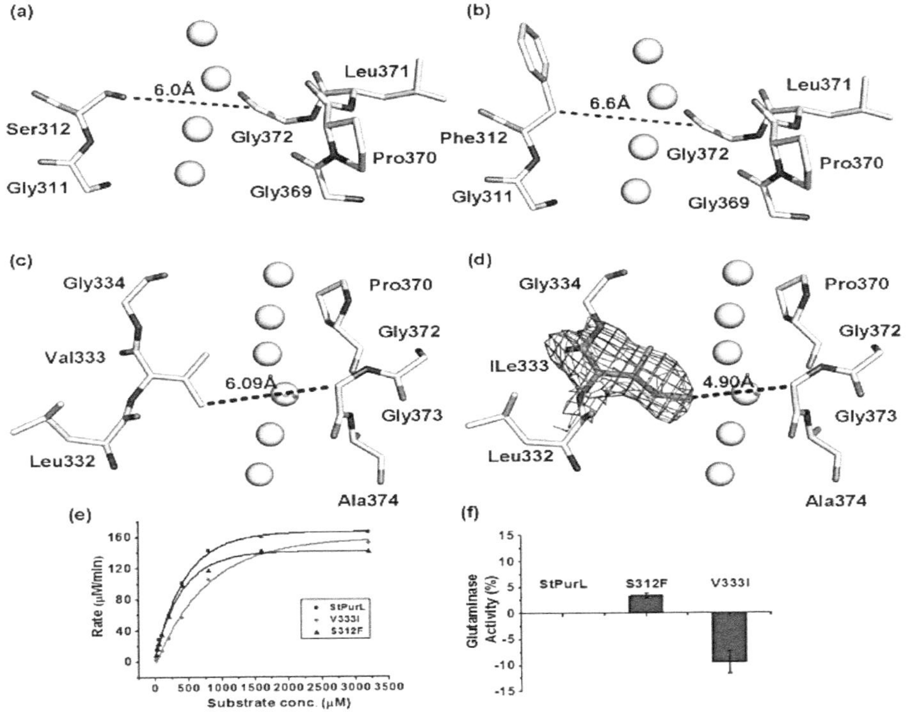

Fig. IV-6.3: Mutations along path 2. (a) WT and (b) molded S312F mutation showing greater path accessibility. (c) WT and (d) crystal structure of V333I mutant demonstrating path constriction. (e) Data of FGAM synthetase activity and (f) percentage ammonia leakage of mutant proteins as compared to WT. Copyright (2014) American Chemical Society.

Table IV-6.2: Kinetic data for FGAM synthetase activity. Copyright (2014) American Chemical Society.

Protein	K_m (mM)	k_{cat} (s^{-1})	k_{cat}/K_m ($M^{-1}s^{-1}$)
StPural	0.27 ± 0.03	1.30 ± 0.08	4.9 ± 10^3
V333I	0.48 ± 0.03	1.28 ± 0.14	2.6 ± 10^3
S312F	0.26 ± 0.01	1.15 ± 0.03	4.4 ± 10^3

IV-7 Role of Ions in Enzymatic Reactions

Small ions such as Mg^{2+} play crucial role in enzymatic reactions as a co-catalyst [Cowan 2002]. The importance will be readily verified by removing such ions from the solution. For example, the hydrolysis reaction of adenosine tri-phosphate (ATP), catalyzed by the ATPase, does not take place without Mg^{2+} in the solution [Kostrewa et al. 1995]. Nevertheless, it has not necessarily been well clarified what is the role such an ion plays in the reaction. There are several reasons why the role has not been well clarified yet, but the most important one is that it is not a trivial task to determine the position of such small ions in an enzyme. The most popular method for determining the positions (or distributions) of water and small ions in a protein is the X-ray crystallography [Kostrewa et al. 1995]. The method provides the structural information of protein, including small ions and water, as an electron distribution (or map) [Cantor and Schimmel 1980]. It has been successful to determine the structure of protein as far as the heavier atoms are concerned, which do not have large spatial and temporal fluctuations. Unfortunately, it is not the case for water molecules and small ions in an enzymatic reaction. They are essentially in diffusive motion inside protein, slow or fast, indicating that the position has large fluctuation. Especially in an enzymatic reaction, those ions and water molecules may change their positions in the active-site dramatically along the reaction pathway. In the case of a hydrolysis reaction, a water molecule as the substrate of the reaction even disappears along the reaction pathway to be converted into other chemical species. In the several sections in this chapter, we have inspected the ability of 3D-RISM/KH to identify the position of small molecules and small ions inside protein. Irisa and coworkers have applied the method to an enzymatic reaction catalyzed by a restriction enzyme, called *Eco*RV, to clarify the role of the Mg^{2+} ion in the reaction mechanism [Onishi et al. 2018].

*Eco*RV is a type II restriction enzyme that recognizes a specific DNA sequence, and selectively cuts double-stranded DNA via hydrolysis [Horton and Perona 2004]. The type II reaction does not require ATP hydrolysis, but reactivity is quite sensitive to solution conditions, such as type and concentration of cations dissolved in the solution. *Eco*RV requires Mg^{2+} for the reaction to be activated. Therefore, it is crucial for elucidating the reaction mechanism to identify the position of Mg^{2+} as well as water molecules as substrates along the reaction pathway. There have been many attempts, experimental as well as the MD simulation, to clarify the reaction mechanism by identifying the position of Mg^{2+} ion, but the results are not conclusive [Kostrewa et al. 1995, Vipond et al. 1995, Groll et al. 1997, Horton and Perona 2001, Horton et al. 2004].

A challenge to clarify the mechanism has been made by Irisa and his coworkers recently, based on the 3D-RISM/KH theory and the molecular dynamics simulation [Onishi et al. 2018]. They have first carried out the 3D-RISM calculation for a structure of *Eco*RV in aqueous solution, which is the structure before the hydrolysis reaction takes place, to find the distribution ($g(\mathbf{r})$) of Mg^{2+} as well as water molecules at the active site of the enzyme and compared the results with a corresponding experimental data due to the X-ray crystallography by Horton et al. [Horton et al. 2004]. The result is shown in Fig. IV-7.1 along with the experimental results.

The positions of Mg^{2+} determined by the X-ray crystallography are depicted in Fig. IV-7.1(a) with the green spheres labeled by I^*, II^* and III^*. The 3D-distribution of Mg^{2+}, or $g(\mathbf{r})$, obtained from 3D-RISM are shown with the web-like picture in Fig. IV-7.1(b) for $g(\mathbf{r}) \geq 42$. The peak position is indicated by $I^\dagger$, $II^\dagger$, $III^\dagger$, an $IV^\dagger$, in which $I^\dagger$ and $III^\dagger$ are major peaks, while $II^\dagger$ and $IV^\dagger$ are local maxima. The positions, $I^\dagger$, $II^\dagger$, and $III^\dagger$ roughly coincide with those from the experiment, I^*, II^* and III^*. However, there is a difference between the results from the two methods, which is the position marked by $IV^\dagger$ in (b). The position is close to the scissile bond of the phosphate, which is supposed to be cleaved by the hydrolysis reaction. The Mg^{2+} position is not seen in (a). According to the 3D-RISM analysis, the peak height of $g(\mathbf{r})$ at $IV^\dagger$ is about a half of those at $I^\dagger$. However, it is a distinct peak indicating that Mg^{2+} stays around the position with rather high probability. It also suggests the possibility that Mg^{2+} ion may migrate into the position with high probability due to the thermal fluctuation.

The migration of an ion is more likely associated with the structural fluctuation of protein, since change in the position of an ion is associated with the change of electrostatic field produced by the ion around the protein. Another interesting finding from the 3D-RISM calculation was the distribution peak of a water molecule at the position right next to $IV^\dagger$, where a water molecule as the nucleophile of the hydrolysis reaction is supposed to be located. So, it may be interesting to perform a *gedanken experiment* to see what happens if Mg^{2+} is placed at the position IV†. Irisa and his coworkers actually carried out such a gedanken experiment using the MD simulation. The results are shown in Fig. IV-7.2, in which the initial structure having the Mg^{2+} ions at I* and IV† in *Eco*RV and the equilibrated structure after the MD simulation are depicted in (a) and (b), respectively.

There are three distinct changes in the structures between the initial one and that induced by the simulation which is referred to as Structure M. The first and the most important change is the rearrangement of the DNA structure. Due to the rearrangement, the distance between OP1 of the scissile phosphate and C_γ of Asp74 decreased about 1.5 A. Secondly, the Mg^{2+} ion

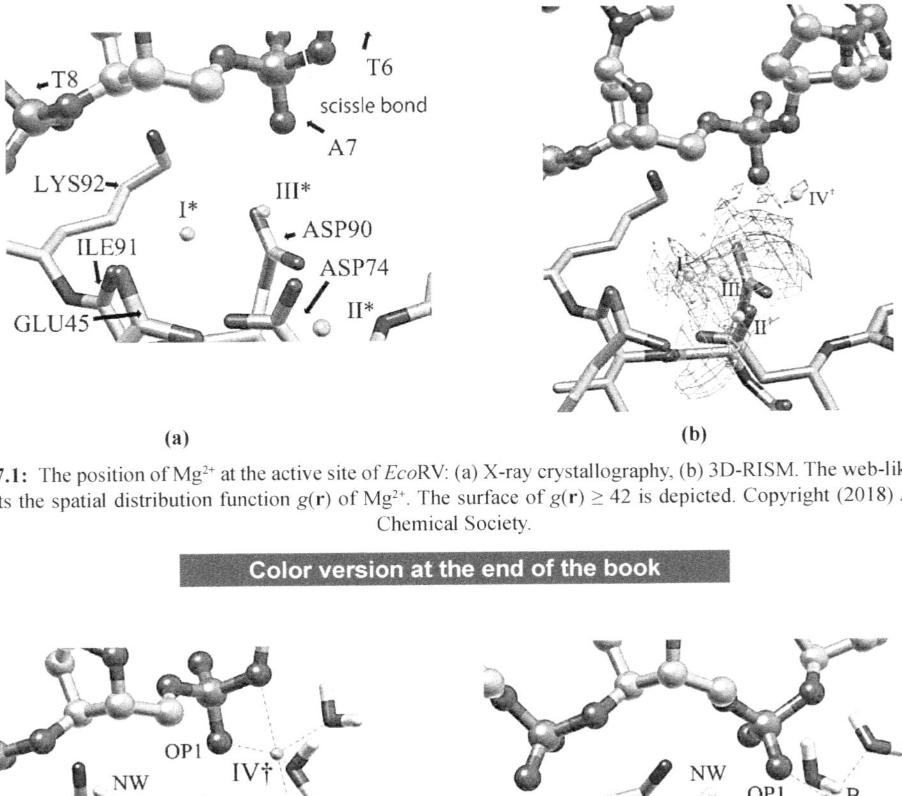

Fig. IV-7.1: The position of Mg^{2+} at the active site of *Eco*RV: (a) X-ray crystallography, (b) 3D-RISM. The web-like surface represents the spatial distribution function $g(\mathbf{r})$ of Mg^{2+}. The surface of $g(\mathbf{r}) \geq 42$ is depicted. Copyright (2018) American Chemical Society.

Color version at the end of the book

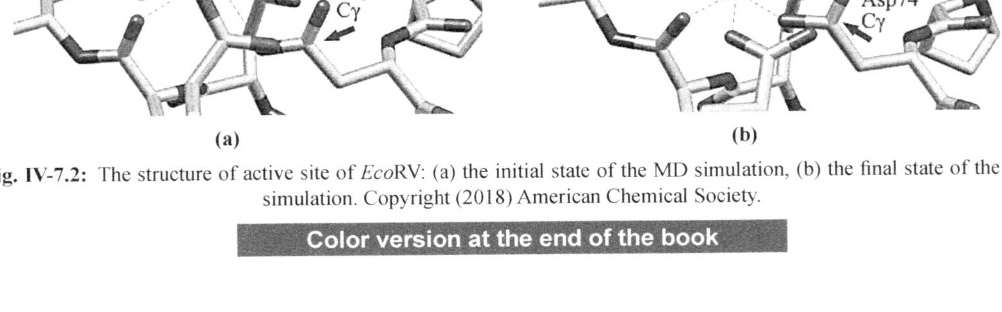

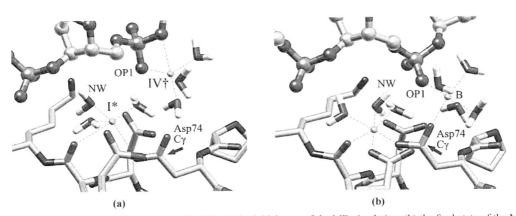

Fig. IV-7.2: The structure of active site of *Eco*RV: (a) the initial state of the MD simulation, (b) the final state of the MD simulation. Copyright (2018) American Chemical Society.

Color version at the end of the book

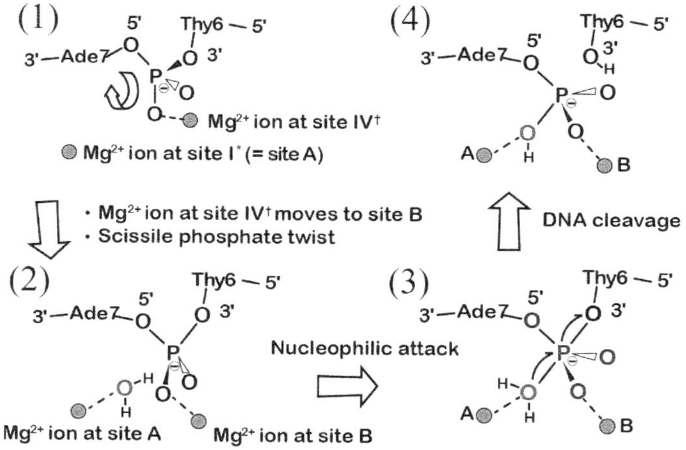

Fig. IV-7-3: Illustration of the reaction mechanism of the DNA cleavage by *Eco*RV, inferred from 3D-RISM and MD. Copyright (2018) American Chemical Society.

initially placed at IV† in the figure (a) moved to the new position labeled by B in the figure (b), while the Mg^{2+} initially located at I* stayed at the same position. The Mg^{2+} ion at IV† moved to the site B, while the scissile phosphate twisted in concerted manner with the move of Mg^{2+}, so that the distance between the phosphate and water-oxygen became ~ 3.5 A. The distance may be close enough for the water molecule to act as the nucleophile of the chemical reaction. So, it will be rational to consider the structure M as the precursor of the chemical reaction. It will be of a great interest to perform the quantum chemistry for the scissile-bond cleavage taking the structure M as an initial configuration.

Although the quantum calculation has not been performed yet, Irisa et al. inferred a mechanism for the hydrolysis reaction catalyzed by the restriction enzyme, EcoRV, which is illustrated in Fig. IV-7.3. The proposed mechanism consists of the four steps: (1) A Mg^{2+} ion which was originally located somewhere around III† migrates into IV† due to fluctuation, while a Mg^{2+} ion stays at site A, which is equivalent to site I* or I†. (Alternatively, it is possible to consider that this state is a result of the spatial or concentration fluctuation of protein solution, instead of a temporal fluctuation.) (2) Mg^{2+} at IV† moves to the site B, while the scissile phosphate twists in concerted manner with the Mg^{2+} move. At the same time, a water molecule comes close to the scissile position for the nucleophilic attack. (3) Nucleophilic attack takes place. (4) The DNA cleavage is completed.

IV-8 3D-RISM-SCF Theory and its Application to the ATP Hydrolysis Reaction

All the biological functions are driven by the energy produced by the ATP hydrolysis reaction that converts ATP into ADP [Meyerhof and Lohmann 1932]. It has long been believed that the large energy originates mainly from the backbone P-O bond in ATP, so called the *high-energy P-O bond* [Boyd and Lipscomb 1969, Hammond et al. 1992, Hill and Morales 1951, Lipmann 1941]. The high-energy bond hypothesis has been challenged by some scientists who have proposed another idea in which the difference in *hydration free energies* between ATP and ADP plays a critical role in determining the reaction energy, not only the high-energy P-O bond in ATP [Akola and Jones 2003, Colvin et al. 1995, George et al. 1970, Hofmann and Zundel 1974]. Thus, there has not been clear consensus regarding the origin of ATP hydrolysis energy. In order to establish a consensus concerning the origin of ATP hydrolysis energy, many experimental and theoretical efforts have been devoted [Colvin et al. 1995, Grigorenko et al. 2006, Hong et al. 2012, Kamerlin and Warshel 2009, Klaehn et al. 2006, Takahashi et al. 2017, Wang et al. 2015, Yamamoto 2010].

Here, we review an analysis of the hydrolysis reaction, carried out based on a revised version of the RISM-SCF theory that was introduced in Chapter II in this book. The revised version of theory employs the 3D-RISM theory in place of the RISM theory in order to calculate the solvent distribution around a solute molecule. So, the theory is referred to as "3D-RISM-SCF." The 3D-RISM-SCF theory is applied to the hydrolysis reaction of ATP and its analogue, or pyrophosphate, in *aqueous solutions*.

Some readers may raise a question if the environment of the reaction is realistic enough, since the actual ATP hydrolysis is taking place in an *enzyme,* not just in *aqueous solutions*. The answer to the question is "yes," as long as the *energy* produced by the hydrolysis reaction is concerned with. By definition, an enzyme is a catalyst that concerns the rate of a reaction, or *kinetics*, and it does not affect the energy difference between reactant and product. On the other hand, what we are concerned with is the difference of energies between reactant and product.

Hydrolysis reaction of pyrophosphate

Here, the 3D-RIMS-SCF study of the hydrolysis free energy of the pyrophosphate, ATP analogue, is reviewed [Hong et al. 2012]. Hong et al. have applied the 3D-RISM-SCF theory to the four types of hydrolysis reaction of the pyrophosphate which have different charged states depending on the protonation level of phosphate oxygen.

The reaction free energy of hydrolysis can be defined as

$$G_{aq} = G_{kin} + E_{gas} + \Delta E_{reorg} + \Delta\mu \tag{IV-8.1}$$

where G_{kin}, E_{gas}, ΔE_{reorg}, and $\Delta\mu$ denote the kinetic free energy, gas phase electronic energy, electronic reorganization energy, and solvation free energy, respectively. The sum of E_{gas} and ΔE_{reorg} corresponds to the solute electronic energy, E_{solute}, in Eq. (II-5.10) in Chapter II. They employed the Kohn-Sham-DFT at the B3LYP/6-31+G(d) for the electronic structure calculation, and the 3D-RISM/KH theory with the OPLS and RESP parameters for the solvation structure calculations (see Section IV-2 for the 3D-RISM/KH theory). The details of the computational condition can be found in original research paper.

The free energy profiles of the reactions are summarized in Table IV-8.1 (Also see Scheme I). In a gas phase, the hydrolysis reaction is almost neutral in the case of reaction A, while the reaction B is exothermic. Reaction C and D are the endothermic reactions and show large negative reaction free energy. These behaviors can be roughly understood by the coulomb repulsion and localization of excess charges. In the case of reaction A, all the reactant and product species are charge neutral, therefore there is no large energy involved. In the case of reaction B, the excess charge on reactant pyrophosphate, $H_3P_2O_7^-$, should be localized in product phosphate, $H_2PO_4^-$, thus the product state becomes unstable than the reactant state. In the cases C and D, the multiple excess charges on the reactant pyrophosphate, $H_2P_2O_7^{2-}$ and $HP_2O_7^{3-}$ in the cases C and D, respectively, can be separated by the reaction. This charge separation drastically reduces the energy loss of Coulomb repulsion in the reactant state, and it

Table IV-8.1: The reaction free energies in a gas phase and an aqueous phase computed by DFT and 3D-RISM-SCF calculations. Solvation effect means the difference between the free energy in aqueous phase and gas phase. Units are in kcal/mol. Copyright (2012) American Chemical Society.

Reaction	Gas phase (DFT)	Aqueous phase (3D-RISM-SCF)	Exp[a)]	Solvation effect
A	–1.7	–8.9	–9.5	–7.2
B	21.3	–6.2	–7.5	–27.5
C	–56.6	–8.1	–7.7	64.7
D	–119.5	–7.7	–7.1	127.2

a) Experimental values from [George et al. 1970]

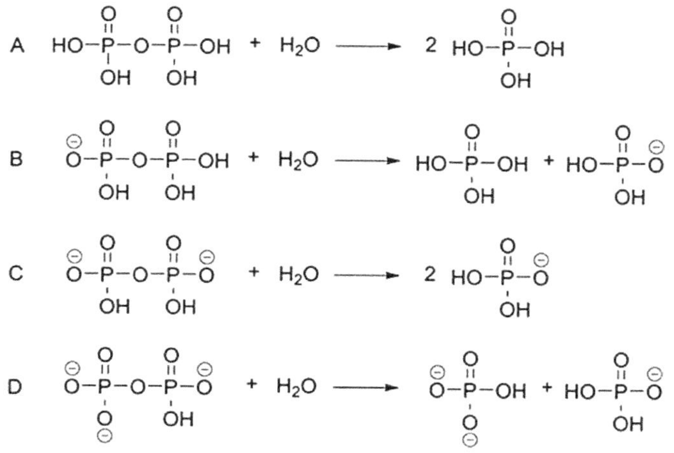

Scheme 1: Schematic description of the hydrolysis reaction of pyrophosphate for the four possible charged states. Copyright (2012) American Chemical Society.

makes these reactions *exothermic*. The large energy gain due to the charge separation may be the reason why P-O bond is called *high energy bond*. However, these results could not explain the experimental observation in aqueous solutions, which indicate that all the four reactions are moderately exothermic, c.a. 8 kcal/mol [George et al. 1970].

In an aqueous phase, the 3D-RISM-SCF gives almost quantitative agreement of the reaction free energy with those of experimental results. The results clearly demonstrate the importance of the solvation effects as a factor to determine the reactivity. In addition, it is interesting to see that the solvation effect on the reaction free energy is negative for the cases, A and B, whereas it is positive for the cases C and D. It indicates that the mechanism of the reaction free energy change due to the solvation is different depending on the charged states. In case A, since both the reactant and product species are charge neutral, the origin of the solvation free energy change may be attributed to the microscopic solvation structural change. In the Table IV8.2, the contribution from solvation to the electronic reorganization energy and the solvation free energy are summarized. Here, the electronic reorganization energy is defined as the difference of the electronic energy of a solute molecule in aqueous phase and gas phase. As can be seen in the table, the solvation free energy change dominates the reaction free energy change. Note that the contribution from the kinetic energy change is rather small, therefore the discussion on the kinetic energy term is not given below.

In the cases of reactions A and B, the products are more stabilized than the reactants by solvation, whereas the reactants are more stabilized in the reactions C and D. This converts the reaction A and B from "thermo neutral" or "endothermic" in the gas phase to "exothermic" in water, whereas the reactions C and D from "highly exothermic" in the gas phase to be "moderately exothermic" in water. In addition, the solvation free energies for the higher charged species have bigger negative values than those for the lower charged species. The excess electrons in more negatively charged pyrophosphates are much more stabilized by the hydration. The stabilization due to hydration leads to quantitative agreement with the experimental data as shown in Table IV-8.2.

To clarify the molecular origin of the solvation effects on the hydrolysis reaction, the solvation structures around the solute species have been investigated. Depicted in Figs. IV-8.1 and IV-8.2 are the spatial and radial distribution functions of water around the solute species. As can be seen in the figures, the hydration patterns are quite different depending on the charged states of the phosphates. When a phosphate has no net charge as H_3PO_4 and $H_4P_2O_7$, the phosphate is primarily hydrated by water-oxygen through hydrogen-bonds. As stated before, the reaction A is almost thermo neutral in the gas phase, therefore the reaction free energy of –8.9 kcal/mol in the aqueous phase mainly comes from the difference in the short-range solute-solvent interactions between the reactants and products, including the energy of cavity

Table IV-8.2: Solvation contributions (electronic reorganization energy ΔE_{reorg} and solvation free energy $\Delta\mu$) to the reaction free energy computed by the 3D-RISM-SCF theory. Units are in kcal/mol. Copyright (2012) American Chemical Society.

Reaction	reactants		products	
	ΔE_{reorg}	$\Delta\mu$	ΔE_{reorg}	$\Delta\mu$
A	8.2	–27.0	8.2	–32.6
B	6.7	–74.4	13.9	–113.7
C	5.8	–224.1	19.6	–194.8
D	9.7	–478.3	21.8	–385.2

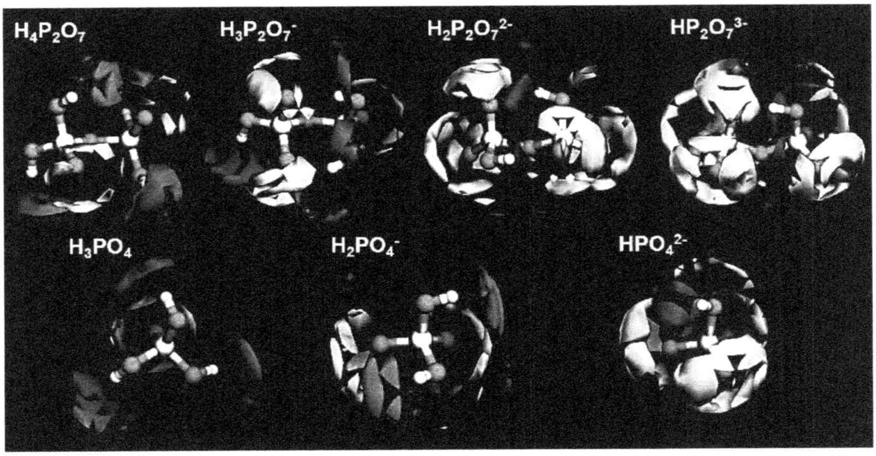

Fig. IV-8.1: Spatial distribution of solvent water around the solute species. Distributions of water oxygen and hydrogen are depicted with dark gray and light gray, respectively. Copyright (2012) American Chemical Society.

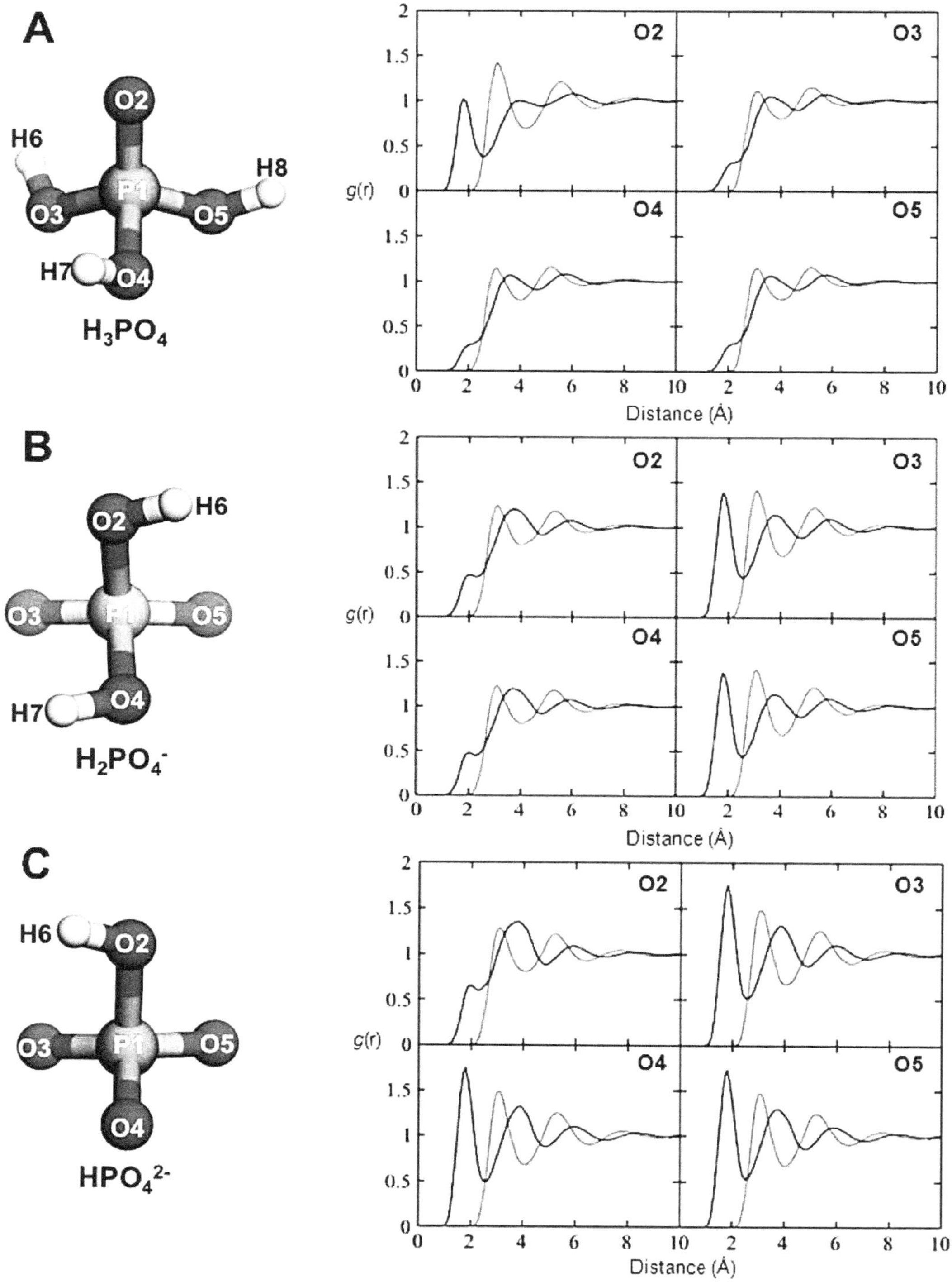

Fig. IV-8.2: Radial distribution function of solvent water around phosphate oxygen of product species. The gray and black curves denote the distribution function of oxygen and hydrogen of water, respectively. Copyright (2012) American Chemical Society.

formation, hydrogen-bond strength, and steric effect. On the other hand, the hydration patterns look quite different in the case in which the phosphates have one negative charge as in $H_2PO_4^-$ and $H_3P_2O_7^-$.

In that case, the distribution of water-hydrogen is observed around the negatively charged oxygen atoms of the phosphates. The trend is especially pronounced in the case of $H_2PO_4^-$, because the excess charge on $H_2PO_4^-$ is more localized than that on $H_3P_2O_7^-$, which makes solute-solvent electrostatic interaction stronger. The difference in the hydration pattern between the reactants and products in reaction B converts the reaction from *endothermic* in the gas phase to *exothermic* in the aqueous phase.

When the phosphates have multiple excess charges as in HPO_4^{2-}, $H_2P_2O_7^{2-}$, and $HP_2O_7^{3-}$, the solvation structures are dramatically changed such that the conspicuous peaks of water-hydrogen are observed around solute oxygen atoms. Because these phosphates have multiple excess charges, their hydration free energies are dominated by the long-range electrostatic interactions. The trend can be explained qualitatively by the simple Born model. According to the model, the solvation free energy is proportional to the square of the net-charge of solute molecule. Based on the model, the highly charged reactant species are greatly stabilized by the solvation than the product state. Consequently, reactions C and D become "moderately exothermic" in water.

Hydrolysis reaction of ATP

The hydrolysis reaction of trivalent anion ATP, which corresponds to the reactant in the ambient condition, is examined by means of the 3D-RISM-SCF theory [Yoshida and Hirata 2018]. The reaction scheme is depicted in Scheme 2. The four types of the calculations are performed namely, B3LYP/6-31+G(d), B3LYP/6-31++G(d,p), M06-2X/6-31+G(d), and M06-2X/6-31++G(d,p) in both gas and aqueous phase. The geometry optimization for all reactant and product species are performed in gas and aqueous phase. It is noted that the geometry optimization in aqueous phase is done by DFT/PCM method whereas the free energy calculations are performed by 3D-RISM-SCF. The initial structure of ATP and ADP are taken from PDB code 5LQZ [Vinothkumar et al. 2016]. GAFF parameter set is employed for ATP, ADP and Phosphate, and SPC/E parameter is used for water molecule.

The reaction free energy in aqueous phase and its components are summarized in Table IV-8.3. The reaction free energies are evaluated by

$$\Delta G = G^{\text{product}} - G^{\text{reactant}} \tag{IV-8.2}$$

where G^{product} and G^{reactant} are the free energy of product and reactant states defined by,

$$G = E_{\text{solute}} + \Delta\mu, \tag{IV-8.3}$$

where E_{solute} and $\Delta\mu$ are the electronic energy of solute and the solvation free energy, respectively. Therefore, the reaction free energy can be splitted into contributions from the electronic energy and solvation free energy,

$$\Delta G = \Delta E_{\text{solute}} + \Delta\Delta\mu, \tag{IV-8.4}$$

where

$$\Delta E_{\text{solute}} = E_{\text{solute}}^{\text{product}} - E_{\text{solute}}^{\text{reactant}} \tag{IV-8.5}$$

and

$$\Delta\Delta\mu = \Delta\mu^{\text{product}} - \Delta\mu^{\text{reactant}} \tag{IV-8.6}$$

Here, the kinetic energy contribution is ignored, since it is known from previous studies that the contribution has only minor effect. All the computed values are in good agreement with the experimental value, –10.7 kcal/mol [George et al. 1970]. The ATP hydrolysis shows the behavior similar to the case of pyrophosphate, that is, the solute electronic energy shows large negative value, while the solvation free energy exhibits large positive value. It can be explained by the same mechanism with the reaction of the multiply charged pyrophosphate, since the reactant ATP^{4-} has multiple excess charges. The electrostatic repulsion decreased and stabilized due to the decomposition into ADP^{3-} and $H_2PO_4^-$. On the other hand, the positive solvation free energy change is caused by the greater stabilization of reactant ATP^{4-} compared to the product ADP^{3-} and $H_2PO_4^-$ (Table IV-8.4). The trend is roughly in accordance with the simple Born model as is the case in pyrophosphates, although the microscopic solute-solvent interaction such as hydrogen bond becomes important for a quantitative estimate, especially, in the real situation of enzymatic reactions.

In Fig. IV-8.3, the spatial distributions of water oxygen and hydrogen around the solute species are shown. The conspicuous peaks of both oxygen and hydrogen are observed around

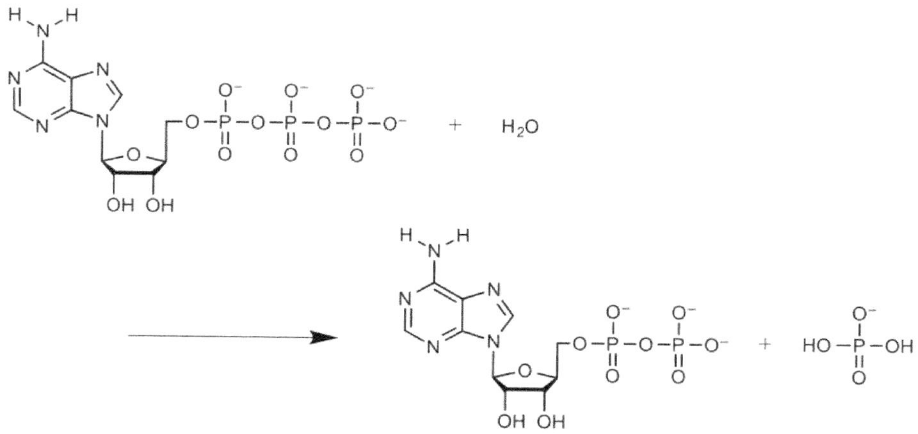

Scheme 2: Schematic description of the hydrolysis reaction of ATP^{4-} into the ADP^{3-}.

Table IV-8.3: The reaction free energies in an aqueous phase and the components computed by DFT/3D-RISM-SCF calculations. Units are in kcal/mol.

Functional/Basis set	ΔG	ΔE_{solute}	$\Delta\Delta\mu$
B3LYP/6-31+G(d)	–10.6	–167.5	156.9
B3LYP/6-31++G(d,p)	–11.0	–167.7	156.7
M06-2X/6-31+G(d)	–9.4	–167.1	157.8
M06-2X/6-31++G(d,p)	–9.6	–166.9	157.3

Table IV-8.4: The solvation free energy of reactant and product state. Units are in kcal/mol.

Functional/Basis set	$\Delta\mu^{reactant}$	$\Delta\mu^{product}$
B3LYP/6-31+G(d)	–726.0	–569.1
B3LYP/6-31++G(d,p)	–725.1	–568.5
M06-2X/6-31+G(d)	–738.4	–580.6
M06-2X/6-31++G(d,p)	–737.5	–580.2

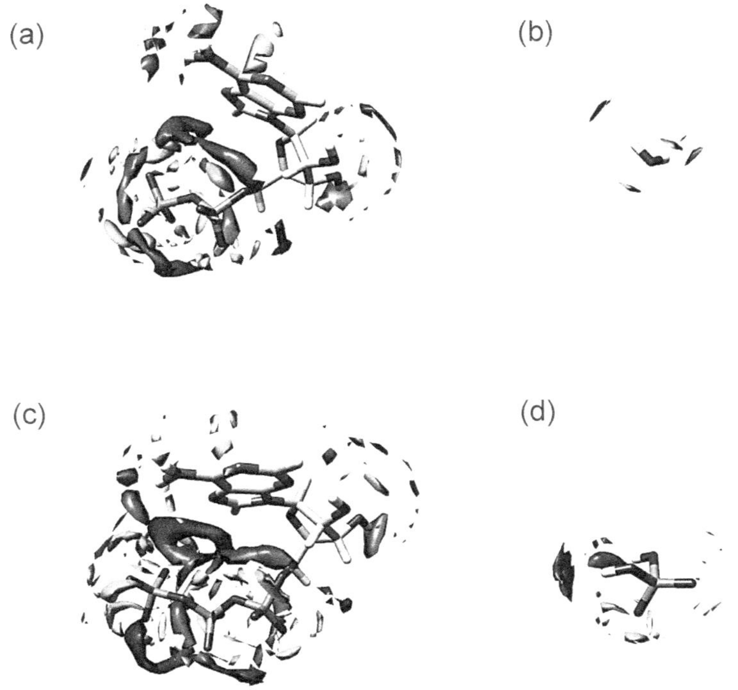

Fig. IV-8.3: Spatial distribution of solvent water around the solute species. (A) ATP, (B) water, (C) ADP, and (D) phosphate. Distributions of water oxygen and hydrogen, g(r) > 4, are depicted with dark gray and light gray, respectively.

the phosphates in ATP^{4-}, ADP^{3-}, and $H_2PO_4^-$. Those peaks are attributed to the strong hydrogen-bonds formed between phosphate and water as seen in Fig. IV-8.2. On the other hand, there are weak peaks around the adenine group of $ATP^{4-}$and ADP^{3-}, which may contribute less to the reaction free energy.

In this chapter, the application of the 3D-RISM-SCF theory to investigate the hydrolysis reaction of ATP and pyrophosphate in aqueous solution was reviewed.

First, the hydrolysis reactions of four different dissociated states of the pyrophosphate, which had been studied experimentally by George, were reviewed (Goerge et al. 1970). The results in gas phase give the reaction free energy ranging from extremely large positive to unrealistically large negative values depending on the dissociated states of the phosphate, which contradict evidently with the experimental results. The 3D-RISM-SCF theory, on the other hand, predicts the experimental results of the reaction free energy almost quantitatively for all the four dissociated states of the phosphates, indicating that the solvent plays crucial roles in determining the energy stabilizing both the reactants and products appropriately. The results also demonstrated the importance of the hydrogen bond between the phosphate and water to make the reaction to be moderately exothermic, about 8 kcal/mol. The study for the hydrolysis reaction of ATP^{4-} into ADP^{3-} by the 3D-RISM-SCF theory was also reviewed. The computed hydration free energy shows good agreement with those of experiment. The mechanism of determining the hydration free energy is essentially the same as the case of pyrophosphate. Therefore, the results given in this part prove the universality of the result for the pyrophosphate. Here, these studies provide unequivocal proof that the origin of the energy produced by the hydrolysis reaction of ATP is the balance of the "hydration" and "electrostatic interaction," not only the "high-energy bond" of the phosphate.

The Literature

A comprehensive description for the molecular biology and biochemistry from a physicochemical viewpoint is provided in the textbook by Cantor and Schimmel [Cantor and Schimmel 1980]. The book covers most of the experimental methodologies used in the molecular biology including NMR and X-ray as well as some theoretical aspects of living phenomena. For the concept of "molecular recognition," the books by Lehn [1990, 1995] will be instructive.

CHAPTER V

Structural Fluctuation and Dynamics of Protein in Aqueous Solutions

V-1 Introduction

Protein molecules in our body are not rigid objects, but they are fluctuating temporarily as well as spatially due to the thermal motion. Let's imagine an experiment in which a test tube contains an aqueous solution of micromolar concentration of protein. There are about $\sim 10^{17}$ protein molecules in the test tube, each of which has different structure or conformation, and they are making a distribution or an ensemble around an average structure with some dispersion or variance in *space*, depending on its thermodynamic condition, or environment (see Chapter I for those words concerning statistics). On the other hand, if one observes one of $\sim 10^{17}$ protein molecules in the test tube for long time, the structure of protein will be fluctuating *temporarily*, and the fluctuating structures will make a distribution around an average structure over the time, with some dispersion. The *spatial* or ensemble average would agree with the *temporal* average by virtue of the *Ergordic theorem*. The purpose of the present chapter is to formulate a statistical mechanics theory to describe the *temporal* fluctuation of protein around its equilibrium structure based on the 3D-RISM/RISM and the generalized Langevin theory explained in the previous chapters. This chapter also treats another problem, with which the structural fluctuation is concerned. It is the conformational change of protein, induced by a perturbation including thermodynamic conditions, as well as chemical perturbations such as amino-acid substitutions or *mutation*. The process is essentially a *relaxation* induced by a perturbation from an equilibrium state to another. There are two aspects in such relaxations: one is static or time independent, while the other is dynamic or the time dependent. This chapter provides basic theories to treat both the aspects of the structural relaxation of protein.

The structural fluctuation of protein plays essential roles in a variety of processes in which a biomolecule performs its intrinsic function [Henzler-Widman and Kern 2007]. For example, so called "gating" mechanisms of ion channels are regulated by the structural fluctuation of amino-acid residues consisting of the gate region of the channel. We saw an example of such gating mechanism in the M2 channel in the Section IV-4.3 of this book [Phongphanphanee et al. 2010]. Molecular recognition such as the formation of an enzyme-substrate complex in an enzymatic reaction is controlled often by structural fluctuation of protein (see section six in Chapter IV for an example) [Tanwar et al. 2015]. A few typical examples of structural fluctuations around a native conformation of protein, related to function, are "breathing" [Makowski et al. 2008], "hinge-bending [Mchaourab et al. 1997], and "arm-rotating" motions [Burghardt et al. 2011]. Those motions are *collective* in nature involving many atoms moving in the same direction. The structural fluctuation associated with protein functions, whether it's large or small, stays around its native conformation, and does not induce global conformational change such as denaturing, with few exceptions exemplified by *intrinsically disordered protein* [Smith et al. 2012]. It is because unless otherwise a protein loses its functional activity. For example, an enzyme in a cell makes large structural change when it is working as a catalyst [Henzler-Widman and Kern 2007]. However, it recovers its original structure upon completing a catalytic cycle, and becomes ready for next reaction cycle.

In actual biological processes, solvent plays vital roles not only in the equilibrium state but also in fluctuation of protein [Feig 2009]. In the previous chapter, we have discussed many examples in which water and ions play vital roles in the molecular recognition due to protein. Here, let us consider roles played by water in *fluctuation* of protein around its native conformation, associated with recognition of a ligand by protein. The process is primarily a thermodynamic process, governed by the free energy difference between the two states before and after the recognition. It is obvious that water plays crucial role in the thermodynamics, since the equilibrium structures are determined by the free energies including the excess chemical potential or the solvation free energy of water. However, it is not the only role of water in the process. Water actually regulates the kinetic pathway of the process as well by controlling the structural fluctuation of amino-acid residues consisting the active site. An example of such processes is a mouth-like motion of amino-acid residues. The open-and-close motion of the mouth is driven not only by the direct force acting among atoms in protein, but by that originated from the solvent induced force which is in turn caused by the fluctuation in the solvation free energy, or the non-equilibrium free energy. Just imagine, what happens if water molecules confined inside the active site stay there as they are when the mouth is closed. It surely induces large increase of the free energy due to packing effect. In an actual biomolecular process, such conformational change around the native state is induced often by some perturbation upon amino-acid residues around the active site, for example, binding of a ligand. It will be rational to consider that the response to such a perturbation should be linear, because the protein recovers its native conformation upon removing the perturbation.

It is not surprising that considerable efforts have been devoted to clarify the conformational fluctuation of protein experimentally as well as theoretically. The structural fluctuation of protein is characterized by a variance-covariance matrix $\langle \Delta \mathbf{R}_\alpha \Delta \mathbf{R}_\beta \rangle$ of atomic coordinates of the molecule in the body-fixed-frame, where $\Delta \mathbf{R}_\alpha$ and $\Delta \mathbf{R}_\beta$ are the displacement of atoms α and β of protein from their equilibrium position, that is

$$\Delta \mathbf{R}_\alpha(t) = \mathbf{R}_\alpha(t) - \langle \mathbf{R}_\alpha \rangle \tag{V-1.1}$$

(See the first chapter of the book, if a reader is not familiar with the variance-covariance matrix) The sum of the diagonal element of the matrix is the mean square displacement of atoms, $\sum_\alpha \langle (\Delta \mathbf{R}_\alpha)^2 \rangle$ that can be measured by the small angle X-ray scattering (SAXS) or the elastic incoherent neutron scattering (EINS). In the case of SAXS, the logarithm of the scattering intensity S from the electrons of all atoms in protein is plotted against the wave vector Q of the

X-ray [Kataoka et al. 1995]. On the other hand, the logarithm of the scattering intensity S from the protons in protein is plotted against the wave vector of neutrons. In both cases, it follows a simple equation,

$$\log S(Q) = -\frac{1}{2}MQ^2 \tag{V-1.2}$$

where Q is small. In the equation, M is the mean square displacement of atoms in protein defined by

$$M = \sum_{\alpha} \left\langle (\Delta \mathbf{R}_{\alpha})^2 \right\rangle \tag{V-1.3}$$

where α denotes all atoms in protein in the case of SAXS, while it denotes all hydrogen atoms in the case EINS [Kataoka et al. 2011]. It is important to realize that the plot of log$S(Q)$ against Q^2 is a straight line with the negative slope with gradient ($M/2$), if the distribution is Gaussian with a single variance. It has been verified experimentally by Kataoka and his coworkers that log$S(Q)$ is linear to Q^2 at least in smaller Q irrespective of the state of protein, native or denatured, but with different gradients (see Fig. V-1.1) [Kataoka et al. 1995]. By plotting log$S(Q)$ against Q^2, one can find the mean square displacement of the protein that characterizes its structural fluctuation: for example, if M is small, the structure of protein is more likely in native conformation, while it can be in the denatured conformation if M is large.

Earliest theoretical attempt to find the structural fluctuation of protein was based on the normal mode analysis of protein [Noguchi and Go 1982]. Suppose that the interaction among atoms in protein is described by a harmonic potential. Then, the deviation (or "fluctuation") of atoms in protein from their equilibrium positions follows the Hook equation of motion

$$M_{\alpha} \frac{d^2 \Delta \mathbf{R}_{\alpha}(t)}{dt^2} = -\sum_{\beta} H_{\alpha\beta} \cdot \Delta \mathbf{R}_{\beta}(t) \tag{V-1.4}$$

$$\Delta \mathbf{R}_{\alpha}(t) = \mathbf{R}_{\alpha}(t) - \left\langle \mathbf{R}_{\alpha} \right\rangle, \tag{V-1.5}$$

where $\mathbf{R}_{\alpha}(t)$, $\langle \mathbf{R}_{a} \rangle$, and M_{a} are the position at time t, the equilibrium position, and the mass, of atom a in protein. $H_{\alpha\beta}$ is the force constant or the Hessian matrix defined by,

$$H_{\alpha\beta} = \frac{\partial^2 U}{\partial \mathbf{R}_{\alpha} \partial \mathbf{R}_{\beta}} \tag{V-1.6}$$

where $U(\mathbf{R}_1, \mathbf{R}_2, ..., \mathbf{R}_N)$ is a harmonic potential, and it has a quadratic form with respect to $\Delta\mathbf{R}$, that is,

$$U(\Delta \mathbf{R}_1, \Delta \mathbf{R}_2, ..., \Delta \mathbf{R}_N) = \frac{1}{2}\sum_{\alpha}\sum_{\beta} H_{\alpha\beta} \Delta \mathbf{R}_{\alpha} \Delta \mathbf{R}_{\beta} \tag{V-1.7}$$

If one considers that the protein atoms are in thermal motion, the distribution of $\Delta\mathbf{R}_{\alpha}$ becomes a "Gaussian,"

$$p(\Delta \mathbf{R}_1, \Delta \mathbf{R}_2, ..., \Delta \mathbf{R}_N) = \sqrt{\frac{A}{(2\pi)^{3N}}} \exp\left[-\frac{1}{2}\sum_{\alpha}\sum_{\beta} A_{\alpha\beta} \Delta \mathbf{R}_{\alpha} \Delta \mathbf{R}_{\beta}\right] \tag{V-1.8}$$

where A_{ab} is defined by,

$$A_{\alpha\beta} = \frac{H_{\alpha\beta}}{kT} \tag{V-1.9}$$

and A is the determinant of the matrix $\{A_{\alpha\beta}\}$. The variance-covariance matrix of the "fluctuation" is related to the Hessian through

$$\left\langle \Delta\mathbf{R}_{\alpha}\Delta\mathbf{R}_{\beta}\right\rangle = \frac{kT}{H_{\alpha\beta}} \tag{V-1.10}$$

The equation implies that the structural "fluctuation" of a protein can be characterized theoretically by calculating the second derivative of the interaction potential with respect to atomic coordinates. The idea has been employed in the early 1990s to characterize the structural fluctuation of protein by means of the computational science. However, the "fluctuation" was essentially a mechanical oscillation. The physical origin of the "Gaussian" distribution of "fluctuation" in this system is not concerned with the central limiting theorem governing the random fluctuation (see Chapter I for the fluctuation-dissipation theorem). Those works have demonstrated the importance of the collective mode in the fluctuation. However, those efforts have not provided a realistic physical insight into the dynamics of actual biological processes, since they are concerned with a protein in "vacuum", which obviously cannot describe the fluctuation conjugated with that of solvent.

The principal component analysis involving diagonalization of the variance-covariance matrix of conformational fluctuation, extracted from the molecular dynamics trajectory of a protein in water, has revealed some important aspects of the concerted fluctuation between a biomolecule and water [Kitao et al. 1991]. The lowest frequency mode of fluctuation around a native conformation exhibits an activated transition from a minimum to another minimum

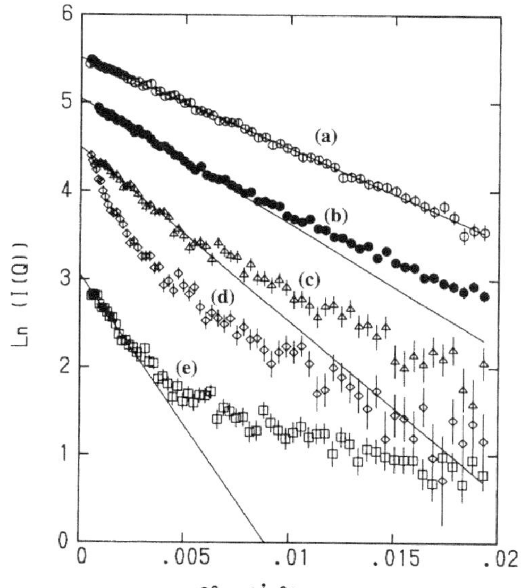

Fig. V-1.1: Guinier plot of scattering factor from the various conformational state of myoglobin: (a) holomyoglobin in native state, (b) apomyglobin in native state, (c) molten globule state, (d) molten globule state, (e) denatured state. (The figure is provided by Kataoka et al. 1995.) Copyright (1995) Academic Press Limited.

in the conformational space, akin to the jump diffusion model of liquids [Eglestaff 1967]. However, the procedure cannot be extended readily to that associated with such a process as ligand binding, because the process is concerned with sampling of large configuration space involving both protein and solvent. It becomes formidable especially when the solvent consists of several chemical components such as the electrolyte solution.

In this chapter, a first-principle approach to treat structural fluctuation of protein concerted with that of solvent is presented, based on the two theoretical frameworks in the statistical mechanics of liquids, presented in earlier chapters in this book, 3D-RISM/KH theory and the generalized Langevin equation (see Chapter III for the generalized Langevin theory). The 3D-RISM/KH theory has proven itself to be capable of predicting the molecular recognition of ligand by protein which has a rigid structure as we saw in Chapter IV. The generalized Langevin equation should be able to describe the fluctuation of a system consisting of protein and solvent around its equilibrium state. Therefore, it is reasonable to expect that the two theories combined together will produce a method which can describe the molecular recognition process by protein, the structure of which is fluctuating. As was described in Chapter III, the key to a successful treatment of dynamics of a system by means of the generalized Langevin theory lies in the choice of *dynamic variables*. Here, four quantities are chosen as dynamic variables in the phase space: the displacement of atom positions in protein from their equilibrium coordinates, the conjugated momentum of those atoms, the fluctuation of the density field of solvent molecules, and their conjugated momentum field or flux.

V-2 Generalized Langevin Equations for a Protein in Solution

Here, a general theory to treat the structural fluctuation of protein in water is formulated. Our main concern is a protein-water system at infinite dilutions. However, the formulation is completely general for any solute-solvent system at infinite dilution. So, in the formulation, we consider a general solute-solvent system. In particular, we consider a solute molecule consisting of N_u atoms immersed in solvent consisting of N molecules, each having n atoms.

V-2.1 Generalized Langevin Theory

Here, main results in Section 3 in Chapter III, derived concerning the generalized Langevin theory, are summarized for convenience. (See Chapter III for more detail on the generalized Langevin theory.)

The generalized Langevin equation describes the time evolution of a dynamic variable $\mathbf{A}(t)$ which is governed by the Liouville operator $i\mathcal{L}$. (See Chapter III)

$$\frac{d\mathbf{A}(t)}{dt} = i\mathcal{L}\mathbf{A}(t) \tag{V-2.1}$$

The formal solution of (V-2.1) is given by

$$\mathbf{A}(t) = \exp(i\mathcal{L}t)\,\mathbf{A}(0) = \exp(i\mathcal{L}t))\mathbf{A} \tag{V-2.2}$$

The projection operator $\mathcal{P}$ is defined as

$$\mathcal{P} \equiv \frac{\mathbf{A}\cdot(\mathbf{A},\ \cdots\ )}{(\mathbf{A},\ \mathbf{A}\ )} \tag{V-2.3}$$

The inner product (**a**, **b**) denotes an average of the canonical distribution $\exp(-H/k_B t)$, or

$$(\mathbf{a},\mathbf{b}) \equiv (\mathbf{a}^*\mathbf{b}) = Z^{-1}\int d\Gamma \mathbf{a}^*\mathbf{b}\exp(-H(\Gamma)/k_B T) \tag{V-2.4}$$

where Z is the partition function (see Chapter I), and Γ denotes all microscopic degrees of freedoms in the system. The operator projects out only the "component" of **A** from the objects (· · ·). Then, obviously $P\,\mathbf{A} = \mathbf{A}$ holds. It also has the idempotent property $P^2 = P$.

After projecting **A**-component out of the microscopic degrees of freedom, the exact time evolution equation for **A**(t) is given by

$$\frac{d\mathbf{A}(t)}{dt} = \mathrm{i}\Omega\mathbf{A}(t) + \int_0^t \mathbf{K}(t-\tau)\mathbf{A}(t)d\tau + \mathbf{f}(t) \tag{V-2.5}$$

where $i\Omega$, $\mathbf{f}(t)$, and $\mathbf{K}(t)$ denote, respectively, the collective frequency, the random force or fluctuating force, and the memory kernel, defined by

$$i\Omega \equiv \frac{(\dot{\mathbf{A}},\mathbf{A})}{(\mathbf{A},\mathbf{A})} \tag{V-2.6}$$

$$\mathbf{f}(t) \equiv \exp(i(1-\mathcal{P})\mathcal{L}t)(1-\mathcal{P})\dot{\mathbf{A}} \tag{V-2.7}$$

$$\mathbf{K}(t) \equiv \frac{(\mathbf{f}(t),\mathbf{f}(0))}{(\mathbf{A},\mathbf{A})} \tag{V-2.8}$$

V-2.2 Generalized Langevin Theory for Protein in Water

The Hamiltonian of the solute-solvent system is then given by

$$H = H_0 + H_1 + H_2 \tag{V-2.9}$$

$$H_0 = \sum_{i=1}^{N}\sum_{a=1}^{n}\left[\frac{\mathbf{p}_i^a\cdot\mathbf{p}_i^a}{2m_a} + \sum_{j\neq i}\sum_{b\neq a}U_0\left(\left|\mathbf{r}_i^a - \mathbf{r}_j^b\right|\right)\right] \quad \text{(solvent)}$$

$$H_1 = \sum_{\alpha=1}^{N_u}\left[\frac{\mathbf{P}_\alpha\cdot\mathbf{P}_\alpha}{2M_\alpha} + \sum_{\beta\neq\alpha}U_1\left(\left|\mathbf{R}_\alpha - \mathbf{R}_\beta\right|\right)\right] \quad \text{(solute)}$$

$$H_2 = \sum_{\alpha=1}^{N_u}\sum_{i=1}^{N}\sum_{a=1}^{n}U_{\text{int}}\left(\left|\mathbf{R}_\alpha - \mathbf{r}_i^a\right|\right) \quad \text{(solute-solvent)}$$

(V-2.10)

where M_α denotes the mass of the αth atom in the solute molecule, and m_a the mass of ath atom in a solvent molecule. (It should be noted that atoms in solvent are labeled by Roman letters instead of Greek letters in order to distinguish from those of solute.) The Hamiltonian of the solvent is denoted by H_0 where $\mathbf{r}_i^a$ and $\mathbf{p}_i^a$ are, respectively, the position and momentum of ath atom in the ith molecule of the solvent, and $U_0(r_{ij}^{ab})$ ($r_{ij}^{ab} \equiv |\mathbf{r}_i^a - \mathbf{r}_j^b|$) is the pair potential energy between them. H_1 is the Hamiltonian of the N_u solute atoms, and $\mathbf{R}_\alpha$ and $\mathbf{P}_\alpha$ are the position and momentum of the αth solute atoms (we preserve the Greek indices for denoting the solute atoms), and $U_{\text{int}}(|\mathbf{R}_\alpha - r_i^a|)$ is the interaction potential energy between the αth solute atom and the ath atom of the ith molecule in the solvent.

The associated Liouville operator $i\mathcal{L}$ is defined by

$$i\mathcal{L} \equiv i\mathcal{L}_0 + i\mathcal{L}_1 \tag{V-2.11}$$

$$i\mathcal{L}_0 \equiv \sum_{i=1}^{N}\sum_{a=1}^{n}\left[\frac{1}{m_\alpha}\mathbf{p}_i^a \cdot \frac{\partial}{\partial \mathbf{r}_i^a} - \sum_{j\neq i}\sum_{b\neq a}\frac{\partial U_0\left(r_{ij}^{ab}\right)}{\partial \mathbf{r}_i^a}\cdot\frac{\partial}{\partial \mathbf{p}_i^a} - \sum_{\alpha=1}^{N_u}\frac{\partial U_{\text{int}}\left(\left|\mathbf{R}_\alpha - \mathbf{r}_i^a\right|\right)}{\partial \mathbf{r}_i^a}\cdot\frac{\partial}{\partial \mathbf{p}_i^a}\right] \tag{V-2.12}$$

$$i\mathcal{L}_1 = \sum_{\alpha=1}^{N_u}\left[\frac{\mathbf{P}_\alpha}{M_\alpha}\cdot\frac{\partial}{\partial \mathbf{R}_\alpha} + \mathbf{F}_\alpha \cdot \frac{\partial}{\partial \mathbf{P}_\alpha}\right] \tag{V-2.13}$$

where $\mathbf{F}_\alpha \equiv \mathbf{F}_\alpha^{(u)} + \mathbf{F}_\alpha^{(v)}$, and $\mathbf{F}_\alpha^{(u)}$ is the force exerted on the αth solute atom by the other solute atoms, $\mathbf{F}_\alpha^{(v)}$ the force exerted on the same solute atom by the solvent molecules. Their explicit expressions are given by

$$\mathbf{F}_\alpha^{(u)} = -\sum_{\beta\neq\alpha}\frac{\partial U_1\left(R_{\alpha\beta}\right)}{\partial \mathbf{R}_\alpha} \tag{V-2.14}$$

$$\mathbf{F}_\alpha^{(v)} = -\sum_{i=1}^{N}\sum_{a=1}^{n}\frac{\partial U_{\text{int}}\left(\left|\mathbf{R}_\alpha - \mathbf{r}_i^a\right|\right)}{\partial \mathbf{R}_\alpha} \tag{V-2.15}$$

Now, we define the dynamic variables onto which all other variables in the phase space are projected. The choice of dynamic variables is crucial for the success of the theory. First of all, the variables should be chosen so that they can represent the essential physics of processes under concern. In the present case, the physics concerns the structural fluctuation of protein and water, and their coupling. Secondly, the variables should be tractable by means of the statistical mechanics. From such considerations, we choose the following variables as the dynamic variable $\mathbf{A}(t)$.

$$\mathbf{A}(t) = \begin{pmatrix} \Delta\mathbf{R}_\alpha(t) \\ \mathbf{P}_\alpha(t) \\ \delta\rho_{\mathbf{k}}^a(t) \\ \mathbf{J}_k^a(t) \end{pmatrix} \tag{V-2.16}$$

Here, $\Delta\mathbf{R}_\alpha(t)$ is the displacement of the position vector $\mathbf{R}_\alpha$ of the αth atom of solute from its equilibrium value, or

$$\Delta\mathbf{R}_\alpha(t) = \mathbf{R}_\alpha(t) - \langle\mathbf{R}_\alpha\rangle \tag{V-2.17}$$

So, it is nothing but the structural fluctuation of a solute molecule, and $\mathbf{P}_\alpha(t) = M_\alpha \Delta\dot{\mathbf{R}}_\alpha(t) = M_\alpha \dot{\mathbf{R}}_\alpha(t)$. $\delta\rho_{\mathbf{k}}^a(t)$ is the Fourier transform of the density fluctuation

$\delta\rho^a(\mathbf{r}, t) \equiv \rho^a(\mathbf{r}, t) - \rho_0^a$ of solvent, where $\rho^a(\mathbf{r}, t)$ is the density field of the ath atom of the solvent, defined by $\rho^a(\mathbf{r}, t) \equiv \sum_i \delta(\mathbf{r} - \mathbf{r}_i^a(t))$, and ρ_0^a is the average number density.

$$\delta\rho^a(\mathbf{r},t) \equiv \sum_i \delta\left(\mathbf{r} - \mathbf{r}_i^a(t)\right) - \rho_0^a \tag{V-2.18}$$

$$\delta\rho_{\mathbf{k}}^a(t) = \int d\mathbf{r} e^{i\mathbf{k}\cdot\mathbf{r}} \delta\rho^a(\mathbf{r},t) = \sum_i \exp\left[i\mathbf{k}\cdot\mathbf{r}_i^a(t)\right] - (2\pi)^3 \rho_0^a \delta(\mathbf{k}) \tag{V-2.19}$$

$\mathbf{J}_{\mathbf{k}}^a(t)$ is the Fourier component of the current of the ath atom in the solvent.

$$\dot{\rho}_{\mathbf{k}}^a(t) = \sum_i i\mathbf{k}\cdot\frac{\mathbf{p}_i^a}{m_a}\exp\left[i\mathbf{k}\cdot\mathbf{r}_i^a(t)\right] \equiv i\mathbf{k}\cdot\mathbf{J}_{\mathbf{k}}^a(t) \tag{V-2.20}$$

A. The correlation matrix

The correlation matrix $\mathbf{C} = (\mathbf{A}, \mathbf{A})$ is given by

$$(\mathbf{A},\mathbf{A}) = \begin{pmatrix} (\Delta\mathbf{R}_\alpha, \Delta\mathbf{R}_\beta) & (\mathbf{P}_\alpha, \Delta\mathbf{R}_\beta) & (\delta\rho_{\mathbf{k}}^a, \Delta\mathbf{R}_\beta) & (\mathbf{J}_{\mathbf{k}}^a, \Delta\mathbf{R}_\beta) \\ (\Delta\mathbf{R}_\alpha, \mathbf{P}_\beta) & (\mathbf{P}_\alpha, \mathbf{P}_\beta) & (\delta\rho_{\mathbf{k}}^a, \mathbf{P}_\beta) & (\mathbf{J}_{\mathbf{k}}^a, \mathbf{P}_\beta) \\ (\Delta\mathbf{R}_\alpha, \delta\rho_{\mathbf{k}}^b) & (\mathbf{P}_\alpha, \delta\rho_{\mathbf{k}}^b) & (\delta\rho_{\mathbf{k}}^a, \delta\rho_{\mathbf{k}}^b) & (\mathbf{J}_{\mathbf{k}}^a, \delta\rho_{\mathbf{k}}^b) \\ (\Delta\mathbf{R}_\alpha, \mathbf{J}_{\mathbf{k}}^b) & (\mathbf{P}_\alpha, \mathbf{J}_{\mathbf{k}}^b) & (\delta\rho_{\mathbf{k}}^a, \mathbf{J}_{\mathbf{k}}^b) & (\mathbf{J}_{\mathbf{k}}^a, \mathbf{J}_{\mathbf{k}}^b) \end{pmatrix} \tag{V-2.21}$$

Among the matrix elements, the following elements vanish,

$$\begin{aligned} &(\Delta\mathbf{R}_\alpha, \mathbf{P}_\beta) = 0, \quad (\Delta\mathbf{R}_\alpha, \mathbf{J}_{\mathbf{k}}^b) = 0, \\ &(\mathbf{P}_\alpha, \Delta\mathbf{R}_\beta) = 0, \quad (\mathbf{P}_\alpha, \delta\rho_{\mathbf{k}}^b) = 0, \quad (\mathbf{P}_\alpha, \mathbf{J}_{\mathbf{k}}^b) = 0, \\ &(\delta\rho_{\mathbf{k}}^a, \mathbf{P}_\beta) = 0, \quad (\delta\rho_{\mathbf{k}}^a, \mathbf{J}_{\mathbf{k}}^b) = 0, \\ &(\mathbf{J}_{\mathbf{k}}^a, \Delta\mathbf{R}_\beta) = 0, \quad (\mathbf{J}_{\mathbf{k}}^a, \mathbf{P}_\beta) = 0, \quad (\mathbf{J}_{\mathbf{k}}^a, \delta\rho_{\mathbf{k}}^b) = 0 \end{aligned} \tag{V-2.22}$$

They vanish because the integrations concerning momentums become zero,

$$\int d\mathbf{p}^{nN} \mathbf{p}_i^a \exp\left(-\beta\sum_i\sum_a \frac{\mathbf{p}_i^a\cdot\mathbf{p}_i^a}{2m_a}\right) = 0 \tag{V-2.23}$$

$$\int d\mathbf{P}^{N_u} \mathbf{P}_\alpha \exp\left(-\beta\sum_\gamma \mathbf{P}_\gamma^2/2M_\gamma\right) = 0 \tag{V-2.24}$$

Now, we look at the nonvanishing elements. The integral for the momentum correlation of solute atoms is readily performed to give,

$$\begin{aligned} (\mathbf{P}_\alpha, \mathbf{P}_\beta) &= \frac{1}{Z_P}\int d\mathbf{P}^{N_u} \mathbf{P}_\alpha \mathbf{P}_\beta \exp\left(-\beta\sum\nolimits_\gamma P_\gamma^2/2M_\gamma\right) \\ &= \frac{1}{Z_P}\int d\mathbf{P}^{N_u} \mathbf{P}_\alpha(-M_\beta k_B T)\frac{\partial}{\partial\mathbf{P}_\beta}\exp\left(-\beta\sum\nolimits_\gamma P_\gamma^2/2M_\gamma\right) = k_B T M_\alpha \mathbf{1}\delta_{\alpha\beta} \end{aligned} \tag{V-2.25}$$

where $Z_P \equiv \int d\mathbf{P}^{N_u} \mathbf{P}_\alpha \mathbf{P}_\beta \exp\left[-\beta \sum_\gamma \mathbf{P}_\gamma^2 / 2M_\gamma\right]$, and **1** is the unit (3 × 3) matrix. Equation (V-2.25) is nothing but the equipartition theorem.

Since the general current-current correlation function ($\mathbf{J}_\mathbf{k}^a$, $\mathbf{J}_\mathbf{k}^b$) will have non-vanishing correlation between the same Cartesian components only, it is sufficient to define the current-current correlation function as

$$J_{ab}(k) \equiv \frac{1}{N}\left\langle \mathbf{J}_{-\mathbf{k}}^a \cdot \mathbf{J}_\mathbf{k}^b \right\rangle \tag{V-2.26}$$

Its calculation is somewhat involved,

$$\begin{aligned} J_{ab}(k) &= \frac{1}{N}\sum_i \sum_j \frac{1}{m_a}\frac{1}{m_b}\left\langle \mathbf{p}_i^a \cdot \mathbf{p}_j^b \exp\left[-i\mathbf{k}\cdot\left(\mathbf{r}_i^a - \mathbf{r}_j^b\right)\right]\right\rangle \\ &= \frac{1}{N}\sum_i \sum_j \frac{1}{m_a}\frac{1}{m_b}\left\langle \mathbf{p}_i^a \cdot \mathbf{p}_j^b\right\rangle\left\langle \exp\left[-i\mathbf{k}\cdot\left(\mathbf{r}_i^a - \mathbf{r}_j^b\right)\right]\right\rangle \\ &= \frac{1}{N}\sum_i \sum_j \left\langle \mathbf{v}_i^a \cdot \mathbf{v}_i^b\right\rangle \delta_{ij}\left\langle \exp\left[-i\mathbf{k}\cdot\left(\mathbf{r}_i^a - \mathbf{r}_j^b\right)\right]\right\rangle \\ &= \frac{1}{N}\sum_i \left\langle \mathbf{v}_i^a \cdot \mathbf{v}_i^b\right\rangle\left\langle \exp\left[-i\mathbf{k}\cdot\left(\mathbf{r}_i^a - \mathbf{r}_i^b\right)\right]\right\rangle \end{aligned} \tag{V-2.27}$$

The remaining elements ($\Delta\mathbf{R}_\alpha, \Delta\mathbf{R}_\beta$), ($\Delta\mathbf{R}_\alpha, \delta\rho_\mathbf{k}^b$), ($\delta\rho_\mathbf{k}^a, \Delta\mathbf{R}_\beta$), and ($\delta\rho_\mathbf{k}^a, \delta\rho_\mathbf{k}^b$) involve the spatial coordinates only. Let us consider them in order. The first one is the displacement correlation matrix among the solute atoms,

$$\mathbf{L}_{\alpha\beta} \equiv \left(\Delta\mathbf{R}_\alpha, \Delta\mathbf{R}_\beta\right) \tag{V-2.28}$$

where $\mathbf{L}_{\alpha\beta}$ is a ($3N_u \times 3N_u$) matrix. The next one is ($\Delta\mathbf{R}_\alpha, \delta\rho_\mathbf{k}^b$). Note that ($\Delta\mathbf{R}_\alpha, \delta\rho_\mathbf{k}^b$) = $\langle \Delta\mathbf{R}_\alpha \delta\rho_\mathbf{k}^b \rangle$ – $(2\pi)^3 \rho_0^b \delta(\mathbf{k})$ = $\langle \Delta\mathbf{R}_\alpha \delta\rho_\mathbf{k}^b \rangle$ since $\langle \Delta\mathbf{R}_\alpha \rangle = 0$. Therefore,

$$\mathbf{B}_\mathbf{k}^{\alpha,b} \equiv \left(\Delta\mathbf{R}, \delta\rho_\mathbf{k}^b\right) = \left\langle \Delta\mathbf{R}_\alpha \rho_\mathbf{k}^b \right\rangle \tag{V-2.29}$$

It has been shown in the paper by Kim and Hirata that the quantity and its transposed one vanishes in the thermodynamic limit [Kim and Hirata 2013].

$$\mathbf{B}_{\alpha,b}(\mathbf{k}) = \mathbf{B}_{a,\beta}(\mathbf{k}) = 0 \tag{V-2.30}$$

The last one is the static structure factor of solvent, that is,

$$\chi_{ab}(\mathbf{k}) \equiv \frac{1}{N}\left(\delta\rho_\mathbf{k}^a, \delta\rho_\mathbf{k}^b\right) = \frac{1}{N}\left(\delta\rho_{-\mathbf{k}}^a, \delta\rho_\mathbf{k}^b\right) \tag{V-2.31}$$

The quantity can be calculated using the RISM theory (Chapter II).

Summing up the above results, we have the following block-diagonal matrix for (**A, A**),

$$(\mathbf{A},\mathbf{A}) = \begin{pmatrix} \mathbf{L}_{\alpha\beta} & \mathbf{O} & \mathbf{0} & \mathbf{0} \\ \mathbf{O} & k_B T M_\alpha \mathbf{1}\delta_{\alpha\beta} & \mathbf{0} & \mathbf{0} \\ \mathbf{0}^T & \mathbf{0}^T & N\chi_{ab}(\mathbf{k}) & \underline{\mathbf{0}} \\ \mathbf{0}^T & \mathbf{0}^T & \underline{\mathbf{0}}^T & NJ_{ab}(k) \end{pmatrix} \quad \text{(V-2.32)}$$

$$Z_P \equiv \int d\mathbf{P}^{N_u} \exp\left(-\beta\sum_\gamma \mathbf{P}_\gamma^2/2M_\gamma\right) \quad \text{(V-2.33)}$$

where **O** denotes the ($3N_u \times 3N$) zero matrix, **0** the ($3N_u \times n$) zero matrix, $\underline{\mathbf{0}}$ the ($n \times n$) zero matrix, and the superscript T the transpose matrix.

B. Inverse of (A, A)

Since the above correlation matrix is block-diagonal, it is trivial to obtain the inverse $(\mathbf{A},\mathbf{A})^{-1}$ as

$$(\mathbf{A},\mathbf{A})^{-1} = \begin{pmatrix} (\mathbf{L}^{-1})_{\alpha\beta} & \mathbf{O} & \mathbf{0} & \mathbf{0} \\ \mathbf{O} & (k_B T M_\alpha)^{-1} \mathbf{1}\delta_{\alpha\beta} & \mathbf{0} & \mathbf{0} \\ \mathbf{0}^T & \mathbf{0}^T & N^{-1}\chi_{ab}(\mathbf{k}) & \underline{\mathbf{0}} \\ \mathbf{0}^T & \mathbf{0}^T & \underline{\mathbf{0}} & N^{-1}J_{ab}^{-1}(k) \end{pmatrix} \quad \text{(V-2.34)}$$

Here, the inverse matrices $(\mathbf{L}^{-1})_{\alpha\beta}$, $\chi_{\alpha\beta}^{-1}(k)$, and $J_{ab}^{-1}(k)$ are defined as

$$\sum_{\gamma=1}^{N_u} \mathbf{L}_{\alpha\gamma}(\mathbf{L}^{-1})_{\gamma\beta} = \mathbf{1}\delta_{\alpha\beta} \qquad \sum_{c=1}^{n} \chi_{ac}(k)\chi_{cb}^{-1}(k) = \delta_{ab} \qquad \sum_{c=1}^{n} J_{ac}(k)J_{cb}^{-1}(k) = \delta_{ab} \quad \text{(V-2.35)}$$

C. The frequency matrix *i*Ω

Here, we find the expression for the frequency matrix $i\Omega$ defined as

$$i\Omega_{\lambda\nu} \equiv \sum_{\lambda'} (\mathbf{A}_{\lambda'}, \dot{\mathbf{A}}_\lambda)\left[(\mathbf{A},\mathbf{A})^{-1}\right]_{\lambda'\nu} \quad \text{(V-2.36)}$$

We first look at the elements of the matrix $(\mathbf{A}, \dot{\mathbf{A}})$,

$$(\mathbf{A},\dot{\mathbf{A}}) = \begin{pmatrix} (\Delta\mathbf{R}_\alpha, \Delta\dot{\mathbf{R}}_\beta) & (\mathbf{P}_\alpha, \Delta\dot{\mathbf{R}}_\beta) & (\delta\rho_\mathbf{k}^a, \Delta\dot{\mathbf{R}}_\beta) & (\mathbf{J}_\mathbf{k}^a, \Delta\dot{\mathbf{R}}_\beta) \\ (\Delta\mathbf{R}_\alpha, \dot{\mathbf{P}}_\beta) & (\mathbf{P}_\alpha, \dot{\mathbf{P}}_\beta) & (\delta\rho_\mathbf{k}^a, \dot{\mathbf{P}}_\beta) & (\mathbf{J}_\mathbf{k}^a, \dot{\mathbf{P}}_\beta) \\ (\Delta\mathbf{R}_\alpha, \dot{\rho}_\mathbf{k}^b) & (\mathbf{P}_\alpha, \dot{\rho}_\mathbf{k}^b) & (\delta\rho_\mathbf{k}^a, \dot{\rho}_\mathbf{k}^b) & (\mathbf{J}_\mathbf{k}^a, \dot{\rho}_\mathbf{k}^b) \\ (\Delta\mathbf{R}_\alpha, \dot{\mathbf{J}}_\mathbf{k}^b) & (\mathbf{P}_\alpha, \dot{\mathbf{J}}_\mathbf{k}^b) & (\delta\rho_\mathbf{k}^a, \dot{\mathbf{J}}_\mathbf{k}^b) & (\mathbf{J}_\mathbf{k}^a, \dot{\mathbf{J}}_\mathbf{k}^b) \end{pmatrix} \quad \text{(V-2.37)}$$

First, we obtain some elements of $\dot{\mathbf{A}}$ using the Liouville operator, Eq. (V-2.5),

$$\Delta\dot{\mathbf{R}}_\alpha = i\mathcal{L}\,\Delta\mathbf{R}_\alpha = \frac{\mathbf{P}_\alpha}{M_\alpha} \tag{V-2.38}$$

$$\dot{\mathbf{P}}_\alpha = i\mathcal{L}\,\mathbf{P}_\alpha = \mathbf{F}_\alpha \tag{V-2.39}$$

$$\dot{\rho}^a_{\mathbf{k}} = i\mathcal{L}\,\rho^a_{\mathbf{k}} = i\mathbf{k}\cdot\mathbf{J}^a_{\mathbf{k}} \tag{V-2.40}$$

$$\dot{\mathbf{J}}^a_{\mathbf{k}} = \sum_i \frac{1}{m_\alpha}\left(\dot{\mathbf{p}}^a_i + \mathbf{p}^a_i i\mathbf{k}\cdot\frac{\mathbf{p}^a_i}{m_a}\right)\exp\left(i\mathbf{k}\cdot\mathbf{r}^a_i\right) \tag{V-2.41}$$

where $\mathbf{F}_\alpha$ is the total force exerted on the αth solute particle by the solvent as well as by other solute atoms. Actually, when we compute the elements involving $\dot{\mathbf{P}}$ or $\dot{\mathbf{J}}^a_{\mathbf{k}}$, it is more convenient to use the integration by parts. It is useful to remember that while $\Delta\dot{\mathbf{R}}$ and $\dot{\rho}^a_{\mathbf{k}}$ involve single momentum ($\mathbf{P}$ or $\mathbf{p}_i$), $\dot{\mathbf{P}}$ and $\dot{\mathbf{J}}^a_{\mathbf{k}}$ involve zero (since $\dot{\mathbf{p}}^a_i$ is the force acting on the ath atom of the ith molecule, which only involves the position of solute atoms and solvent molecules), or two momentums (two $\mathbf{p}_i$). Using this fact, we can identify the vanishing elements,

$$\begin{aligned}
&\left(\Delta\mathbf{R}_\alpha,\Delta\dot{\mathbf{R}}_\beta\right)=0, && \left(\delta\rho^a_{\mathbf{k}},\Delta\dot{\mathbf{R}}_\beta\right)=0, && \left(\mathbf{J}^a_{\mathbf{k}},\Delta\dot{\mathbf{R}}_\beta\right)=0,\\
&\left(\mathbf{P}_\alpha,\dot{\mathbf{P}}_\beta\right)=0, && \left(\delta\rho^a_{\mathbf{k}},\dot{\mathbf{P}}_\beta\right)=0, && \left(\mathbf{J}^a_{\mathbf{k}},\dot{\mathbf{P}}_\beta\right)=0,\\
&\left(\Delta\mathbf{R}_\alpha,\dot{\rho}^b_{\mathbf{k}}\right)=0, && \left(\mathbf{P}_\alpha,\dot{\rho}^b_{\mathbf{k}}\right)=0, && \left(\delta\rho^a_{\mathbf{k}},\dot{\rho}^b_{\mathbf{k}}\right)=0\\
&\left(\Delta\mathbf{R}_\alpha,\dot{\mathbf{J}}^b_{\mathbf{k}}\right)=0, && \left(\mathbf{P}_\alpha,\dot{\mathbf{J}}^b_{\mathbf{k}}\right)=0, && \left(\mathbf{J}^a_{\mathbf{k}},\dot{\mathbf{J}}^b_{\mathbf{k}}\right)=0
\end{aligned} \tag{V-2.42}$$

The nonvanishing elements are

$$\begin{aligned}
&\left(\Delta\mathbf{R}_\alpha,\dot{\mathbf{P}}_\beta\right)=-\frac{1}{M}(\mathbf{P}_\alpha,\mathbf{P}_\beta)=-k_BT\mathbf{1}\delta_{\alpha\beta}\\
&\left(\mathbf{P}_\alpha,\Delta\dot{\mathbf{R}}_\beta\right)=\frac{1}{M}(\mathbf{P}_\alpha,\mathbf{P}_\beta)=k_BT\mathbf{1}\delta_{\alpha\beta}\\
&\left(\delta\rho^a_{\mathbf{k}},\dot{\mathbf{J}}^b_{\mathbf{k}}\right)=\left(\delta\dot{\rho}^a_{\mathbf{k}},\mathbf{J}^b_{\mathbf{k}}\right)=i\mathbf{k}\cdot\left(\mathbf{J}^a_{\mathbf{k}},\mathbf{J}^b_{\mathbf{k}}\right)=iN\mathbf{k}\,J_{ab}(k)\\
&\left(\mathbf{J}^a_{\mathbf{k}},\dot{\rho}^b_{\mathbf{k}}\right)=\left(\mathbf{J}^a_{\mathbf{k}},\mathbf{J}^b_{\mathbf{k}}\right)\cdot i\mathbf{k}=iN\mathbf{k}\,J_{ab}(k)
\end{aligned} \tag{V.2-43}$$

Taking all these into account, we obtain

$$i\Omega=\begin{pmatrix}
\mathbf{O} & k_BT\mathbf{1}\delta_{\alpha\beta} & \mathbf{0} & \mathbf{0}\\
-k_BT\mathbf{1}\delta_{\alpha\beta} & \mathbf{O} & \mathbf{0} & \mathbf{0}\\
\mathbf{0}^T & \mathbf{0}^T & \underline{\mathbf{0}} & iN\mathbf{k}J_{ab}(k)\\
\mathbf{0}^T & \mathbf{0}^T & iN\mathbf{k}J_{ab}(k) & \underline{\mathbf{0}}
\end{pmatrix}\cdot(\mathbf{A},\mathbf{A})^{-1} \tag{V-2.44}$$

Using the inverse correlation matrix (V-2.34), we get $i\Omega$ as

$$i\Omega = \begin{pmatrix} \mathbf{O} & \frac{1}{M_\alpha}\mathbf{1}\delta_{\alpha\beta} & \mathbf{0} & \mathbf{0} \\ -k_B T\left(\mathbf{L}^{-1}\right)_{\alpha\beta} & \mathbf{O} & \mathbf{0} & \mathbf{0} \\ \mathbf{0}^T & \mathbf{0}^T & \underline{\mathbf{0}} & i\mathbf{k}\delta_{ab} \\ \mathbf{0}^T & \mathbf{0}^T & i\mathbf{k}\sum_{c=1}^{n} J_{ac}(k)\chi_{cb}^{-1}(k) & \underline{\mathbf{0}} \end{pmatrix} \qquad \text{(V-2.45)}$$

D. The reversible part

From Eq. (V-2.5), the reversible part of the Langevin equation is given by $i\Omega\cdot\mathbf{A}(t)$. Using Eq. (V-2.45), one finds

$$i\Omega\cdot\mathbf{A}(t) = \begin{pmatrix} \mathbf{P}_\alpha(t)/M_\alpha \\ -k_B T\left(\mathbf{L}^{-1}\right)_{\alpha\beta}\cdot\Delta\mathbf{R}_\beta(t) \\ i\mathbf{k}\cdot\mathbf{J}_{\mathbf{k}}^\alpha(t) \\ i\mathbf{k}\sum_{c=1}^{n} J_{ac}(k)\chi_{cb}^{-1}(k)\delta\rho_{\mathbf{k}}^b(t) \end{pmatrix} \qquad \text{(V-2.46)}$$

E. The fluctuating force

The fluctuating force at $t = 0$ from Eq. (V-2.7) is given by

$$\mathbf{f} = (1-\mathcal{P})\dot{\mathbf{A}} = \dot{\mathbf{A}} - i\Omega\cdot\mathbf{A} \qquad \text{(V-2.47)}$$

where we used $P\dot{\mathbf{A}} = (\mathbf{A}, \dot{\mathbf{A}})\cdot(\mathbf{A}, \mathbf{A})^{-1}\mathbf{A} = i\Omega\cdot\mathbf{A}$. We first identify

$$\dot{\mathbf{A}} = \begin{pmatrix} \Delta\dot{\mathbf{R}}_\alpha \\ \dot{\mathbf{P}}_\alpha \\ \dot{\rho}_{\mathbf{k}}^a \\ \dot{\mathbf{j}}_{\mathbf{k}}^a \end{pmatrix} = \begin{pmatrix} \mathbf{P}_\alpha/\mathbf{M}_\alpha \\ \mathbf{F}_\alpha \\ i\mathbf{k}\cdot\mathbf{j} \\ \dot{\mathbf{j}}_{\mathbf{k}}^a \end{pmatrix} \qquad \text{(V-2.48)}$$

and

$$i\Omega\cdot\mathbf{A} = \begin{pmatrix} \mathbf{P}_\alpha/\mathbf{M}_\alpha \\ -k_B T\left(\mathbf{L}^{-1}\right)_{\alpha\beta}\cdot\Delta\mathbf{R}_\beta \\ i\mathbf{k}\cdot\mathbf{j} \\ i\mathbf{k}\sum_{c=1}^{n} J_{ac}(k)\chi_{cb}^{-1}(k)\delta\rho_{\mathbf{k}}^b \end{pmatrix} \qquad \text{(V-2.49)}$$

which are obtained by setting $t = 0$ in Eq. (V-2.46). Using the two results, we obtain the expression for the fluctuating force as

$$\mathbf{f} = \begin{pmatrix} \mathbf{0} \\ \mathbf{W}_\alpha \\ 0 \\ \Xi^a_\mathbf{k} \end{pmatrix} \quad \text{and} \quad \mathbf{f}(t) = e^{it(1-\mathcal{P})\mathcal{L}} \begin{pmatrix} \mathbf{0} \\ \mathbf{W}_\alpha \\ 0 \\ \Xi^a_\mathbf{k} \end{pmatrix} \tag{V-2.50}$$

where

$$\mathbf{W}_\alpha \equiv \mathbf{F}_\alpha + k_B T \sum_\beta (\mathbf{L}^{-1})_{\alpha\beta} \cdot \Delta \mathbf{R}_\beta \tag{V-2.51}$$

$$\Xi^a_\mathbf{k} = \dot{\mathbf{J}}^a_\mathbf{k} - i\mathbf{k} \sum_{b,c} J_{ac}(k) \chi^{-1}_{cb}(k) \delta\rho^b_\mathbf{k} \tag{V-2.52}$$

F. The memory matrix

The memory function matrix $\mathbf{K}$(t) is calculated as

$$\mathbf{K}(t) \equiv (\mathbf{f}, \mathbf{f}(t)) \left[(\mathbf{A}, \mathbf{A})^{-1} \right]$$

$$= \begin{pmatrix} \mathbf{O} & \mathbf{O} & \mathbf{0} & \mathbf{0} \\ \mathbf{O} & \frac{1}{k_B T M_\alpha} (\mathbf{W}_\alpha, \mathbf{W}_\beta(t)) & \mathbf{0} & \frac{1}{N} \sum_b J^{-1}_{ab}(k) \left(\Xi^b_\mathbf{k}, \mathbf{W}_\beta(t) \right) \\ \mathbf{0}^T & \mathbf{0}^T & \mathbf{0} & \mathbf{0} \\ \mathbf{0}^T & \frac{1}{k_B T M_\alpha} (\mathbf{W}_\alpha, \Xi^b_\mathbf{k}(t)) & \mathbf{0} & \frac{1}{N} \sum_c J^{-1}_{ac}(k) \left(\Xi^c_\mathbf{k}, \Xi^b_\mathbf{k}(t) \right) \end{pmatrix} \tag{V-2.53}$$

where $\mathbf{W}(t)$ and $\Xi^a_\mathbf{k}$ are defined as

$$\mathbf{W}(t) \equiv e^{it(1-\mathcal{P})\mathcal{L}} \mathbf{W} \quad \text{and} \quad \Xi^a_\mathbf{k}(t) \equiv e^{it(1-\mathcal{P})\mathcal{L}} \Xi^a_\mathbf{k} \tag{V-2.54}$$

In Eq. (V-2.53), the two terms exhibit explicit dependence on N. In the thermodynamic limit in which N is taken to be infinite while the number of atoms N_u in a solute remains finite, the term $N^{-1}\sum_b J^{-1}_{ab}(k)(\Xi^b_\mathbf{k}, \mathbf{W}(t))$ vanishes since the ensemble average $(\Xi^b_\mathbf{k}, \mathbf{W}(t))$ remains finite. The other term $N^{-1}\sum_c J^{-1}_{ac}(k)(\Xi^c_\mathbf{k}, \Xi^b_\mathbf{k}(t))$ will not vanish since the ensemble average $(\Xi^c_\mathbf{k}, \Xi^b_\mathbf{k}(t))$ is proportional to N. Therefore, only the latter term survives in the thermodynamic limit. It can be also proved that

$$(\mathbf{W}_\alpha, \Xi^b_\mathbf{k}(t)) = 0 \tag{V-2.55}$$

Therefore, the final expression of the memory matrix is given by

$$\mathbf{K}(t) = \begin{pmatrix} \mathbf{O} & \mathbf{O} & \mathbf{0} & \mathbf{0} \\ \mathbf{O} & \frac{1}{k_B T M_\alpha} (\mathbf{W}_\alpha, \mathbf{W}_\beta(t)) & \mathbf{0} & \mathbf{0} \\ \mathbf{0}^T & \mathbf{0}^T & \mathbf{0} & \mathbf{0} \\ \mathbf{0}^T & \mathbf{0}^T & \mathbf{0} & \frac{1}{N} \sum_c J^{-1}_{ac}(k) \left(\Xi^c_\mathbf{k}, \Xi^b_\mathbf{k}(t) \right) \end{pmatrix} \tag{V.2.56}$$

G. The generalized Lagenvin equation for a protein in solution

Combining all the results derived in the subsections **A** through **G**, we are now ready to write down explicit formulas for the structural dynamics of a protein conjugated with the density fluctuation of solvent.

$$\frac{d\Delta\mathbf{R}_\alpha(t)}{dt} = \frac{\mathbf{P}_\alpha(t)}{M_\alpha} \tag{V-2.57}$$

$$\frac{d\mathbf{P}_\alpha(t)}{dt^2} = -k_B T \sum_\beta \left(\mathbf{L}^{-1}\right)_{\alpha\beta} \cdot \Delta\mathbf{R}_\beta(t) - \int_0^t ds \sum_\beta \Gamma_{\alpha\beta}(t-s) \cdot \frac{\mathbf{P}_\beta(s)}{M_\beta} + \mathbf{W}_\alpha(t) \tag{V-2.58}$$

$$\frac{d\delta\rho_{\mathbf{k}}^a}{dt} = i\mathbf{k} \cdot \mathbf{J}_{\mathbf{k}}^a(t) \tag{V-2.59}$$

$$\frac{d\mathbf{J}_{\mathbf{k}}^a(t)}{dt} = i\mathbf{k} \sum_{b,c} J_{ac}(\mathbf{k}) \chi_{cb}^{-1}(\mathbf{k}) \delta\rho_{\mathbf{k}}^b(t) - \frac{1}{N} \sum J_{ac}^{-1}(k) \int_0^t ds M_{\mathbf{k}}^{bc}(t-s) \cdot \mathbf{J}_k^b(s) + i\mathbf{k} \cdot \Xi_{\mathbf{k}}^a(t) \tag{V-2.60}$$

In the above set of dynamic equations for the solute and solvent molecules, the random forces take the following forms:

$$\mathbf{W}_\alpha(t) = e^{it(1-\mathcal{P})\mathcal{L}} \left(\mathbf{F}_\alpha + k_B \sum_\beta (\mathbf{L}^{-1})_{\alpha\beta} \cdot \Delta\mathbf{R}_\beta \right) \tag{V-2.61}$$

$$\Xi_{\mathbf{k}}^a(t) = e^{it(1-\mathcal{P})\mathcal{L}} \left(\mathbf{j}_{\mathbf{k}}^a - i\mathbf{k} \sum_{b,c} J_{ac}(k) \chi_{cb}^{-1}(k) \delta\rho_{\mathbf{k}}^b \right) \tag{V-2.62}$$

The memory functions in Eq. (V-2.58) are given by the time correlations of the random forces,

$$\Gamma_{\alpha\beta}(t) = \frac{1}{k_B T} \left\langle \mathbf{W}_\alpha(t) \mathbf{W}_\beta(0) \right\rangle, \quad \mathbf{M}_{\mathbf{k}}^{bc}(t) = \left\langle \Xi_{\mathbf{k}}^b(t) \Xi_{-\mathbf{k}}^c(0) \right\rangle \tag{V-2.63}$$

V-2.3 Physics of the Generalized Langevin Equations for Protein in Water

A. Solvent dynamics

When the fluid is far from protein, or in bulk, where perturbation from protein vanishes, the last two expressions concerning solvent, Eq. (V-2.59) and Eq. (V-2.60), reduce to the equations for pure-water dynamics, Eq. (III-4.24) and Eq. (III-4.25), derived in Chapter III.

The equation is further simplified to produce the site-site Smoluchowski-Vlasov (SSSV) equation if one makes the memory function local in time as well as in space. The equation can be analytically solved by means of the Laplace transform to produce the van Hove or space-time correlation function of water, with the *input* of the site-site pair correlation functions of the solvent obtained from the RISM theory. The theory has been successfully applied to a variety of solvent relaxation processes induced by an abrupt change in the electronic structure of a solute molecule, or solvation dynamics, as has been described in Chapter III. So, the two equations concerning solvent, Eqs. (V-2.59) and (V-2.60), can be regarded as a generalization of the theories developed for pure water to that subject to the field exerted from protein atoms. There are several remarks to be made with respect to the generalization. Firstly, the translational invariance of the system is no longer valid. Therefore, the equations should be solved in three-dimensional Cartesian-space. Secondly, the factor $\chi_{ab}(\mathbf{r}, \mathbf{r}') = N^{-1} \langle \delta\rho^a(\mathbf{r})\delta\rho^b(\mathbf{r}') \rangle$ appearing in the equation is a two body density correlation function, but subject to the "external force" due to protein. Such a theory has not been well developed yet. Therefore, it may be practical to adopt the superposition approximation $\langle \delta\rho^a(\mathbf{r})\delta\rho^b(\mathbf{r}') \rangle \approx \langle \delta\rho^a(\mathbf{r}) \rangle \langle \delta\rho^b(\mathbf{r}') \rangle$ to this case. Then, $\langle \delta\rho^a(\mathbf{r}) \rangle$ can be readily evaluated from the 3D-RISM theory.

A number of possible applications of the dynamic equations for the solvent are conceivable. An interesting example is the current-current correlation function of water and ions in a molecular channel, which is concerned with many observables including the permeability of water and ions across the cell membrane [Phongphanphanee et al. 2007, 2008, 2010a, 2010b]. The equation for the correlation function can be readily obtained by coupling the two equations for solvent with the aid of the RISM and 3D-RISM/KH theories. The derivation of the equation and the method of solving the equations are similar to the case of uniform liquids described in Chapter III, except for that the density distribution function is not a function of scalar distance, but depends on the positional vector.

B. Solute dynamics

The two equations concerning solute dynamics, Eqs. (V-2.57) and (V-2.58) are combined together to result in

$$M_\alpha \frac{d^2 \Delta\mathbf{R}_\alpha(t)}{dt^2} = -k_B T \sum_\beta \left(\mathbf{L}^{-1}\right)_{\alpha\beta} \cdot \Delta\mathbf{R}_\beta(t) - \int_0^t ds \sum_\beta \Gamma_{\alpha\beta}(t-s) \cdot \frac{d\Delta\mathbf{R}_\beta(s)}{ds} + W_\alpha(t) \quad \text{(V-2.64)}$$

The equation is regarded as a generalization of the equation for a coupled set of Langevin-oscillators, first examined by Wang and Uhlenbeck, to a realistic model of protein in water [Wang and Uhlenbeck 1945]. Wang and Uhlenbeck proposed a model in which a coupled set of oscillators consisting of spherical beads is immersed in a viscous liquid, and applied the Langevin theory to the oscillators. Later on, Lamm and Szabo performed a normal mode analysis on the Wang-Uhlenbeck oscillators, assuming a phenomenological friction term [Lamm and Szabo 1986]. Kottlam and Case, and Ansari 1999 applied the Langevin-mode equation of Lamm and Szabo to proteins [Kottlam and Case 1990, Ansari 1999]. The same method was also applied to the dynamics of DNA [Shih and Georghiou 2000] and RNA [Zacharias 2000] in solvents.

There are several comments to be made on the Eq. (V-2.64). Firstly, the equation does not include a term related to the force which originates from the first derivative of the free energy surface with respect to the position. The force acting on an atom of protein comes from three contributions: one which is proportional to the displacement of the atom from its equilibrium position (second term on the left hand side in Eq. (V-2.64)), and the friction term proportional to the velocities of the atoms, and that due to the random force (the term on the right hand side in Eq. (V-2.65)). The physical origin as to why the equation does not include the first derivative of the free energy lies in our treatment based on the generalized Langevin theory. The whole idea of the generalized Langevin theory is to project all the degrees of freedom in the phase space onto few dynamic variables under concern. The projection is carried out using a projection operator, defined by Eqs. (V-2.3) and (V-2.4), in terms of an ensemble average of two variables which are fluctuating around an equilibrium in the phase space. Obviously by definition, the ensemble average of the displacement of atoms in protein should be zero in equilibrium.

Such a force as the first derivative of the free energy, which may cause complete shift of the equilibrium, is not included in the treatment. The situation is somewhat analogous to the case of a harmonic oscillator, in which an oscillator swings back and forth around a minimum of the harmonic potential. Only force acting on the system is the restoring force proportional to the displacement from the potential minimum. In our case, too, the only force acting on the protein atoms is the one which restores atom positions from fluctuating to equilibrium ones. However, there is an essential difference in physics between the two systems. The equilibrium position of a harmonic oscillator is the minimum of mechanical potential energy, while that of protein in water is the minimum in the thermodynamic potential or the free energy, which is concerned not only with energy but also with the entropy both of protein and of water. So, in the case of protein in water, the stochastic character of the dynamics is attributed not only to the random force term, but also to the conformational fluctuation of protein around its equilibrium state, induced by solvent, while the stochastic character results just from the random force term in the case of the coupled harmonic oscillators treated by Wang and Uhlenbeck.

C. Variance-covariance matrix and structural fluctuation

The argument above suggests interesting physics implied in Eq. (II-2.64), and its application to biological functions. If one ignores the friction and random force terms in Eq. (II-2.64), one gets

$$M_\alpha \frac{d^2 \Delta \mathbf{R}_\alpha(t)}{dt^2} = -k_B T \sum_\beta \left(\mathbf{L}^{-1}\right)_{\alpha\beta} \cdot \Delta \mathbf{R}_\beta(t) \qquad \text{(V-2.65)}$$

This equation can be viewed as a coupled set of "harmonic oscillators", whose "Hessian" matrix is given by $k_B T(\mathbf{L}^{-1})_{\alpha\beta}$. Considering Eq. (V-2.29), the "Hessian" matrix is related to the variance-covariance matrix of the positional fluctuation by

$$k_B T \mathbf{L}^{-1} \equiv k_B T \langle \Delta \mathbf{R} \Delta \mathbf{R} \rangle^{-1} \qquad \text{(V-2.66)}$$

The observation strongly suggests that the dynamics described by Eq. (V-2.64) is that of fluctuation around a minimum of the free energy surface consisting not only of the interactions among atoms in the protein, but of the solvation free energy. In this respect, the configuration corresponding to the free energy minimum is not just one but an ensemble of distinguishable configurations concerning protein and solvent, which can be converted among each other due to the thermal noise. The free energy surface can be given by

$$F(\{\mathbf{R}\}) = U(\{\mathbf{R}\}) + \Delta\mu(\{\mathbf{R}\}) \qquad \text{(V-2.67)}$$

where $U(\{\Delta \mathbf{R}\})$ is the interaction potential energy among atoms in a protein, and $\Delta\mu(\{\Delta \mathbf{R}\})$ is the solvation free energy of protein whose conformation is $\{\Delta \mathbf{R}\}$ (see Chapter IV).

The above consideration further suggests a method to evaluate the variance-covariance matrix, which characterizes structural fluctuation of protein, based on the 3D-RISM theory. The variance-covariance matrix is closely related to the "Hessian" matrix with Eq. (V-2.66). However, the "Hessian" matrix here should not be the second derivative of the potential surface with respect to the protein coordinates in *vacuum*, or Eq. (V-1.6), but it should be the second derivative of the *free energy surface* including *solvation free energy*. The consideration leads naturally to the following expression for the Hessian matrix for protein in water.

$$k_B T \left(\mathbf{L}^{-1} \right)_{\alpha\beta} = \frac{\partial^2 F\left(\{\mathbf{R}\}\right)}{\partial \Delta \mathbf{R}_\alpha \partial \Delta \mathbf{R}_\beta} \tag{V-2.68}$$

Since the free energy surface $F(\{\mathbf{R}\})$ can be obtained by solving the 3D-RISM/RISM equation, Eq. (V-2.68) provides a way to evaluate the variance-covariance matrix.

The variance-covariance matrix is by itself quite informative for characterizing the structural fluctuation of protein around its native state in atomic detail. As an example, let us consider a hinge-bending motion of protein. The variance-covariance matrix should have a structure in which a block of elements $\langle \Delta \mathbf{R}_\alpha \Delta \mathbf{R}_\beta \rangle$ for atom pairs, α, β, belonging to the two sides of the hinge-axis, have the negative sign because the direction of the displacements $\Delta \mathbf{R}_\alpha$ and $\Delta \mathbf{R}_\beta$ is opposite.

The Eq. (V-2.65) also suggests an important physics concerning the structural fluctuation of protein. The equation implies that the restoring force acting on atoms in a fluctuated state of protein around its native state is *linear* to the displacement of atoms. In another word, a working protein restores its native structure *reversibly* after having conformational change in order to perform its function, for example, enzymatic reaction.

The physics deduced above also derives another important conclusion concerning the structural distribution of protein in water. By integrating Eq. (V-2.68) with respect to $\Delta \mathbf{R}_\alpha$ and $\Delta \mathbf{R}_\beta$, one finds the free energy of protein in an equilibrium conformation as

$$F\left(\{\Delta \mathbf{R}\}\right) = \frac{1}{2} k_B T \sum_{\alpha,\beta} \Delta \mathbf{R}_\alpha \cdot \left(\mathbf{L}^{-1} \right)_{\alpha,\beta} \cdot \Delta \mathbf{R}_\beta \tag{V-2.69}$$

Then, one gets the following expression concerning the structural distribution of protein in water.

$$\begin{aligned} w(\{\Delta \mathbf{R}\}) &= C \exp\left[-F\left(\{\Delta \mathbf{R}\}\right) / k_B T \right] \\ &= \sqrt{\frac{\|A\|}{(2\pi)^{3N}}} \exp\left[-\frac{1}{2} \sum_\alpha \sum_\beta \Delta \mathbf{R}_\alpha A_{\alpha\beta} \Delta \mathbf{R}_\beta \right] \end{aligned} \tag{V-2.70}$$

where the matrix $A_{\alpha\beta}$ is defined by

$$A_{\alpha\beta} = \frac{\partial^2 F\left(\{\mathbf{R}\}\right) / k_B T}{\partial \Delta \mathbf{R}_\alpha \partial \Delta \mathbf{R}_\beta} \tag{V-2.71}$$

and $\|A\|$ is the determinant of the matrix. Equation (V-2.70) is saying that the structural distribution of protein is *Gaussian*. This is a rather surprising conclusion deduced from the generalized Langevin theory. The reader may ask a question how the distribution of such complicated structures of protein can be Gaussian. Following is the answer.

There are two physical origins in the Gaussian distribution. The first one is the harmonic character of the potential surface in the local structure: bond stretching, bond-angle bending, etc. When the oscillation is small, the potential surface can be regarded as harmonic, and accordingly the distribution of bond length and angle is Gaussian. However, such a Gaussian character limited to small oscillation is not relevant to the large structural-fluctuation governing functions of protein. The Gaussian distribution governing the global fluctuation of protein has its origin in a mathematical theorem called "central limiting theorem." (See Chapter I for the central limiting theorem.) Roughly speaking, the theorem is stated as follows. A process consisting of many probabilistic variables, fluctuating randomly, makes a normal (or Gaussian) distribution, if there are no variables that make extraordinary large deviation from majority. A Brownian particle that makes a random walk in solvent is an example of such a process as we have seen in Chapter I. The probability distribution of integrated displacement of many random walks becomes a Gaussian distribution, and the half-width (or variance) of the distribution is related to the diffusion constant. Another problem in which the central limiting theorem plays a crucial role is the conformation of a polymer, which is also described in Chapter I. In this case, too, the distribution of end-to-end distance that characterizes the global conformation of a polymer becomes essentially a Gaussian distribution, and the half width of the distribution is related to the dimension of a polymer. Our protein in solution has the degrees of freedom, $\sim 10^{23}$, including those of water. Namely, the huge degrees of freedom of the system is the deep origin of the Gaussian distribution.

V-2.4 Structural Change due to a Perturbation

The linearity of the structural fluctuation of protein, clarified above, gives a great advantage to investigate structural change induced by a perturbation, mechanical, thermodynamical, or chemical, applied to the protein. For example, if a substrate molecule, or ligand binds to an active site of an enzyme, structure of the protein will undergo some change. Such a structural change can be described by the linear response theory as follows (see Chapter III for the linear response theory).

In the presence of a small perturbation due to, for example, ligand binding, the free energy associated with the perturbation can be described by

$$F\left(\{\Delta\mathbf{R}\}\right)=\frac{1}{2}k_BT\sum_{\alpha,\beta}\Delta\mathbf{R}_\alpha\cdot\left(L^{-1}\right)_{\alpha,\beta}\cdot\Delta\mathbf{R}_\beta-\sum_{\alpha}\Delta\mathbf{R}_\alpha\cdot\mathbf{f}_\alpha \tag{V-2.72}$$

where $\mathbf{f}_\alpha$ is the force acting on the α-th protein atom due to the perturbation. The first term in (V-2.72) is the free energy of unperturbed system that is given in Eq. (V-2.69), and the second term is the free energy change due to a perturbation. Then, the conformational change caused by the perturbation can be determined by the variational principle

$$\frac{\partial F\left(\{\Delta\mathbf{R}\}\right)}{\partial\Delta\mathbf{R}_\alpha}=0 \tag{V-2.73}$$

With Eq. (V-2.72), Eq. (V-2.73) gives

$$\left\langle\Delta\mathbf{R}_\alpha\right\rangle_1=\left(k_BT_0\right)^{-1}\sum_{\beta}\left\langle\Delta\mathbf{R}_\alpha\Delta\mathbf{R}_\beta\right\rangle_0\cdot\mathbf{f}_\beta \tag{V-2.74}$$

where $\langle\cdots\rangle_1$ and $\langle\cdots\rangle_0$ indicate the presence and absence of the perturbation, respectively. Therefore, Eq. (V-2.74) combined with Eq. (V-2.68) provides a theoretical basis for analyzing the conformational relaxation of protein in water due to a perturbation such as ligand binding. The Eq. (V-2.74) is first derived by Ikeguchi et al. based on the linear response theory [Ikeguchi et al. 2005].

The Eq. (V-2.74) is also a generalized equation which provides molecular basis for the phenomenological Rouse-Zimm model of the polymer dynamics, with a proper account of the variance-covariance matrix, the diagonal terms of which correspond to the mean square displacement of each atom in equilibrium states [Doi and Edwards 1986]. This suggests that the theory can be applied not only to the native conformation of protein but also to characterizing the denatured or random-coil state. However, the application requires special care of the ensemble average to evaluate the variance-covariance matrix, since the average, by definition, should be taken over virtually an infinite number of conformations randomly appearing in the solution. Nevertheless, a practical method to evaluate the variance-covariance matrix for the random-coil state of protein can be suggested based on the 3D-RISM/RISM theory as follows. First, produce some small number of conformations for protein in water by means of a generalized ensemble technique such as the replica-exchange algorithm. Second, evaluate the second derivative of the free energy surface of each conformation based on Eq. (V-2.69), and take the average of the results over the conformations, which will give rise to the variance-covariance matrix for the sampled conformational space. Third, add more conformations to the sample to take the average. Repeat the procedure until the convergence is attained. Our implication is that the convergence will be attained rather quickly, because the variance-covariance matrix for each conformation, obtained from Eq. (V-2.68), is already an average over a large number of conformations in the free energy surface. The converged variance-covariance matrix can be compared with observable quantities which characterize a random coil state of protein, such as the gyration radius and the distribution of end-to-end distance.

V-3 Mean Square Displacement Measured by the Scattering Experiment

In the preceding section, we have formulated a theory to describe the structural fluctuation of protein in solution. The equation suggests strongly that the *global* fluctuation of protein structure around the native state is *linear, or harmonic, and* the distribution of fluctuated structures is Gaussian. The Gaussian aspect of the global fluctuation is consistent with the Guinier plot obtained from the small angle X-ray scattering, which exhibits linearity when it is plotted against a square of the scattering angle or the wave vector (see Fig. V-1.1). The same behavior is seen in the results from the elastic incoherent neutron scattering (EINS). Shown in Fig. V-3.1 are the results obtained by Kataoka and his coworkers [Nakagawa et al. 2004, Nakagawa and Kataoka 2010]. In the left panel, the structure factor $\ln S(Q, \omega = 0)$ is plotted against the wave vector Q for different temperatures (filled squares). The plots exhibit linearity in the range of small wave vector (< 15 A^{-2}). From the slope of the plots, the mean square displacement M was obtained for the different temperatures. The results are plotted against temperature in the right panel of Fig. V-3.1.

The plots show a characteristic behavior in which the slope of the temperature dependence changes abruptly at ~ 230 K. It has also been clarified that the onset of the abrupt change in gradient largely depends on the hydration level of a sample: no or less gradient change in the low humidity sample, while phenomenal gradient change in a fully hydrated sample [Kataoka and Nakagawa 2011]. There have been many hypotheses and/or models proposed in order to explain such behaviors, but a consensus is yet to be reached: "glass transition," [Doster et al. 1989] "alpha to beta transition," [Chen et al. 2008] "harmonic to anharmonic transition," [Nakagawa et al. 2008] "solvent mediated transition," [Nakagawa et al. 2008] and so forth. A cause of such lack in consensus seems to be attributed to the fact that those models and

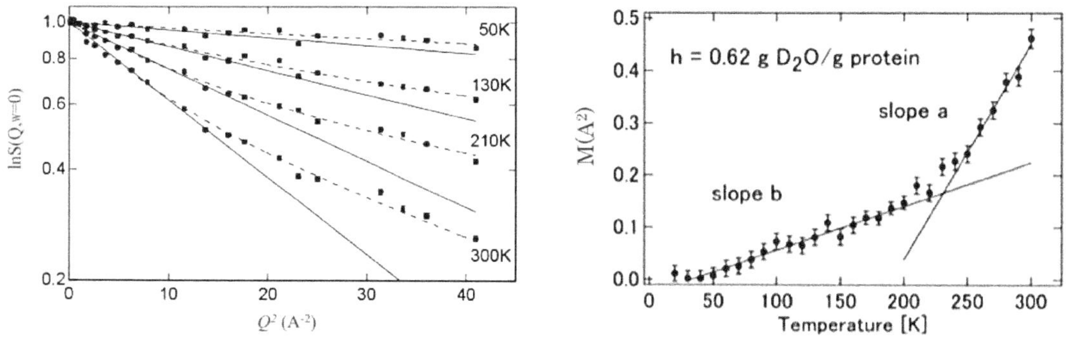

Fig. V-3.1: Left: the structure factor $\ln S(Q, \omega = 0)$ from EINS. Right: the mean square displacement against temperature. (Figure provided by Kataoka.) Copyright (2004) and (2010) Physical Society of Japan.

hypotheses are proposed based on an analogy with the phenomenological models, elasticity in the other field of science, i.e., *glass transition*, not based on a microscopic theory concerning the structural fluctuation of protein in solution. Here, we look into the problem based on the theory described in the Section V-2 [Hirata 2018].

V-3.1 The Mean Square Displacement (MSD) of Protein

The mean square displacement of protein is defined by Eq. (V-1.1) in the Section V-1 as,

$$M = \sum_{\alpha} \left\langle \Delta \mathbf{R}_{\alpha}^{2} \right\rangle$$

where α labels all atoms in protein in the case of SAXS, while it denotes hydrogen atoms in the case of EINS. The probability distribution of the displacement of hydrogen atoms is defined based on Eq. (V-2.70) by

$$w\left(\{\Delta \mathbf{R}\}\right) == \left(2\pi L_{\alpha\alpha}\right)^{-3n/2} \exp\left[-\frac{1}{2}\sum_{\alpha}^{n} \frac{\Delta \mathbf{R}_{\alpha}^{2}}{L_{\alpha\alpha}}\right] \qquad \text{(V-3.1)}$$

where n denotes the number of atoms (or hydrogen atoms) in protein. The structure factor ($S(Q, \omega = 0)$) can be defined by the Fourier transform of $w(\{\Delta \mathbf{R}\})$ into the momentum space. However, the transformation is not straightforward because the real space vector $\{\Delta \mathbf{R}\}$ is a vector in the $3n$-dimensional space, while the wave vector measured by the scattering experiments is a vector in the three dimensional Euclidian space. Therefore, we define a displacement vector in the three dimensional coordinate space by

$$\Delta \mathbf{R} = \sum_{\alpha} \Delta \mathbf{R}_{\alpha} \qquad \text{(V-3.2)}$$

and its differential volume element by

$$d\Delta \mathbf{R} = \prod_{\alpha} \Delta \mathbf{R}_{\alpha} \qquad \text{(V-3.3)}$$

Then, the structure factor may be defined by the Fourier transform of $w(\{\Delta \mathbf{R}\})$,

$$S\left(Q, \omega = 0\right) \equiv \int w\left(\{\Delta \mathbf{R}\}\right) \exp\left(-i\mathbf{Q} \cdot \Delta \mathbf{R}\right) d\Delta \mathbf{R} \qquad \text{(V-3.4)}$$

Considering the definitions Eqs. (V-3.2) and (V-3.3), Eq. (V-3.4) can be rewritten as

$$S(Q,\omega=0)=\prod_{\alpha}^{n}\left(2\pi L_{\alpha\alpha}\right)^{-3/2}\int\exp\left[-\frac{\Delta R_{\alpha}^{2}}{2L_{\alpha\alpha}}\right]\exp\left(-i\mathbf{Q}\cdot\Delta\mathbf{R}_{\alpha}\right)d\Delta\mathbf{R}_{\alpha} \qquad \text{(V-3.5)}$$

The Fourier transform of the Gaussian function can be readily performed to give,

$$\begin{aligned}S(Q,\omega=0)&=\prod_{\alpha}^{n}\left(2\pi L_{\alpha\alpha}\right)^{-3/2}\int\exp\left[-\frac{\Delta R_{\alpha}^{2}}{2L_{\alpha\alpha}}\right]\exp\left(-i\mathbf{Q}\cdot\Delta\mathbf{R}_{\alpha}\right)d\Delta\mathbf{R}_{\alpha}\\&=\prod_{\alpha}^{n}\exp\left(-\frac{L_{\alpha\alpha}}{2}Q^{2}\right)\end{aligned} \qquad \text{(V-3.6)}$$

or

$$\log S(Q,\omega=0)=-\frac{1}{2}\left(\sum_{\alpha=1}^{n}L_{\alpha\alpha}\right)Q^{2}=-\frac{1}{2}MQ^{2} \qquad \text{(V-3.7)}$$

That is, $\log S(Q,\,\omega=0)$ is proportional to Q^2, and the proportional constant is the mean square displacement M.

The Eq. (V-3.7) indicates that $S(Q,\,\omega=0)$ follows a Gaussian distribution in accordance with both the scattering experiments, SAXS and EINS. However, it should be noted that the statement is true only in the small Q region, where the global fluctuation of all atoms in the protein is involved. As Q gets greater, $S(Q,\,\omega=0)$ probes more and more the local fluctuation of atoms, for example, the bimodal distribution between two dihedral angles, gauche and trans, in an amino acid. The local fluctuation is dominated by the direct interaction among the atoms, while the global fluctuation is governed by the central limiting theorem.

V-3.2 Temperature Dependence of MSD

Since the mean square displacement increases with temperature, the gradient of the plot $\log S(Q,\,\omega=0)$ against Q^2 should increase with increasing temperature. The physics deduced from the equation is depicted schematically in Fig. V-3.2. The general behavior of the theoretical plots is in harmony with the experimental data plotted in Fig. V-3.1 (left).

The variance-covariance matrix of the structural fluctuation are related to the free energy surface of the protein by Eq. (V-2.68) as

$$L_{\alpha\beta}\equiv\left\langle\Delta\mathbf{R}_{\alpha}\Delta\mathbf{R}_{\beta}\right\rangle=k_{B}T\left(\frac{\partial^{2}F\left(\{\Delta\mathbf{R}\}\right)}{\partial\Delta\mathbf{R}_{\alpha}\partial\Delta\mathbf{R}_{\beta}}\right)^{-1} \qquad \text{(V-3.8)}$$

where $F(\{\Delta\mathbf{R}\})$ denotes the free energy surface of protein, defined by Eq. (V-2.67). From Eqs. (V-3.7) and (V-3.8), the mean square displacement can be related to the second derivative of the free energy surface with respect to the displacement as

$$M=\sum_{\alpha}L_{\alpha\alpha}=k_{B}T\sum_{\alpha}\left(\frac{\partial^{2}F\left(\{\Delta\mathbf{R}\}\right)}{\partial\Delta\mathbf{R}_{\alpha}^{2}}\right)^{-1} \qquad \text{(V-3.9)}$$

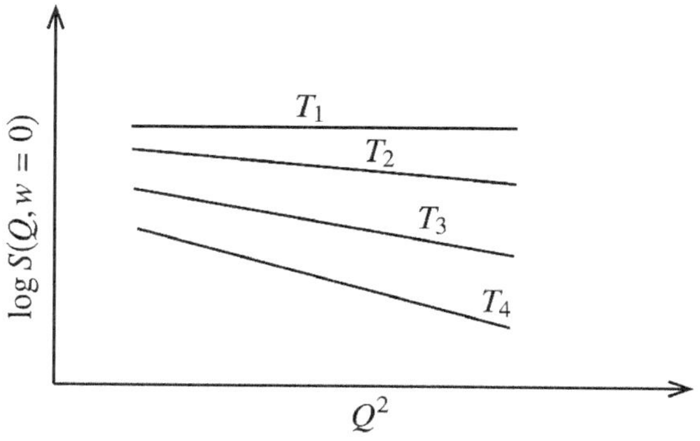

Fig. V-3.2: Structure factor plotted against Q^2 ($T_1 < T_2 < T_3 < T_4$).

Considering the Eq. (V-2.67), one can split the second derivative into two contributions as follows

$$K_{\alpha\alpha}^{E} = \frac{\partial^2 U(\{\Delta\mathbf{R}\})}{\partial \Delta\mathbf{R}_\alpha^2} \qquad \text{(V-3.10)}$$

$$K_{\alpha\alpha}^{W} = \frac{\partial^2 \Delta\mu(\{\Delta\mathbf{R}\})}{\partial \Delta\mathbf{R}_\alpha^2} \qquad \text{(V-3.11)}$$

Then, Eq. (V-3.9) can be written in a simple form as

$$M = k_B \sum_\alpha \left(\frac{1}{K_{\alpha\alpha}^{E} + K_{\alpha\alpha}^{W}} \right) T \qquad \text{(V-3.12)}$$

In the Eq. (V-3.10), $K_{\alpha\alpha}^{E}$ just depends on the interaction potential energy among atoms in protein, and it does not have any dependence on temperature. On the other hand, $K_{\alpha\alpha}^{W}$ is the second derivative of the solvation free energy with respect to the protein coordinates, thereby it surely depends on temperature. $K_{\alpha\alpha}^{W}$ may make the contribution to the force constant in the same or opposite direction with $K_{\alpha\alpha}^{E}$ depending on temperature. In the lower temperature, where water is in the state of *ice* or *amorphous solid*, the sign of $K_{\alpha\alpha}^{W}$ may be the same as $K_{\alpha\alpha}^{E}$, because the hydrogen-bond network of water is likely to restrict the fluctuation of protein. On the other hand, in the higher temperature, where water is in the liquid state, the sign of $K_{\alpha\alpha}^{W}$ is most likely to be opposite to that of $K_{\alpha\alpha}^{E}$, because the hydrogen-bonds among the back-bone atoms in the secondary structure, for example, will be loosened or even broken by interactions with water molecules around the atoms. Therefore, the proportional constant should become larger at higher temperature in which water molecules are activated to be *solvent*.

The temperature dependence of MSD that can be deduced from Eq. (V-3.12) is illustrated schematically in Fig. V-3.3(a), which faithfully reproduces the general behavior of the experimental observations depicted in Fig.V-3.3(b). In Fig. V-3.3(a), the dotted line corresponds to a *dry* sample in which water is supposed to have no significant effect on the fluctuation or elasticity, or $K_{\alpha\alpha}^{W} \approx 0$. Therefore, the MSD is determined essentially by the thermal excitation of the vibrational modes in harmonic potential wells with the "force constant," $K_{\alpha\alpha}^{E}$. Of course, the

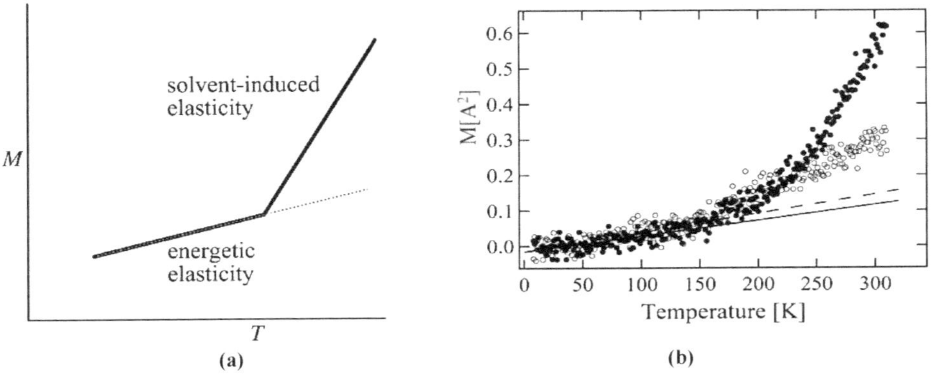

Fig. V-3.3: The mean square displacement of protein plotted against temperature. (a) Theoretical prediction: dashed line, dry sample; solid line, hydrated. Copyright (2018) Elsevir. (b) Experimental results: •, wet sample; ○, dry sample. Copyright (2011) Taylor and Francis.

slope of the line will be different from protein to protein, since the interaction energy among atoms of each protein is different. The temperature dependence of MSD for a *hydrated* sample is also depicted schematically with the solid line in Fig. V-3.3(a). In contrast to the case of a dry sample, the temperature dependence of MSD should change at certain temperature, due to the onset of the *solvent-induced elasticity* $K^W_{\alpha\alpha}$ stated above that is activated by the excitation of the diffusive motion, or fluidity, of water molecules. The temperature dependence of MSD after the onset of $K^W_{\alpha\alpha}$ may not necessarily be *linear*, but may have some curvature in general, since the solvation free energy is dependent on temperature. It should also be noted that the case depicted here is concerned with the two *ideal* cases: a completely *dry* sample and a *fully hydrated* sample. The temperature dependence of MSD will have a further variety depending on the hydration level, as has been clarified by the experiments [Kataoka and Nakagawa 2011].

The physics concerning the temperature dependence of MSD, deduced from the theory of structural fluctuation, can be summarized as follows. In the low temperature regime, the fluctuation is harmonic or linear within the harmonic *potential* well. In the higher temperature, the fluctuation is also harmonic or linear, but the "harmonicity" is originated from the *central limiting theorem*, not from the harmonicity of the *potential energy surface*. Therefore, the transition from the lower to higher gradients in the temperature dependence is not due to that between the harmonic and non-harmonic fluctuations, but between the two *harmonic* fluctuations with different physical origins, the harmonicity in the potential energy surface and due to the *central limiting theorem*, or the *solvent-induced elasticity*. The phenomena can be simply interpreted as the transition from the *energetic elasticity* to the *solvent-induced elasticity* [Hirata 2018].

The concept developed here may not be specific just to protein solution, but may be applied to more general class of polymer materials. There is a phenomenon that may have a physical origin, the same with or at least similar to the aqueous solution of protein, that is, the *soft contact-lens*. The soft contact-lens is made of synthetic polymers, called "hydro gel", that contain a large amount of water, 38 to 75 weight percent. The lens can be distorted in almost any shape by applying pressure with fingers, but it restores its original shape by removing the pressure, when it is *in aqueous solution*. Namely, it is an *elastic* deformation. On the other hand, the lens is broken into pieces by applying pressure, and never recovers its original shape, when it is completely *dry*. That is a *plastic* deformation. So, it is obvious that elasticity of the contact lens is induced by water or solvent. The phenomenon should not be confused with *wetting* of a hydrophobic polymer, such as polyethylene. If one applies a large distorting force on such a polymer in water, it may be broken, or undergoes plastic deformation. It is because water molecules in such a case are not participating as a part of the structural unit of polymer, but just concerned with wetting of polymer on the surface. On the other hand, in the case of a contact lens, water molecules are apparently participating in the polymer as a building unit by making the hydrogen-bonds with hydrophilic moieties of the polymer, just like in protein.

V-4 Structural Change of Proteins Induced by Thermodynamic Perturbation

The structure of a protein is subject to its thermodynamic environment such as temperature, pressure, solution conditions including denaturant, pH, and so on. A change of a thermodynamic condition can be regarded as a perturbation to protein conformation. Concerning pressure as a perturbation, there is an insightful experiment carried out by Akasaka and coworkers [Akasaka 2003, 2006]. By applying pressure on protein solutions, they have measured the NMR signals that probe the fluctuated states of protein, ranging from partially denatured to fully denatured, depending on the magnitude of pressure. When they have removed the pressure, the protein reversibly recovered its native conformation irrespective of whether it is partially or fully denatured. Based on their observation, Akasaka and his coworkers have deduced a picture concerning the conformational change of protein, which is completely different from the conventional one that has been practiced in both experimental and theoretical studies for long time. The conventional picture considers a *single* molecule of protein, and has focused their analysis on such a property as the "activation barrier" of the native-to-denatured transition of the structure. On the other hand, Akasaka sees a state of protein as an *ensemble consisting of* many molecules that distributes around the most propense structure corresponding to the thermodynamic condition. So, when the thermodynamic condition is *native*, a majority of molecules is surely in the native conformation, but small amount of molecules may take a "random-coil" conformation, and *vice versa*. The conformational change, say from native to denatured, is viewed as *a shift of the distribution from that having its peak at the native conformation to that at denatured one [Hirata and Akasaka 2015].*

In the present section, we introduce a theory for the structural change of protein induced by thermodynamic perturbations that correspond to the picture deduced by Akasaka et al., based on the theories described in the previous sections. When the thermodynamic perturbation is small, one may apply the linear response theory (Eq. (V-2.74)) developed in the Section V-2. (Here, we put the equation again for convenience.)

$$\langle \Delta \mathbf{R}_\alpha \rangle_1 = (k_B T_0)^{-1} \sum_\beta \langle \Delta \mathbf{R}_\alpha \Delta \mathbf{R}_\beta \rangle_0 \cdot \mathbf{f}_\beta \tag{V-4.1}$$

V-4.1 Structural Change Induced by Pressure

We define the force acting on an atom of protein in solution by the derivative of the Helmholtz free energy of the system with respect to the coordinates of the atom.

$$\mathbf{f}_\alpha = -\left(\frac{\partial F_s}{\partial \mathbf{R}_\alpha} \right), \tag{V-4.2}$$

where $\mathbf{R}_\alpha$ and F_s denote the α-th atom in protein and the Helmholtz free energy of the entire system, respectively. (Note that the notation F_s is used to distinguish it from the free energy surface of the protein defined by $F(\{\Delta\mathbf{R}\})$ In order to clarify the physical meaning of $\mathbf{f}_\alpha$, we define the free energy of the entire system in terms of the statistical mechanics by the configurational integral as follows

$$F_s = -kT\log\int\cdots\int\exp\left[-\frac{1}{kT}U\left(\{\mathbf{R}\},\{\mathbf{r}\}\right)\right]\prod_{\alpha=1}^{N_u} d\mathbf{R}_\alpha \prod_{i=1}^{N}\prod_{a=1}^{n} d\mathbf{r}_i^a, \tag{V-4.3}$$

where $\{\mathbf{R}\}$ and $\{\mathbf{r}\}$ denote sets of coordinates of atoms in protein and water, respectively, $U(\{\mathbf{R}\},\{\mathbf{r}\})$ is the interaction energy among the atoms in the system, $\mathbf{R}_\alpha$ denotes the coordinate of the α-th atom in protein, and $\mathbf{r}_i^a$ labels the a-th atom in the i-th water molecule. Taking the derivative of F_s with respect to the position of an atom in protein, we get the following expression.

$$-\left(\frac{\partial F_s}{\partial \mathbf{R}_\alpha}\right) = \frac{\int\cdots\int\left(-\frac{\partial U}{\partial \mathbf{R}_\alpha}\right)\exp\left[-\frac{1}{kT}U\left(\{\mathbf{R}\},\{\mathbf{r}\}\right)\right]\prod_{\alpha=1}^{N_u} d\mathbf{R}_\alpha \prod_{i=1}^{N}\prod_{a=1}^{n} d\mathbf{r}_i^a}{\int\cdots\int\exp\left[-\frac{1}{kT}U\left(\{\mathbf{R}\},\{\mathbf{r}\}\right)\right]\prod_{\alpha=1}^{N_u} d\mathbf{R}_\alpha \prod_{i=1}^{N}\prod_{a=1}^{n} d\mathbf{r}_i^a}. \tag{V-4.4}$$

The expression indicates that $\mathbf{f}_\alpha$ is a *mean* force acting on the atom α due to all other atoms in protein and in solvent, averaged over the statistical ensemble.

It is the thermodynamic work due to pressure that induces the structural change of protein. With this consideration in mind, we decouple the derivative into two steps as follows.

$$\left(\frac{\partial F_s}{\partial \mathbf{R}_\alpha}\right) = \left(\frac{\partial F_s}{\partial V}\right)_T \left(\frac{\partial V}{\partial \mathbf{R}_\alpha}\right)_{P,T}, \tag{V-4.5}$$

where V denotes the volume of the system. The first factor in the right-hand side is related to the pressure in terms of the standard thermodynamics,

$$-\left(\frac{\partial F_s}{\partial V}\right)_T = P, \tag{V-4.6}$$

while the second factor is identified as

$$\left(\frac{\partial V}{\partial \mathbf{R}_\alpha}\right)_{P,T} = \left(\frac{\partial \Delta\bar{V}}{\partial \mathbf{R}_\alpha}\right)_{P,T}, \tag{V-4.7}$$

where $\Delta\bar{V}$ is the partial molar volume of a protein in the solution. It results from the heuristic consideration that the change in the volume due to the structural change is only concerned with the partial molar volume. Substituting Eqs. (V.4-5) ~ (V.4-7) into Eq. (V-4.1) gives rise to the following equation.

$$\left\langle \Delta\mathbf{R}_\alpha \right\rangle = \left(k_B T_0\right)^{-1}\sum_\beta \left\langle \Delta\mathbf{R}_\alpha \Delta\mathbf{R}_\beta \right\rangle_0 \left(\frac{\partial \Delta\bar{V}}{\partial \mathbf{R}_\beta}\right)_{P,T} P. \tag{V-4.8}$$

The equation provides a physical explanation concerning how the hydrostatic pressure applied to the system induces the structural change in protein. According to the equations, the structural response to pressure is realized through the successive response functions. The first one is the first derivative of the partial molar volume with respect to the atomic coordinates of protein. The second is the variance-covariance matrix of the fluctuation in the unperturbed system. When pressure is applied, the displacement of the atom labeled by β occurs so that the partial molar volume decreases in accordance with the Le Chateirer's law. Then, the displacement or conformational change is propagated to other atom labeled by α through the variance-covariance matrix. Summing the perturbation over all atoms β gives rise to the displacement of atom α from the equilibrium conformation.

V-4.2 Structural Change Induced by Denaturant

The structural fluctuation is induced by adding the denaturant such as urea into the system. It is obviously the change in the chemical potential that drives the structural change due to denaturant. The thermodynamic potential which includes the chemical potential as independent variables is (PV), and its change caused by adding denaturant is defined by

$$d(PV) = SdT + \sum_i N_i d\mu_i + PdV, \tag{V-4.9}$$

where N_i and μ_i are the number of molecules and the chemical potential of the species i included in the solution. We define the thermodynamic perturbation due to the denaturant by,

$$\mathbf{f}_\alpha = -\left(\frac{\partial(PV)}{\partial \mathbf{R}_\alpha}\right)_{T,V} = -\sum_i N_i \left(\frac{\partial \mu_i}{\partial \mathbf{R}_\alpha}\right)_{T,V}. \tag{V-4.10}$$

Combining Eq. (V-4.10) with Eq. (V-4.1), one finds

$$\langle \Delta \mathbf{R}_\alpha \rangle = -(k_B T_0)^{-1} \sum_\beta \langle \Delta \mathbf{R}_\alpha \Delta \mathbf{R}_\beta \rangle_0 \sum_i \left(\frac{\partial \mu_i}{\partial \mathbf{R}_\beta}\right) N_i. \tag{V-4.11}$$

According to the equation, the perturbation of the denaturant on the structure of protein is created first through a response of the structure to the change in the chemical potential of the species included in solution. Then, the structural response is propagated over the entire molecule through the variance-covariance matrix.

V-4.3 Structural Change Induced by Heat

The deformation due to heat is the most well-studied property of protein. In this case, the change in the entropy is the perturbation that induces the structural change of protein, and it is the energy that contains the entropy as the independent variable. The energy is expressed by,

$$dE = TdS - pdV. \tag{V-4.12}$$

Then, the perturbation produced on the protein structure by heat is defined as,

$$\mathbf{f}_\alpha = -\frac{dE}{d\mathbf{R}_\alpha} = -\left(\frac{dE}{dS}\right)_V \left(\frac{dS}{d\mathbf{R}_\alpha}\right) = -T\left(\frac{d\Delta \bar{S}}{d\mathbf{R}_\alpha}\right), \tag{V-4.13}$$

where $\Delta \bar{S}$ denotes the partial molar entropy of protein, which is also a functional of protein structure. Substituting Eq. (V-4.13) into Eq. (V-4.1), one finds

$$\left\langle \Delta \mathbf{R}_{\alpha} \right\rangle = -\left(k_B T_0\right)^{-1} \sum_{\beta} \left\langle \Delta \mathbf{R}_{\alpha} \Delta \mathbf{R}_{\beta} \right\rangle_0 \left(\frac{d\Delta \bar{S}}{d\mathbf{R}_{\beta}} \right) T, \tag{V-4.14}$$

where T_0 denotes the temperature concerned with the reference system. The expression tells us how the structural fluctuation of protein due to heat or temperature is coupled with the change in the partial molar entropy of protein. The change in the partial molar entropy presumably consists of two major contributions: one from solvent, i.e., the hydration, the other from the *structural entropy*.

The last statement concerning the structural entropy suggests us a caution to be taken in the actual application of Eq. (V-4.14). If the structural entropy increases with rising temperature, the variance-covariance matrix should also increase, since the quantity is a measure of variation or dispersion of protein structure. Therefore, the response of protein structure to temperature is not considered in general as a linear process. Nevertheless, the theory may be applied to the non-linear regime, employing the idea of *analytical continuation*, which is suggested in Section IV.

V-4.4 Correspondence between the Theory and the Pressure NMR Experiment

The theory presented above can be applied to provide a physical explanation for the experimental finding concerning the pressure NMR by Akasaka and coworkers [Akasaka et al. 1999, Akasaka 2003, 2006].

In Fig. V-4.1, depicted *conceptually* is the free energy surface of protein for two different pressures, which can be deduced from the general features of Eqs. (V-2.71) and (V-2.74): both the surfaces should be quadratic with the same curvature, but with different equilibrium structures. In the figure, the x-axis represents the structure of protein which is denoted by $\{\mathbf{R}\}$, and the y-axis is the free energy surface, $F(\{\mathbf{R}\})$, of the structure. The curves marked by P_1 and P_2 ($P_1 < P_2$) describe the surface corresponding to the two pressures, and the minimum of each curve represents the equilibrium conformation corresponding to the respective pressure. The figure illustrates that the conformation is *thermally* fluctuating around an equilibrium structure, producing a Gaussian distribution.

The vertical arrow indicates that the pressure is abruptly increased from P_1 to P_2 so that the equilibrium structure in the state denoted by P_1 becomes a fluctuated or non-equilibrium conformation in the P_2 state, as is indicated by the end point of the arrow landing on the P_2 surface. Then, the conformation relaxes to the equilibrium point along the P_2 surface corresponding to the new thermodynamic condition, which is illustrated by the winding arrow. The fluctuated conformations are distributed around the new minimum to contain a conformational ensemble. The equilibrium structure in the P_2 surface corresponds to a fluctuated conformation in the P_1 surface, which is indicated by the vertical dashed-line.

The picture drawn here is in complete accordance with the hypothesis proposed by Akaska and his coworkers based on the high pressure NMR. The hypothesis states that different conformations induced by pressure are the *fluctuating* states of the protein when the solution is in the native condition, and that pressure would not create a new state, but simply creates a new population distribution among the existing states. In order to show the consistency of the two concepts, let us suppose that P_1 in Fig. V-4.1 is the ambient pressure, thereby the P_1 surface represents the conformational fluctuation taking place around the native structure, while the surface labeled by P_2 depicts the surface of higher pressure. The minimum of the P_2 surface represents the equilibrium conformation at the pressure P_2, but the structure corresponds

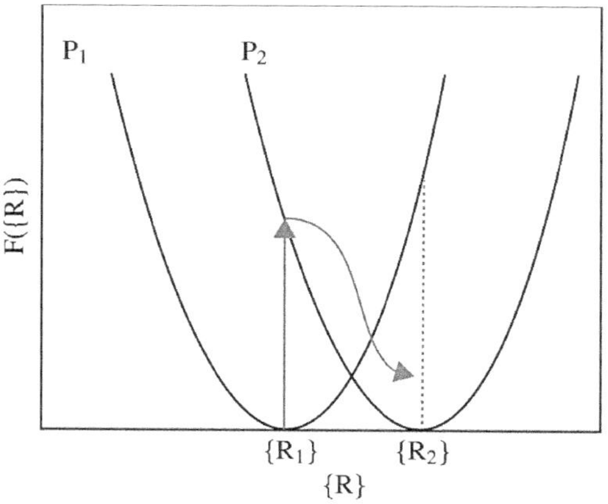

Fig. V-4.1: Schematic picture of the free energy surface of a protein in solution, drawn against the structure of protein: {**R**} represents the structure of protein; the labels P_1 and P_2 indicate that the surfaces correspond to the two pressures; $\{R_1\}$ and $\{R_2\}$ represent the equilibrium structure corresponding to the two pressures. Copyright (2015) American Institute of Physics.

to a fluctuated state on the P_I surface, which is indicated by the vertical dashed line. Since the argument applies to any higher pressure within the linear regime, the conformation corresponding to the completely denatured state, produced by pressure, should exist in the ambient condition as a fluctuated state, though in much lower probability.

Basically, the same argument applies to the reverse process, or, "refolding." Suppose that the P_2 surface in Fig. V-4.1 represents the unfolded state of a protein. If one lowers the pressure to the ambient pressure P_I, abruptly, as is practiced in the ordinary experiment, the system jumps up vertically to a fluctuated state of the P_I surface, and relaxes to the native conformation along the surface. The relaxation process is nothing but the refolding.

It would be worthwhile to make a further comment on Fig. V-4.1. The surface drawn in Fig. V-4.1 is just a conceptual illustration presenting the global nature of the free energy surface. An actual free energy surface is not monotonical, but has a lot of local minima corresponding to conformational sub states that have a hierarchical structure. The rate of fluctuation and relaxation is determined by the detailed structure of the free energy surface. For example, if the surface has two major sub states, i.e., the native and denatured states, then the rate will be determined by the height of the barrier which separates the two sub states.

An important implication of the new theory is that the fluctuated states of protein can be reproduced theoretically, which are identical in its physics to those found by Akasaka and coworkers [Akasaka et al. 1999, Akasaka 2003, 2006]. It has been already shown in Section V-2 that the variance-covariance matrix in Eq. (Fig. V-4.1) is related to the second derivative of the free energy surface with respect to the atomic coordinates of protein (see Eq. (V-2.70)). The free energy surface, which is an explicit function of the atomic coordinates, can be calculated from the 3D-RSM/RISM theory (see Chapter IV). The first derivative of the partial molar volume can also be calculated based on the 3D-RISM/RISM theory combined with the site-site Kirkwood-Buff theory (see Chapter II and Chapter IV) [Imai et al. 2004].

$$\Delta\bar{V}(\{\mathbf{R}\}) = k_B T \chi_T^0 \left(1 - \rho \sum_{\gamma} \int_V c_\gamma(\mathbf{r};\{\mathbf{R}\}) d\mathbf{r}\right) \tag{V-4.15}$$

In the equation, $c_\gamma(\mathbf{r}; \{\mathbf{R}\})$ is the direct correlation function of the water molecules around protein, which is a functional of the protein structure $\{\mathbf{R}\}$.

V-4.5 Extending the Theory to a Large Fluctuation

It is likely that the variance-covariance matrix in Eq. (V-4.1) changes significantly due to a perturbation, even though the distribution still remains Gaussian or approximately Gaussian. One such example is the thermal denaturation processes described by Eq. (V-4.14). The variance-covariance matrix for the denatured state is apparently much greater than that of the native state, since the conformational fluctuation for the denatured state should be greater than the native state due to the increased structural entropy. The linear response theory, Eq. (V-4.1), in its original form may not be applied to those processes. However, we may be able to extend the theory to such a case by means of the *analytical continuation*. The well-regarded mathematical theorem tells us how an analytical function can be expressed in a series, i.e., a Taylor series. A Taylor series can be truncated at the first term, if the term is well within a range called the *convergence radius*. Shifting the center of the expansion into the new position of the function, expanding the function into a series around the position, and truncating the function at the first term, one may be able to find the next value of the function. Iterating the procedure successively, one may be able to get the entire function. The procedure is valid for any analytic function that does not include a singularity, and we believe that the free energy surface of a protein in an ordinary condition does not include such a singularity.

We divide the perturbation and the response into several steps for which the linear response theory is valid. We refer to such a process as a "quasi-linear" process. The process can be written as

$$\langle \Delta\mathbf{R}_\alpha \rangle_{j+1} = \left(k_B T_j\right)^{-1} \sum_\beta \langle \Delta\mathbf{R}_\alpha \Delta\mathbf{R}_\beta \rangle_j \cdot \mathbf{f}_\beta^j \qquad \text{(V-4.16)}$$

where j denotes the each quasi-linear step, starting from zero. In the equation, the variance-covariance matrix at the $(j+1)$-th step should be calculated based on the protein coordinates in the j-th step. With careful scheduling for increasing the order parameter, we may be able to get successive curves similar to Fig. V-4.1, but with different values for the curvature, or the variance-covariance matrix.

V-5 Structural Fluctuation of Protein and Anfinsen's Thermodynamic Hypothesis

One of the most important findings in the field of biophysical chemistry is undoubtedly that made by C. Anfinsen. He found experimentally that an enzyme called Ribonuclease deactivated by adding urea in the solution reversibly restores its activity by removing the denaturant [Anfinsen 1973]. Based on his experimental finding, and with the assumption concerning the structure-activity relationship, Anfinsen proposed his famous hypothesis concerning the protein folding, which is casually referred to as Anfinsen's "dogma." Anfinsen described his hypothesis in the paper published in *Science* right after he was awarded the Nobel Prize in 1972. Let us quote the entire statement Anfinsen has written in order to explain his hypothesis. *"This hypothesis states that three-dimensional structure of a native protein in its normal physiological milieu (solvent, pH, ionic strength, presence of other components such as metal ions or prosthetic groups, temperature, and others) is the one in which the Gibbs free energy of the whole system is the lowest; that is, the native conformation is determined by the totality of interatomic interactions and hence by the amino acid sequence, in a given environment."* The statement is essentially different from the most popular understanding of Anfinsen's dogma, represented by *Wikipedia*, which says "Anfinsen's dogma is a postulate in molecular biology that, at least for small globular proteins, the native structure is determined *only by the protein's amino-acid sequence*." This is a misunderstanding of Anfinsen's hypothesis. It is important to realize that Anfinsen identified the native conformation of protein as the state in which the *Gibbs free energy* of the *whole* system including solvent environment is the lowest. He never said "the native structure is determined *only by the protein's amino-acid sequence*."

After Anfinsen proposed his hypothesis, Levinthal asked a serious question on the hypothesis based on a *gedanken experiment* concerning the folding mechanism [Levinthal 1968]. The comment is referred to as "Levinthal paradox." Just imagine the folding of a protein consisting of a hundred amino-acids. The protein has three dihedral angles at each peptide bond that creates $\sim 3^{100}$ possible conformations. Suppose the conformational change for each dihedral angle takes about nano-seconds, or $\sim 10^{-9}$ seconds, then, it takes $\sim 3^{100} \times 10^{-9} \approx 10^{40}$ seconds, a cosmological time, for searching a native conformation. Nonetheless, a protein folds into a native conformation from a random-coil state in short time, like a second. Why and how? That is the Levinthal paradox. It should be noted that Anfinsen's dogma is concerned with the *thermodynamic* stability of native and denatured *states* of protein, while Levinthal's paradox concerns the *kinetics* or rate of the folding process of a *single* protein molecule.

The protein solution Anfinsen has studied in his experiment contains a number of protein molecules, $\sim 10^{20}$, in the test tube, the structures of which are fluctuating due to thermal motion. The fluctuated state of protein makes a distribution or an ensemble around the most propensive conformation corresponding to the thermodynamic condition. According to many experimental studies, the difference of free energy between the native and denatured states is marginal, or few kcal/mol. Such experimental observation suggests strongly that there are some non-native conformations, even completely denatured states, in the solution of native condition, and *vice versa*. The pressure NMR studies by Akasaka et al. have proven unambiguously that the change of structural distribution with increasing pressure is continuous, indicating that there is no abrupt change from the native to denatured conformations concerning *each molecule* of protein [Akasaka 2006]. Rather, induced by pressure is the change in the *structural distribution* of protein from the native to denatured states, or its peak position and variance. Such a viewpoint may give a solution to the Levinthal paradox.

Let's imagine a process to apply pressure on a 1.0 mM protein solution in an ambient condition. The system includes as many as $\sim 10^{20}$ protein molecules, structures of which likely form a distribution having a peak around the native conformation. The foot of the distribution consists of partially or completely denatured conformations, depending on the distance or deviation from the peak. If one applies pressure on the system, the peak position will shift toward the denatured conformation, and the degree of peak shift, or the degree of *denaturation*, will be greater and greater as the pressure increases, in harmony with the experiment by Akasaka [Akasaka 2006]. Now, we conjecture that a shift of the peak from one to neighboring positions along the conformational space won't take much time, because the structures between the neighboring peaks won't be so different from each other, just few in terms of the dihedral angles. So, integrating such small shifts of the distribution peak to complete the native to denatured conformational change will not take too much time as well, at least not as much as the *cosmological* time estimated by Levinthal.

V-5.1 Realization of the Protein Folding based on the Generalized Langevin Theory

It is the concept of ensemble or structural distribution that may resolve the Levinthal paradox. Then, what type of distribution should be attributed to an ensemble of protein molecules which are subject to structural fluctuation? The natural answer to the question deduced from the theoretical and experimental considerations given in the previous sections is the *Gaussian* distribution (see Eq. V-2.70). Then, the folding or unfolding process of protein can be viewed as the change of the Gaussian distribution due to a change in thermodynamic conditions. Such changes in the Gaussian distribution due to a thermodynamic perturbation can be treated by the linear or non-linear response theory described in the Section V-4. Recently, Hirata and his coworkers developed a new concept and a theory for the protein folding based on the response theories [Hirata et al. 2018]. The concept sees the protein folding (or unfolding) process as *a shift of the Gaussian distribution formed by fluctuated structures.* The shift of the distribution is likely to be associated with a change in the variance-covariance matrix in general, depending on the thermodynamic perturbation. The unfolded state of protein very likely has larger variance-covariance matrix compared to that of the native state. In that respect, the unfolding or refolding is not a linear process. Nevertheless, it is considered that the distribution itself remains Gaussian in both folded and unfolded states, and in all intermediate states, but with different variance-covariance matrices. An experimental evidence for the statement can be found in the article written by Kataoka and his coworkers [Kataoka et al. 1995]. In the article, they have presented a Guinier plot for a myoglobin in various thermodynamic states: holo-myoglobin in native state, apo-myoglobin in native state, trichloroacetate-stabilized molten-globule state, and acid-unfolded state (see Fig. V-1.1). According to their experiment, all the plots for different thermodynamic conditions stay linear with negative slope up to the wave number Q less than ~ 0.07 A^{-1}, which corresponds to ~ 90 A in real space. The results suggest strongly that the fluctuation of *global* structure of protein is Gaussian, although that of more local structures inside protein is not necessarily so. (For example, a dihedral angle in an amino acid residue can take a bimodal or trimodal distribution, gauche and trans, and not a Gaussian distribution.) The results also indicate that the slope of the linear plot, or the variance-covariance matrix, for the unfolded state is greater than that of the native state. The findings support our view that sees the protein folding (or unfolding) process as *a shift of the Gaussian distribution, the average structure and the variance-covariance matrix of which varies according to changes in thermodynamic conditions*.

The concept stated above can be formulated by the linear response theory extended to a non-linear regime in the previous Section (V-4), or Eq. (V-4.17), in which the response function varies with the thermodynamic perturbation. The extension was made by means of the analytical continuation. We divide the perturbation as well as the response into several steps in such a way that each step lies within a convergence radius, where the linear response theory is valid. The equation is shown below again for convenience

$$\langle \Delta \mathbf{R}_\alpha \rangle_{j+1} = \left(k_B T_j\right)^{-1} \sum_\beta \langle \Delta \mathbf{R}_\alpha \Delta \mathbf{R}_\beta \rangle_j \cdot \mathbf{f}_\beta^j \qquad \text{(V-5.1)}$$

where j starts from zero. In the equation, the variance-covariance matrix at the (j+1)-th step should be calculated based on the protein coordinates in the j-th step. (Although temperature has suffix j, or T_j, it should stay constant unless temperature is the thermodynamic perturbation.)

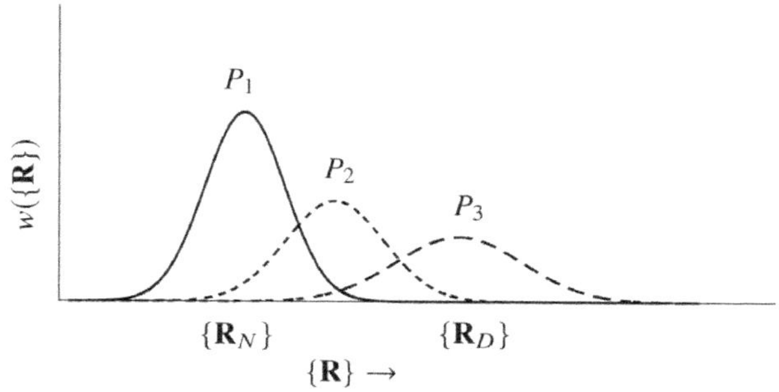

Fig. V-5.1: Illustration of the shift of Gaussian distribution of protein structure along the pathway from native to denatured states, or vice versa. It should be noted that the figure is just a schematic presentation. (A real distribution is composed of many Gaussian functions with different half-widths.)

With careful scheduling for increasing (or decreasing) the perturbation, so that it stays within a convergence radius, we may be able to get successive Gaussian distributions for varying thermodynamic conditions with corresponding $\langle \Delta \mathbf{R}_\alpha \rangle$ and $\langle \Delta \mathbf{R}_\alpha \Delta \mathbf{R}_\beta \rangle$. Such a process is schematically illustrated in Fig. V-5.1.

In the figure, $w(\{\mathbf{R}\})$ denotes the probability distribution of the structure of protein, represented by $\{\mathbf{R}\}$, and $\{\mathbf{R}\}_N$ and $\{\mathbf{R}\}_D$ represent the structure of native and denatured states, respectively. The figure describes the protein folding (or unfolding) as a process in which the Gaussian distribution (the average structure and the variance covariance matrix) shifts reversibly according to a thermodynamic perturbation described by P_i.

V-5.2 Characterization of Folding Mechanism

Given all the parameters characterizing the Gaussian distribution, we may be able to answer the long-standing questions concerning the mechanism of protein folding: if it is a two states model or not, if it is not, how many intermediate states exit, and what are the structure of intermediate states. By changing a thermodynamic parameter, we can get a series of Gaussian distributions, as illustrated in Fig. V-5.2.

The distribution for the native condition is most likely to be a single distribution around the native structure. (Here, it should be noted that the word "single Gaussian distribution" does not mean a *single* Gaussian distribution with a *single* variance. It really means a single Gaussian distribution composed of *many* Gaussian distributions with different variance and covariance.) However, the distributions of a non-native condition in general have multiple peaks, or average conformations. If the distribution peak is one, the process can be identified as a two state model (Fig. V-5.2(a)). On the other hand, if the distribution shows multiple peaks, the process may be a multiple-step transition passing through several intermediate states, each step of which has different average structure and variance-covariance matrix (Fig. V-5.2(b)). If the average structure changes continuously, then any of the multi-state mechanisms does not apply. The mechanism may be different from protein to protein. It is also very likely that the mechanism depends on thermodynamic parameters, temperature, pressure, or denaturant, to induce structural changes.

The appearance of multi-state mechanisms implies existence of a gap in the distribution or a barrier in the free energy surface, between any of the two states. Such a gap or a barrier may have several physical origins. One of the well-known origin is the *frustration* due to the potential

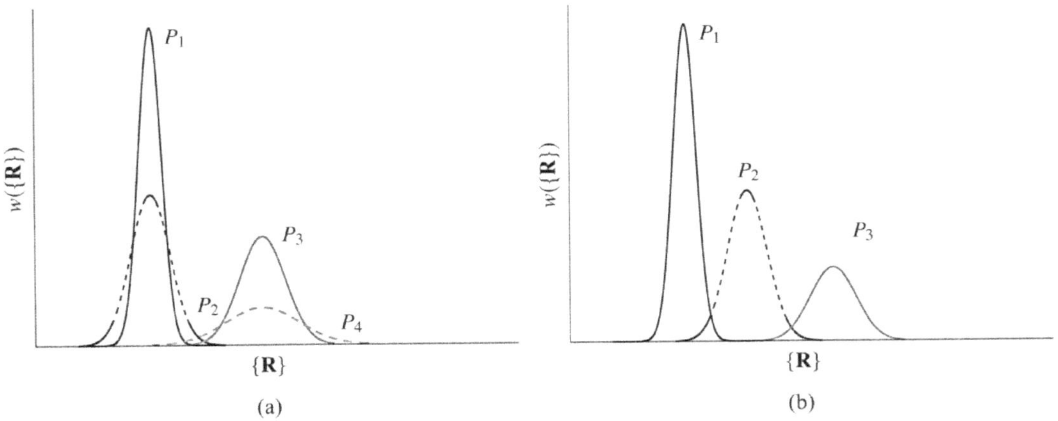

Fig. V-5.2: Schematic representation of folding mechanisms in terms of Gaussian distributions.

energy and the structural entropy of protein itself: the potential energy of the native state is lower than that of the denatured state, while the native state has much less entropy than the denatured state. It is surely an important cause of the gap or barrier. However, there is another important cause that may produce the gap or barrier. It is a *frustration* between the structure of protein and the configuration of solvent. In the process of unfolding and refolding, the initial structures in both ends of the process are more or less well-characterized by the hydrogen-bond among backbone atoms. The native structure has the secondary structures, alpha-helices or beta-sheets, constructed by hydrogen-bonds between a pair of *back-bone* atoms, and by the hydrophobic contact among alkyl and aromatic groups of amino-acid side-chains in the core region. On the other hand, in the denatured state, many of the hydrogen-bond donors and acceptors in the *back-bone* are likely to make hydrogen-bonds with water molecules, while the alkyl groups in the side chain may be making hydrophobic contact with each other. It means that *there should be global and extensive reorganization of hydrogen-bonds among back-bone atoms of protein and water molecules in the process of unfolding or refolding.* Such a hydrogen-bond rearrangement is essentially a random search process, and any specific conformation during the search will be unstable with short life-time. For example, in the unfolding process, some of water molecules should be sneaking into the core region to make hydrogen-bonds with backbone atoms which were making intramolecular hydrogen-bonds to form alpha-helices or beta-sheets. Conversely, in the refolding process, the backbone atoms making hydrogen-bonds with water molecules should release water molecules to make intramolecular hydrogen-bonds to form the secondary structures. On the other side, an extensive rearrangement of hydrogen-bond network among water molecules should take place. In the middle of such a rearrangement of hydrogen-bonds, potential bond-donors and acceptors in the backbone region may make contact with a hydrophobic group, or/and hydrophobic groups may be exposed to water. We hypothesize that the gap in the distribution or the barrier in the free energy surface originate in such a manner. If the hypothesis is correct, it means that any of methods ignoring the *molecularity* of water fails to describe the mechanism of folding and unfolding because it may not account for the exchange of hydrogen-bonds with water molecules.

An experimental method to probe the state of transition has been proposed by Akasaka and his coworkers, based on the pressure NMR [Akasaka 2006]. The idea is to apply pressure to *stabilize* a state which otherwise is a *fluctuated state* in the ambient condition. If the state is stabilized, the reorganization of hydrogen-bonds involving atoms in protein and water molecules may be probed by some empirical methods, i.e., the NMR chemical shift. Alternatively, the state can be probed theoretically by means of the 3D-RISM method, provided the conformation of protein is at high pressure. Imai et al. have calculated the distribution of water molecules around Ubiquitin at high pressure (300 MPa) based on the 3D-RISM method, and found that water molecules are sneaking into the core region consisting of hydrophobic residues [Imai et al. 2007]. Of course, such a solvation state is apparently unstable in the ambient condition.

Although we have emphasized the global reorganization of hydrogen-bonds of *backbone* atoms with water molecules in the process of folding and unfolding, it is just a general aspect of the story. It is the *specificity* in the structure that makes protein as a functional molecule: how many alpha-helices and beta-sheets are included, how those secondary structures are arranged in the conformation, what is the structure of active site where water molecules and ions are located in the protein, and so on. It is the side-chains that are responsible to the specificity of protein. The problem should be solved for each protein, since it is specific to each protein. However, it may be safe to say that any theoretical attempt ignoring the *molecularity* of water will fail to explain the specificity of protein conformation because it is also determined largely by the reorganization of hydrogen-bonds between water and protein and among water molecules in characteristic manner to each protein.

V-5.3 Comments on Concurrent Models of Protein Folding

Motivated by Anfinsen's hypothesis, many experimental and theoretical ideas and models have been proposed to explain the protein folding [Arai and Kuwajima 2000]. Among such ideas and models, the latest and the most highlighted one is the *energy landscape model* in general, or the *funnel model* in particular [Bryngelson et al. 1995]. The model has been applied successfully to many biophysical processes related to the protein folding [Panchenko et al. 1997, Clementi et al. 2000, Koga and Takada 2001, Shen et al. 2005]. So, it may be worthwhile to comment on the model. The model has some similarity to the theory proposed by Hirata and coworkers in the sense that it handles distribution of protein in the configurational space [Hirata 2018]. However, the model is entirely different from ours in its level of description. The reason is explained below.

The theory behind the funnel model is embodied in the following two equations for the free energy surface along an order parameter "n" that is a *measure of similarity to the native conformation*; according to the authors: $n = 1$, most similar to the native conformation, and $n = 0$, least similar to the native structure [Bryngelson et al. 1995].

$$P(E) = \frac{1}{\sqrt{2\pi E^2(n)}} \exp\left(-\frac{\left(E - \langle E(n)\rangle\right)^2}{2\Delta E^2(n)} \right) \tag{V-5.2}$$

$$F(n) = \langle E(n)\rangle - \frac{\Delta E(n)^2}{2k_B T} - TS_0(n) \tag{V-5.3}$$

The first equation, Eq. (V-5.2), implies that the authors assume a Gaussian fluctuation for the energy $E(n)$ around the structure described by n, the first and second moments of which are denoted by $\langle E(n)\rangle$ and $\Delta E^2(n)$, respectively. According to authors' explanation, $E(n)$ is not just the conformational energy of protein, but it implicitly includes the solvation free energy such as the hydrophobic interaction. Therefore, the order parameter n in fact represents a thermodynamic state including those of solvent, not a *structure* of protein. A Gaussian fluctuation for the thermodynamic variables is reasonable assumption. But, it is nothing new. More than half a century ago, Landau created the theory of thermodynamic fluctuation based on the assumption that relates the fluctuation to the second order derivative of the free energy [Landau and Lifshits 1958]. On the other hand, the theory described in the preceding sections is based on the Gaussian fluctuation of the *structure* (atomic coordinates of protein), not *thermodynamic variables*. The Gaussian distribution was *derived*, not *assumed*, from the generalized Langevin analysis of the phase space consisting of all the degrees of freedom including protein and solution at atomic level.

The second equation, Eq. (V-5.3), gives the free energy of a state measured by the order parameter n, where $S_0(n)$ denotes the structural entropy of protein. $S_0(n)$ is related to the number of possible structures $\Omega(n)$ consistent with n, by $S_0(n) = k_B \ln\Omega(n)$: $\Omega(n)$ depends on the detail of the structural description of the model. As an approximation, the authors have suggested a primitive expression, $\Omega(n) = \gamma^{*N(1-n)}$, where γ^* denotes the number of possible configurations per a peptide unit, and N is the number of peptides in the protein. The second term in the equation, or $\Delta E(n)^2/2k_BT$, is a contribution to the entropy from the thermodynamic fluctuation described by Eq. (V-5.2). The stability of a state measured by n is determined essentially by the balance between $\langle E(n)\rangle$ and the structural entropy $S_0(n)$: $\langle E(n)\rangle$ is dominative in the state $n = 1$, while $S_0(n)$ dominates the state $n = 0$. The most essential hypothesis concerning the model is an energetic predominance of the state $n = 1$, or the native state, over the contribution from the conformational entropy. The hypothesis is made apparently in order to be consistent with the thermodynamic hypothesis by Anfinsen in the process of thermal denaturation. However, it does not provide much structural insight into Anfinsen's hypothesis, since it does not have atomic scale information concerning the protein structure. The model does not have predictive ability either. In order to *predict* the native structure of protein, the model has to *assume* the native structure, since the definition of similarity measure "n" requires structural information concerning the native state.

The most serious problem of the model is that it does not have independent variables to control the thermodynamic condition such as temperature, pressure, pH, ionic strength, and so on. In the case of temperature, Eq. (V-5.3) apparently includes the temperature dependence in the second and third terms. However, it is just apparent, because the quantity $\langle E(n)\rangle$ and $\Delta E(n)^2$ should be also dependent on the temperature according to their definition. The most striking phenomenon, in which the dominance of the structural entropy in non-native conformation breaks down, is the so-called "cold denaturation process." It suggests that the model breaks down even for the temperature dependence. The model does not include pressure and composition of the system as independent variables *at all*. How a model can reproduce Anfinsen's thermodynamic hypothesis without providing the thermodynamic variables? A coarse-grained simulation based on the model may not solve the problem, because such simulations inherit all the essential feature of the model, described above.

Recently, interesting results concerning the folding kinetics have been presented by Englander and his coworkers, based on the hydrogen-exchange mass-spectroscopy (HXMS) experiment [Englande and Mayne 2017]. In the experiment, a protein unfolded by denaturant is diluted into folding conditions. After various elapse times of folding, the sample is subject to brief D-to-H HX pulse to label partially folded transient intermediates. The protein sample is quenched to halt further exchange, and fragmented into many peptides. The many peptides are roughly separated by HPLC, and analyzed by the high resolution MS. In this way, the method is able to realize the time resolved change of mass concerning the fragments, called "foldon", that represent the structural units of protein. By analyzing many proteins with the method, the authors have deduced a general model of protein folding, which sees the folding as a linear, stepwise pathway. While the general model contradicts seriously the energy landscape model, briefly reviewed above, it is quite consistent with the concept proposed in this chapter. In fact, the mass distribution of each *foldon* unit, Fig. 2 in the paper by Englander, looks like *Gaussian*, and the first moment (peak position) and the second moment (variance) of the Gaussian distribution change along the time evolution of protein folding [Englande and Mayne 2017]. There is some apparent difference between their experimental results and the theoretical concept described in this book. Their experimental results focus on some specific segment of protein, or "foldon", in the resolution of amino-acids, while the concept here concerns the structure of entire protein in atomic resolution. However, the problem may not be essential, because the information with atomic resolution can be always *reduced* or *coarse-grained* into the amino-acid resolution. In near future, the theory will be able to resolve the structure of "foldon" in atomic scale along the folding pathway.

In the present section, a new concept and a theory of protein folding were presented based on the generalized Langevin equation that describes the thermodynamic hypothesis by Anfinsen, or so called "Anfinsen's dogma." The concept sees the protein folding (or unfolding) process as *a shift of the Gaussian distribution of fluctuated structures, the first moment (average structure) and the second moment (variance-covariance matrix) of which vary depending on changes of thermodynamic conditions*. The idea is not just "a pie in the sky." The concept can be realized by calculating the free energy surface of the protein in aqueous solution, as well as its first and second derivatives with respect to atomic coordinates.

V-6 A Molecular Theory of the Structural Dynamics of Protein Induced by a Perturbation

Structural dynamics of protein induced by a perturbation has been a major concern in many fields of life sciences, since it is intimately related to function of the biomolecule [Terazima 2015]. For example, a photo excitation of retinal bound to bacterio-rhodopsin induces a structural dynamics of the protein, and triggers successive signal transduction process for light sensing [Koua and Kandori 2016]. A variety of experimental techniques to explore structural dynamics of protein has been flourished in the last two decades, which include the time-resolved X-ray crystalography [Kim et al. 2016], NMR [Sugase et al. 2007], so-called "single molecule measurement" [Yanagida et al. 2000, Noji et al. 1997], a variety of spectroscopy including time resolved Raman spectroscopy [Gao et al. 2006], transient grating spectroscopy [Nishihara et al. 2004], and so on. Among the techniques, the transient grating spectroscopy developed by Terazima and his coworkers is of particular interest, because it probes time-resolved thermodynamic as well as transport properties of protein solution along the course of structural change induced by a photo-excitation [Nishihara et al. 2004, Terajima 2015]. For example, they could have probed the partial molar volume of protein, a thermodynamic quantity, at each step of the photo-dissociation process of carbon mono-oxide from myogrobin, induced by transient grating of light [Nishihara et al. 2004]. So, it is essentially an irreversible thermodynamics of protein measured by a method of spectroscopy. Unfortunately, since the technique does not provide structural information of protein in atomic scale, it is not easy to examine a relationship between structure and phenomenological properties during a course of the structural change. On the other hand, methods such as the time-resolved X-ray crystallography, which provides the structural information, may not be able to probe irreversible thermodynamics. So, it is highly desirable to develop a theory to bridge these methods for comprehensive understanding of life phenomena. The present section is concerned with a theoretical approach to trace the structural change triggered by some perturbation applied to a protein solution, based on the linear response theory (see Chapter III for the linear response theory). The approach is quite general so that it can be applied to any type of perturbation, mechanical, thermodynamic, photo excitation, etc., by devising a force or perturbation acting on atoms of protein (see Section V-4 for the thermodynamic perturbation).

The linear response theory originated by R. Kubo [Kubo 1957] has been developed further by H.L. Friedman and his coworkers to explore dynamics of a variety of liquid systems, including electrolyte solutions [Friedman 1985]. Among the ideas they have developed in order to apply the theory to liquid systems, a concept named a "surrogate Hamiltonian" deserves a special attention. The problem they have formulated with the linear response theory is so called "solvation dynamics" or "dynamic Stokes shift", which has been highlighted by the community of the non-linear spectroscopy (see Chapter III for the solvation thermodynamics). The problem is to describe the relaxation of solvent polarization around a solute after an abrupt change of the electronic state of the solute is induced by photo-excitation. In conventional set up of the linear response theory, a perturbed part of Hamiltonian is described by direct interactions between solute and solvent. However, such a set-up faces a serious difficulty because it will leave the most important many-body problem concerning the solute-solvent interaction *unsolved*. In order to overcome the problem, Friedman and his coworkers employed the Ornstein-Zernike direct correlation function for solute-solvent interactions, which effectively renormalizes the many-body interactions between solute and solvent into an interaction between a pair of atoms [Friedman et al. 1995]. Since the methods to evaluate the direct correlation function between solute and polar solvent have been well developed in the field of liquid state theory, the problem has become tractable with the use of renormalized interaction (see Chapter II for the direct correlation function). They referred to the renormalized interaction as "surrogate Hamiltonian." Soundness of the surrogate Hamiltonian has been verified by a number of applications to the solvation dynamics [Friedman et al. 1995].

In the study to develop a theory for the structural dynamics of protein, Hirata employed the same basic idea developed by Friedman's group, by replacing the physics of the solvation dynamics by that of the structural dynamics of protein. In order for the replacement of physics, Hirata made changes in the formulation in two respects: (1) a way to divide an entire system into the unperturbed (or reference) and perturbed systems, and (2) definition of the renormalized (or surrogate) Hamiltonian. Apart from the difference in physics, the development in the present chapter is highly isomorphic to that of Friedman's group.

V-6.1 Time Dependent Linear Response Theory of Protein Structure

Hamiltonian:

We define the Hamiltonian of protein in water by

$$H(t) = H^{(0)} + H^{(1)}(t) \tag{V-6.1}$$

where $H^{(0)}$ is the Hamiltonian of an unperturbed system, which consists of that of water, H_{ww}, protein, H_{pp}, and the atom-atom interactions between water and protein, H_{pw}, that is,

$$H^{(0)} = H_{ww} + H_{pp} + H_{pw} \tag{V-6.2}$$

Each term of the unperturbed Hamiltonian is defined by the following equations as in the Section V-2 [16].

$$H_{ww} = \sum_{i=1}^{N}\sum_{a=1}^{n}\left[\frac{\mathbf{p}_i^a \cdot \mathbf{p}_i^a}{2m_a} + \sum_{j\neq i}\sum_{b\neq a} U_{ww}\left(\left|\mathbf{r}_i^a - \mathbf{r}_j^b\right|\right)\right] \tag{V-6.3}$$

$$H_{pp} = \sum_{\alpha=1}^{N_u}\left[\frac{\mathbf{P}_\alpha \cdot \mathbf{P}_\alpha}{2M_\alpha} + \sum_{\beta\neq\alpha} U_{pp}\left(\left|\mathbf{R}_\alpha - \mathbf{R}_\beta\right|\right)\right] \tag{V-6.4}$$

$$H_{pw} = \sum_{\alpha=1}^{N_u}\sum_{i=1}^{N}\sum_{a=1}^{n} U_{pw}\left(\left|\mathbf{R}_\alpha - \mathbf{r}_i^a\right|\right) \tag{V-6.5}$$

$H^{(1)}(t)$ denotes a perturbation applied to the system. Here, a photo-activated structural change of protein is taken as an illustrative example of the perturbation. Many proteins related to photo-activated functions have a common feature, including a photo-sensing unit or chromophore, retinal, coumarin, and so on, as a part of the structure. A structural change of a chromophore, such as isomerization and change of charge distribution, caused by photo-excitation, induces a successive change in the structure of backbone as well as amino-acid residues of protein. Since the initial photo-chemical reaction in a chromophore is very fast compared to the succeeding structural change of protein, it would be reasonable to assume that the perturbation due to photo-excitation is an abrupt process, described by a *step-function*. The perturbation will cause abrupt change not only on direct interactions among atoms in protein including a chromophore, but those between the protein-chromophore complex and water. In order to take account for such perturbation associated with water, a strategy devised by Friedman and his coworkers in the field of so called "solvation dynamics" or the dynamic Stokes shift is adopted [Friedman et al. 1995]. In the case of solvation dynamics, such a situation was considered by so called "surrogate Hamiltonian," a renormalized Hamiltonian, which takes not only direct interactions between solute and solvent but also the many-body interactions among solvent molecules into consideration in a renormalized way. They have chosen the Ornstein-Zernike direct correlation function between solute and solvent as a renormalized potential [Friedman et al. 1995].

In the present study concerning structural dynamics of protein, the free energy surface of protein is chosen as a renormalized interaction. So, the renormalized interaction is defined by

$$F(\{\mathbf{R}\}) = U(\{\mathbf{R}\}) + \Delta\mu(\{\mathbf{R}\}) \tag{V-6.6}$$

where $\{\mathbf{R}\}$ represents the Cartesian coordinates of all atoms in protein, $U(\{\mathbf{R}\})$ and $\Delta\mu(\{\mathbf{R}\})$ denote the direct interaction energy among atoms in protein and the solvation free energy, respectively. All the interactions among water molecules as well as protein-water interactions are renormalized into $\Delta\mu(\{\mathbf{R}\})$. It should be noted that analytical expressions of $\Delta\mu(\{\mathbf{R}\})$ have been given as a function of $\{\mathbf{R}\}$ by means of the integral equation theory of molecular liquids, i.e., the 3D-RISM/KH theory (see Chapter IV for the solvation free energy of protein).

Description of perturbation:

The equilibrium distribution function for structure of a protein-chromophore complex in water may be defined by

$$p_{eq}^{(D)}(\Gamma) \propto \exp\left\{-\beta F_{eq}^{D}(\Gamma)\right\} \qquad \text{(V-6.7)}$$

where Γ is a phase variable, $F_{eq}^{D}(\Gamma)$ is the free energy of the system, and $\beta \equiv 1/k_BT$. The free energy is further decoupled into that of protein without a chromophore, or of an "apo-protein," and that associated with a chromophore such that,

$$F_{eq}^{D}(\Gamma) = F_{eq}^{0}(\Gamma) + \delta F_{eq}^{D}(\Gamma) \qquad \text{(V-6.8)}$$

where $F_{eq}^{0}(\Gamma)$ and $\delta F_{eq}^{D}(\Gamma)$ are the free energy of an "apo-state" of protein and the free energy change due to coupling of a chromophore. In consistence with the linear response theory, we express Eq. (7) by a linearlized form with respect to $\delta F_{eq}^{D}(\Gamma)$.

$$p_{eq}^{(D)}(\Gamma); p_{eq}^{(0)}(\Gamma)\left[1 - \beta\delta F_{eq}^{D}(\Gamma)\right] \qquad \text{(V-6.9)}$$

In all above equations, superscript "D" distinguishes if the system is in the state before or after photo-excitation. We use "P" (precursor) and "S" (successor) for the states before and after a photo-excitation, respectively [Friedman et al. 1995].

Now, we describe the perturbation ($H^{(1)}(t)$ in Eq. (V-6.1)) in terms of the renormalized interaction, taking the photo-induced structural relaxation as an example, which is regarded as an abrupt process. Suppose that the system is photo-excited at $t = 0$, then $H^{(1)}(t)$ can be expressed as follows,

$$H^{(1)}(t) = \theta(t)\left\{\delta F^{S}(\{\mathbf{R}\}) - \delta F^{P}(\{\mathbf{R}\})\right\} \qquad \text{(V-6.10)}$$

where $\delta F^{S}(\{\mathbf{R}\})$ and $\delta F^{P}(\{\mathbf{R}\})$ are the renormalized interaction between a chromophore and protein in equilibrium before and after the photo-excitation, respectively. (Here, the Cartesian coordinates $\{\mathbf{R}\}$ of atoms in protein are used instead of the phase variable Γ in order to express direct dependence of the free energy upon the spatial coordinates.) $\theta(t)$ is a Heaviside step function defined by,

$$\theta(t) = \begin{cases} 0 & t < 0 \\ 1 & t \geq 0 \end{cases} \qquad \text{(V-6.11)}$$

Then, the renormalized force acting on an atom of protein from other atoms in protein and water molecules can be written as

$$\mathbf{f}_{\beta}(t) = -\frac{\partial}{\partial \mathbf{R}_{\beta}} H^{(1)}(\{\mathbf{R}\};t) \qquad \text{(V-6.12)}$$

where $\mathbf{R}_{\beta}$ denotes the coordinate of an atom β of protein. Then, the perturbed Hamiltonian can be written as,

$$H^{(1)}(t) = -\sum_{\alpha} \Delta\mathbf{R}_{\alpha} \cdot \mathbf{f}_{\alpha}(t) \qquad \text{(V-6.13)}$$

where $\Delta\mathbf{R}_{\alpha}$ is the displacement of atom a, coupled to the renormalized force $\mathbf{f}_{\alpha}(t)$.

Linear response theory:

A structural response to such a perturbation defined by Eq. (V-6.13) can be described based on the linear response theory in a manner similar to the work proposed by H. Friedman in case of so called "dynamic Stokes shift" or "solvation dynamics," just by transferring the physics of solvation dynamics to structural dynamics of protein. Time evolution of the non-equilibrium distribution function $p(\Gamma, t)$ of the system with a Hamiltonian defined by Eq. (V-6.1) is described by the Liouville equation,

$$\left(\frac{\partial}{\partial t} + \hat{L}\right) p(\Gamma, t) = 0 \qquad \text{(V-6.14)}$$

where $\hat{L} \equiv H^{\times} = i\mathcal{L}$ is a Liouville operator corresponding to the Hamiltonian defined by Eq. (V-6.1). Following the spirit of the linear response theory, we divide the Liouville operator as well as the distribution function into two parts, perturbed and unperturbed parts,

$$p(\Gamma, t) = p^{(0)}(\Gamma, t) + p^{(1)}(\Gamma, t) \qquad \text{(V-6.15)}$$

$$\hat{L} = \hat{L}^{(0)} + \hat{L}^{(1)} \qquad \text{(V-6.16)}$$

where the suffices "0" and "1" indicate unperturbed and perturbed parts of the variables, respectively, and $\hat{L}^{(0)}$ and $\hat{L}^{(1)}$ are Liouville operators corresponding to the unperturbed and perturbed Hamiltonians defined by Eq. (V-6.2) and Eq. (V-6.13), respectively. Substituting Eqs. (V-6.15) and (V-6.16) into Eq. (V-6.14), and taking up to the first order of $\hat{L}^{(1)}$ and $p^{(1)}$, we get

$$\left(\frac{\partial}{\partial t} + \hat{L}^{(0)}\right) p^{(0)}(\Gamma, t) = 0 \qquad \text{(V-6.17)}$$

$$\left(\frac{\partial}{\partial t} + \hat{L}^{(0)}\right) p^{(1)}(\Gamma, t) = -\hat{L}^{(1)} p^{(0)}(\Gamma, t) \qquad \text{(V-6.18)}$$

Formal solutions of those equations can be obtained as,

$$p^{(0)}(\Gamma, t) = \exp\left(-t\hat{L}^{(0)}\right) p^{(0)}(\Gamma, 0) \qquad \text{(V-6.19)}$$

$$p^{(1)}(\Gamma, t) = \exp\left(-t\hat{L}^{(0)}\right) p^{(1)}(\Gamma, 0) - \int_0^t \exp\left\{(t' - t)\hat{L}^{(0)}\right\} \hat{L}^{(1)} p^{(0)}(\Gamma, t')\, dt' \qquad \text{(V-6.20)}$$

The initial condition separates into two parts, the unperturbed and perturbed parts, according to the order in the renormalized interactions as,

$$p^{(0)}(\Gamma, 0) = p_{eq}^{(0)}(\Gamma) \qquad \text{(V-6.21)}$$

$$p^{(1)}(\Gamma, 0) = -\beta p_{eq}^{(0)}(\Gamma)\, \delta F^{P}(\{\mathbf{R}\}) \qquad \text{(V-6.22)}$$

Substituting Eq. (V-6.21) into Eq. (V-6.19) leads to

$$p^{(0)}(\Gamma, t) = \exp\left(-t\hat{L}^{(0)}\right) p_{eq}{}^{(0)}(\Gamma) = p_{eq}{}^{(0)}(\Gamma) \qquad \text{(V-6.23)}$$

Taking those relations into consideration, Eq. (V-6.20) can be written as

$$p^{(1)}(\Gamma,t) = -\beta \exp\left(-t\hat{L}^{(0)}\right) p_{eq}{}^{(0)}(\Gamma) H^{(1)}(0) - \int_0^t \exp\left\{(t'-t)\hat{L}^{(0)}\right\}\hat{L}^{(1)} p_{eq}{}^{(0)}(\Gamma) dt' \quad \text{(V-6.24)}$$

Considering the obvious relation, $\hat{L}^0 p_{eq}^{(0)}(\Gamma) = 0$, the first term on the right hand side of Eq. (V-6.24) becomes

$$-\beta \exp\left(-t\hat{L}^{(0)}\right) p_{eq}{}^{(0)}(\Gamma) H^{(1)}(0) = -\beta p_{eq}{}^{(0)}(\Gamma) H^{(1)}(0) \quad \text{(V-6.25)}$$

The integral in the second term on the right hand side of Eq. (V-6.24) can be carried out readily to give rise to

$$\int_0^t \exp\left((t'-t)\hat{L}^{(0)}\right)\hat{L}^{(1)} p^{(0)}(\Gamma,t')dt' = \beta p_{eq}^{(0)}(\Gamma)\left(H^{(1)} - H^{(1)}(-t)\right) \quad \text{(V-6.26)}$$

where the following relation was applied, which can be readily verified.

$$\hat{L}^{(1)} p_{eq}^{(0)}(\Gamma) = \beta p_{eq}^{(0)}(\Gamma)\hat{L}^{(0)} H^{(1)}(0) \quad \text{(V-6.27)}$$

From Eqs. (V-6.24), (V-6.25), and (V-6.26), the perturbed part of the distribution function is obtained as

$$p^{(1)}(\Gamma,t) = -\beta p_{eq}^{(0)}(\Gamma)\delta F^S(\{\mathbf{R}\}) + \beta p_{eq}{}^{(0)}(\Gamma)\Delta F(-t) \quad \text{(V-6.28)}$$

where $\Delta F(-t)$ is defined with the time displacement operator $\exp(-\hat{L}^{(0)}t)$ by,

$$\Delta F(-t) = \exp\left(-\hat{L}^{(0)}t\right)\Delta F(\{\mathbf{R}\}) \quad \text{(V-6.29)}$$

$$\Delta F(\{\mathbf{R}\}) \equiv \left[\delta F^S(\{\mathbf{R}\}) - \delta F^P(\{\mathbf{R}\})\right] \quad \text{(V-6.30)}$$

From Eqs. (V-6.15), (V-6.21), and (V-6.28), we get the time evolution of the non-equilibrium probability distribution function as

$$p(\Gamma,t) = p_{eq}^{(S)}(\Gamma) + \beta p_{eq}{}^{(0)}(\Gamma)\Delta F(-t) \quad \text{(V-6.31)}$$

where $p_{eq}^{(S)}(\Gamma)$ is defined by Eq. (V-6.9) with D = S, or

$$p_{eq}^{(S)}(\Gamma) \equiv p_{eq}^{(0)}(\Gamma)\left[1 - \beta\delta F^S(\{\mathbf{R}\})\right]$$

Structural relaxation of protein induced by a perturbation:

Now, let us choose the displacement of atom positions of protein from their equilibrium state as a dynamic variable to take an ensemble average with the non-equilibrium probability distribution in phase space.

$$\begin{aligned} \langle \Delta\mathbf{R}_\alpha(t)\rangle &= \int \exp\left(t\hat{L}^{(0)}\right)\Delta\mathbf{R}_\alpha(\Gamma)p(\Gamma;t)d\Gamma \\ &= \langle \Delta\mathbf{R}_\alpha\rangle_{eq}^{(S)} + \beta\int p_{eq}^{(0)}(\Gamma)\Delta\mathbf{R}_\alpha(\Gamma;t)\Delta F(-t)d\Gamma \end{aligned} \quad \text{(V-6.32)}$$

Substituting Eq. (V-6.13) in $\Delta F(-t)$, the second term becomes,

$$\begin{aligned}\beta\int p_{eq}^{(0)}(\Gamma)\Delta\mathbf{R}_{\alpha}(\Gamma;t)\Delta F(-t)d\Gamma &= \beta\int p_{eq}^{(0)}(\Gamma)\Delta\mathbf{R}_{\alpha}(\Gamma;t)\left(-\sum_{\beta}\Delta\mathbf{R}_{\beta}\cdot\mathbf{f}_{\beta}(0)\right)d\Gamma \\ &= -\beta\sum_{\beta}\left\langle\Delta\mathbf{R}_{\alpha}(t)\Delta\mathbf{R}_{\beta}\right\rangle_{eq}^{(0)}\cdot\mathbf{f}_{\beta}(0)\end{aligned} \tag{V-6.33}$$

where $\langle\cdots\rangle_{eq}^{0}$ represents an ensemble average over the equilibrium reference system. Finally, we get a linear-response expression for the structural dynamics of protein induced by a perturbation as

$$\left\langle\Delta\mathbf{R}_{\alpha}(t)\right\rangle = \left\langle\Delta\mathbf{R}_{\alpha}\right\rangle_{eq}^{(S)} - \beta\sum_{\beta}\left\langle\Delta\mathbf{R}_{\alpha}(t)\Delta\mathbf{R}_{\beta}\right\rangle_{eq}^{(0)}\cdot\mathbf{f}_{\beta}(0) \tag{V-6.34}$$

At $t = \infty$, it is easy to see that the equation satisfies the condition,

$$\left\langle\Delta\mathbf{R}_{\alpha}(\infty)\right\rangle = \left\langle\Delta\mathbf{R}_{\alpha}\right\rangle_{eq}^{S} \tag{V-6.35}$$

At $t = 0$, the Eq. (V-6.34) reduces to the linear response expression for the static perturbation, derived by Ikeguchi et al. [Ikeguchi et al. 2005]

$$\left\langle\Delta\mathbf{R}_{\alpha}(0)\right\rangle = \left\langle\Delta\mathbf{R}_{\alpha}\right\rangle_{eq}^{(S)} - \beta\sum_{\beta}\left\langle\Delta\mathbf{R}_{\alpha}(0)\Delta\mathbf{R}_{\beta}\right\rangle_{eq}^{(0)}\cdot\mathbf{f}_{\beta}(0) \tag{V-6.36}$$

In the above expression, $\langle\Delta\mathbf{R}_{\alpha}(0)\rangle$ and $\langle\Delta\mathbf{R}_{\alpha}\rangle_{eq}^{(S)}$, should be interpreted as the structural change due to a perturbation before and after the *static* perturbation $\mathbf{f}_{\beta}(0)$ being applied. Such a consideration sets up the initial condition as

$$\left\langle\Delta\mathbf{R}_{\alpha}(0)\right\rangle = \left\langle\Delta\mathbf{R}_{\alpha}\right\rangle_{eq}^{P} \tag{V-6.37}$$

where $\langle\Delta\mathbf{R}_{\alpha}\rangle_{eq}^{P}$ is structural change induced in the protein due to binding of a chromophore (or $\delta F^{P}(\{\mathbf{R}\})$ in Eq. (V-6.8)) before photo-excitation. Equation (V-6.36) also provides a method to calculate $\langle\Delta\mathbf{R}_{\alpha}\rangle_{eq}^{(S)}$.

So far, we have described everything in terms of fluctuation, or displacement of atoms from their equilibrium position (Eq. (V-6.34)). However, it will be easier to understand the physics by expressing equations in terms of atom position (or structure) by itself. For that purpose, we define,

$$\mathbf{R}_{\alpha}(t) = \mathbf{R}_{\alpha,eq}^{(0)} + \left\langle\Delta\mathbf{R}_{\alpha}(t)\right\rangle \tag{V-6.38}$$

where $\mathbf{R}_{\alpha,eq}^{(0)}$ is the position of atom α of an unperturbed protein in equilibrium, and $\mathbf{R}_{\alpha}(t)$ is the position at time t after the perturbation is applied. Substituting Eq. (V-6.34) into (V-6.38), one gets

$$\mathbf{R}_{\alpha}(t) = \mathbf{R}_{\alpha}(t=\infty) - \beta\sum_{\beta}\left\langle\Delta\mathbf{R}_{\alpha}(t)\Delta\mathbf{R}_{\beta}\right\rangle_{eq}^{(0)}\cdot\mathbf{f}_{\beta}(0) \tag{V-6.39}$$

where $\mathbf{R}_{\alpha}(t = \infty)$ is defined by

$$\mathbf{R}_{\alpha}(t=\infty) \equiv \mathbf{R}_{\alpha,eq}^{(0)} + \left\langle\Delta\mathbf{R}_{\alpha}\right\rangle_{eq}^{(S)} \tag{V-6.40}$$

Equation (V-6.39) describes the time evolution of the structural relaxation of protein to a final conformation after a perturbation being applied, starting from an initial structure $\mathbf{R}^{(0)}_{\alpha,eq}$. Here, it is worthwhile to make a comment on Eq. (V-6.39). The equation looks as if it is a deterministic process. However, it is not true, because all the variables included in the equation are an average over the ensemble. The position of an atom should be interpreted as the one averaged over fluctuating structures.

The new equation can be applied to a structural dynamics of protein induced by any time-dependent perturbation, provided that the general force $\mathbf{f}_\gamma(t)$ is modeled appropriately [Hirata and Akasaka 2015]. However, an actual implementation of the theory requires information of the time evolution of the structural fluctuation for unperturbed protein, or the variance-covariance matrix $\langle \Delta\mathbf{R}_\alpha(t)\Delta\mathbf{R}_\gamma(0)\rangle$. Apparently, it is not a trivial task for a molecular dynamics (MD) simulation, because the calculation of the variance-covariance matrix itself requires a time average over a trajectory in place of an ensemble average in the configuration space. The procedure loses the temporal information of the variance-covariance matrix. Only a legitimate way to realize the time dependence is to perform an ensemble average over many MD trajectories, starting from a different initial configuration in the phase space. In the following section, a possible way to evaluate time evolution of the variance-covariance matrix is presented based on the generalized Langevin theory and the 3D-RISM/KH theory.

V-6.2 Time Evolution of the Variance-covariance Matrix

In order to derive an equation to describe time evolution of the variance-covariance matrix, we start from the generalized Langevin equation for the structural dynamics of protein we have derived previously for the same Hamiltonian with H_0 described in Eqs. (V-6.2)–(V-6.5) with a slight change in the notation (see Section V-2 in this chapter).

$$\frac{d\Delta\mathbf{R}_\alpha(t)}{dt} = \Delta\mathbf{V}_\alpha(t) \tag{V-6.41}$$

$$\frac{d\mathbf{V}_\alpha(t)}{dt} = -\frac{k_BT}{M_\alpha}\sum_\beta\left(\mathbf{L}^{-1}\right)_{\alpha\beta}\cdot\Delta\mathbf{R}_\beta(t) - \int_0^t ds\sum_\beta\frac{1}{M_\alpha}\Gamma_{\alpha\beta}(t-s)\cdot\Delta\mathbf{V}_\beta(t) + \mathbf{W}_\alpha(t) \tag{V-6.42}$$

where $\mathbf{L}$ is the variance-covariance matrix of structural fluctuation in equilibrium, defined by

$$L_{\alpha\beta} = \left\langle \Delta\mathbf{R}\Delta\mathbf{R}\right\rangle_{\alpha\beta}, \tag{V-6.43}$$

$\Gamma_{\alpha\beta}(t)$ is the friction kernel, and $\mathbf{W}_\alpha(t)$ is the random force. The variance-covariance matrix is related to the second derivative of the free energy surface $F(\{\mathbf{R}\})$ of protein with respect to the positional deviation as

$$k_BT\left(\left\langle \Delta\mathbf{R}\Delta\mathbf{R}\right\rangle^{-1}\right)_{\alpha\beta} = \frac{\partial^2 F\left(\{\mathbf{R}\}\right)}{\partial\mathbf{R}_\alpha\partial\mathbf{R}_\beta} \tag{V-6.44}$$

Multiplying Eqs. (V-6.41) and (V-6.42) by $\Delta\mathbf{R}_\gamma(0)$, and taking the ensemble average, one gets

$$\frac{d\left\langle \Delta\mathbf{R}_\alpha(t)\Delta\mathbf{R}_\gamma(0)\right\rangle}{dt} = \left\langle \Delta\mathbf{V}_\alpha(t)\Delta\mathbf{R}_\gamma(0)\right\rangle \tag{V-6.45}$$

$$\frac{d\left\langle \mathbf{V}_\alpha(t)\Delta\mathbf{R}_\gamma(0)\right\rangle}{dt} = -\frac{k_BT}{M_\alpha}\sum_\beta\left(\mathbf{L}^{-1}\right)_{\alpha\beta}\cdot\left\langle \Delta\mathbf{R}_\beta(t)\Delta\mathbf{R}_\gamma(0)\right\rangle - \int_0^t ds\sum_\beta\frac{1}{M_\alpha}\Gamma_{\alpha\beta}(t-s)\cdot\left\langle \Delta\mathbf{V}_\beta(t)\Delta\mathbf{R}_\gamma(0)\right\rangle$$
$$+\left\langle \mathbf{W}_\alpha(t)\Delta\mathbf{R}_\gamma(0)\right\rangle \tag{V-6.46}$$

The last term in the second equation vanishes by definition, because $\Delta\mathbf{R}_r(0)$ is one of the dynamic variables, which is orthogonal to the random force $\mathbf{W}_\alpha(t)$, that is,

$$\left\langle \mathbf{W}_\alpha(t)\Delta\mathbf{R}_\gamma(0)\right\rangle = 0 \tag{V-6.47}$$

We further assume that the friction coefficient is local in time, or,

$$\Gamma_{\alpha\beta}(t-s) = \Gamma_{\alpha\beta}\delta(t-s) \tag{V-6.48}$$

Taking all those into consideration, the second equation becomes

$$\frac{d\left\langle \mathbf{V}_\alpha(t)\Delta\mathbf{R}_\gamma(0)\right\rangle}{dt} = -\sum_\beta K_{\alpha\beta}\cdot\left\langle \Delta\mathbf{R}_\beta(t)\Delta\mathbf{R}_\gamma(0)\right\rangle - \sum_\beta\varsigma_{\alpha\beta}\cdot\left\langle \Delta\mathbf{V}_\beta(t)\Delta\mathbf{R}_\gamma(0)\right\rangle \tag{V-6.49}$$

where K_{ab} and $\varsigma_{\alpha\beta}$ are defined as follows,

$$K_{\alpha\beta} \equiv \frac{k_BT}{M_\alpha}\left(\mathbf{L}^{-1}\right)_{\alpha\beta}, \quad \varsigma_{\alpha\beta} \equiv \frac{\Gamma_{\alpha\beta}}{M_\alpha} \tag{V-6.50}$$

Equations (V-6.45) and (V-6.49) are written in a matrix form as

$$\frac{d}{dt}\begin{pmatrix}\mathbf{C}(t)\\ \dot{\mathbf{C}}(t)\end{pmatrix} = \begin{pmatrix}\mathbf{0} & \mathbf{1}\\ -\mathbf{K} & -\varsigma\end{pmatrix}\begin{pmatrix}\mathbf{C}(t)\\ \dot{\mathbf{C}}(t)\end{pmatrix} \tag{V-6.51}$$

where $\mathbf{C}(t)$ and $\dot{\mathbf{C}}(t)$ are defined by

$$\mathbf{C}_{\alpha\gamma}(t) \equiv \left\langle \Delta\mathbf{R}_\alpha(t)\Delta\mathbf{R}_\gamma(0)\right\rangle, \quad \dot{\mathbf{C}}_{\alpha\gamma}(t) \equiv \left\langle \Delta\mathbf{V}_\alpha(t)\Delta\mathbf{R}_\gamma(0)\right\rangle \tag{V-6.52}$$

The formal solution of Eq. (V-6.51) is

$$\mathbf{C}(t) = \exp\left(\mathbf{A}t\right)\mathbf{C}(\infty) \tag{V-6.53}$$

where $\mathbf{C}(\infty) = \mathbf{C}(0)$ in the equilibrium, and the matrix $\mathbf{A}$ is defined by

$$\mathbf{A} = \begin{pmatrix}\mathbf{0} & \mathbf{1}\\ -\mathbf{K} & -\varsigma\end{pmatrix} \tag{V-6.54}$$

The problem is thus reduced to an eigen value problem of the matrix **A**. The equation is formally identical to that solved by several people in the field of the statistical mechanics [Wang and Uhlenbeck 1945, Lamm and Szabo 1986]. The only difference is the first term on the right-hand-side which defines the restoring force proportional to the displacement of atoms. In the earlier works, the force constant or the Hessian has been defined by the second derivative of the direct interaction potential of atoms with respect to the position of atoms [Wang and Uhlenbeck 1945, Lamm and Szabo 1986], while it is defined in Eq. (V-6.44) by corresponding second derivative of the *free energy* which includes not only the direct interaction but the *solvation free energy*. Nevertheless, those equations are formally identical, thereby all the procedure to solve the equation can be imported from the former works to the present problem. In particular, the seminal paper concerning the Langevin Mode analysis, written by Lamm and Szabo, is quite instructive, which provides a detailed procedure to solve the eigen value problem [Lamm and Szabo 1986].

The Eq. (V-6.54) includes two physical parameters, $\mathbf{K}$ and ς, which should be provided in advance. $\mathbf{K}$ is related to the second derivative of the free energy surface of protein with respect to its atomic coordinates (Eq. (V-6.44)), and it can be calculated based on the 3D-RISM-KH theory. On the other hand, it is a highly non-trivial problem to calculate ς for each atom in protein, because solvent environment of an atom in protein is extremely inhomogeneous. For example, there are few or no solvent molecules around an atom located at deep inside protein, while an atom located at surface is well exposed to solvent. Therefore, in order to get a friction coefficient of each atom in protein, the dynamics of solvent under such an inhomogeneous field of protein should be solved. An equation governing such solvent dynamics has been derived in Section V-2 in this book as Eqs. (V-2.59) and (V-2.60) [Kim and Hirata 2013]. However, there is still a distance to get final solution. Considering such a situation, a tentative solution for the problem is proposed in the following section.

Friction coefficient

The source of friction that is proportional to the velocity of a particle is, by definition, "dissipation" or/and "loss" of momentum and energy of the particle due to collisions and other interactions with particles in environment. In the earlier treatments based on the conventional Langevin theory, a phenomenological expression represented by the Oseen tensor have been employed for the friction matrix [Lamb and Szabo 1986, Ermak and McCammon 1978]. The diagonal elements of the Oseen tensor are the Stokes friction acting on each atom of protein due to solvent, and the off-diagonal elements, referred to as the *hydrodynamic interaction*, describe the force acting on an atom from the other atoms of protein, propagated through solvent. Here, it is worthwhile to emphasize that the both elements of the friction matrix, by definition, originate from the fluctuation of solvent in the *momentum* space, not in the *positional* space or the density fluctuation. Suppose that a protein molecule is isolated in vacuum. An atom in the molecule is interacting with the other atoms in the molecule. However, those forces are not considered as "friction," because the momentum and energy of the atom are not *dissipated*, but just *transferred* to the other atoms or vibrational modes. Such energy and momentum transfer associated with the intra-molecular interaction are completely taken care of by the direct interaction terms in our equation. The friction should be exerted on each individual atom of protein, and it should be different from atom to atom depending on the position where the atom is located in the protein. We emphasize again that the friction on each atom of protein originates entirely from solvent molecules.

We also emphasize that the hydrodynamic interaction should be distinguished from that originated from the potential of mean force, although both are solvent-mediated forces. Actually, in the theory presented here, the two contributions are clearly separated. In Eq. (V-6.42), the first term is related to the forces originated from the potential of mean force, while the hydrodynamic interactions are included in the friction-kernel matrix as the off-diagonal elements. Taking the separation of solvent-mediated forces into consideration, we assume that the hydrodynamic interaction between atoms in protein can be negligibly small. It is because the momentum fluctuation of water molecules around an atom, caused by the motion of the atom, is considered to be "thermalized" or to dissipate quickly before the force is propagated to the other atoms in protein. Based on this assumption, we focus our attention on the diagonal terms of the friction matrix.

Based on the arguments above, an approximation can be proposed for the friction $\zeta_{\alpha\alpha}$ exerted on each atom of protein as follows [Hirata and Kim 2016],

$$\zeta_{\alpha\alpha} = f_\alpha \zeta_{\alpha\alpha,bulk} \tag{V-6.55}$$

in which $\zeta_{\alpha\alpha,bulk}$ is the friction exerted on atom a in bulk solvent, and f_α is the fraction of the atom contacting with solvent. We define f_α by the following equation based on the radial distribution function (RDF) of solvent around the atom,

$$f_\alpha = \frac{g_w(\sigma; protein)}{g_w(\sigma; bulk)} \tag{V-6.56}$$

where $g_w(\sigma;bulk)$ is the RDF of water in bulk at the contact separation between the atom and a solvent molecule, and $g_w(\sigma;protein)$ is that corresponding to the atom in protein. If the atom is at surface of protein, f_α is close to unity, because the atom is well exposed to solvent, and $g_w(\sigma;protein)$ will become close to $g_w(\sigma;bulk)$. On the other hand, $g_w(\sigma;protein)$ will become small or zero if the atom is buried inside the protein, since there are no or few water molecules around such an atom. Both $g_w(\sigma;bulk)$ and $g_w(\sigma;protein)$ can be readily calculated based on the 3D-RISM/KH theory by making an appropriate definition for the contact separation s between the protein atom and solvent. A typical choice of σ can be the position of the first peak in RDF in the bulk solvent.

It is a non-trivial problem to determine $\zeta_{\alpha\alpha,bulk}$ by experimental means, because an atom in protein has a partial charge in general, which is not the case if an atom is isolated by itself in solution: a charged atom in solution exists just in the form of an "ion" which of course has full charges, such as monovalent and divalent ions. Such an atom with a partial charge in solution is just an "imaginary" atom. Then, how can one estimate the friction of such an imaginary atom in solution? In the following, we propose a recipe to determine the friction of an imaginary atom of solution based on the site-site mode coupling theory for the dynamics of ions developed by Chong and Hirata [Chong and Hirata 1998c, 1999c] (see Chapter III for the Chong-Hirata treatment of ion dynamics). Alternatively, the quantity can be estimated by the standard MD simulation for an ion with hypothetical *partial-charges* in solution [Koneshan et al. 1988]. Based on the model, a protocol to determine the friction on an atom of protein in water is proposed as follows [Hirata and Kim 2016].

1. Assume that $\zeta_{\alpha\beta}(\alpha \neq \beta) = 0$.
2. Make a table of the friction $\zeta_{\alpha\alpha,bulk}$ exerted on the hypothetical atoms in bulk water using the Chong-Hirata theory (or MD) for ion dynamics described above. In the course of calculation, the radial distribution function $g_w(r;bulk)$ of water around each hypothetical atom is also produced.
3. Perform the 3D-RISM/KH calculation for the each step of protein dynamics to determine the 3D-density distribution $g_w(\mathbf{r};protein)$ of water around and inside protein, from which the radial distribution function $g_w(r;protein)$ is readily obtained.
4. Combining those steps above, one can calculate the friction $\zeta_{\alpha\alpha}$ of atoms in protein due to Eqs. (V.6-55) and (V.6-56).

In this section, we have looked at a theoretical approach to trace the structural change triggered by some perturbation applied to a protein solution, based on the linear response theory. Application of the linear response theory to the structural fluctuation and relaxation of protein may require some explanation. The idea is entirely based on an ordinary observation of the way how protein, for example an enzyme, performs its function. When an enzyme is working in our body, it should change its structure due to a perturbation originated from a reaction of substrate molecules. If not, the reaction will not take place, because the structural restriction will produce a high activation barrier for the reaction. The reaction is completed by releasing product molecules from the active-site, and the enzyme restores its original structure completely. That process makes an enzyme ready for a next cycle. It is the reason why enzyme is called "biocatalyst." In general, a protein in action, deformed by a perturbation, recovers its original structure completely after the perturbation is removed. [Akasaka 2006] Such a behavior of a protein working in our body matches with a general definition of "linearity" of a structural response of a material to a perturbation. Some readers may further ask a question "which law of physics the *linearity* originates from?" The answer to the question is the *central limiting theorem* [Chandrasekhar 1943, Kubo et al. 1992]. The theorem states that the probability distribution of a randomly modulating variable becomes a normal distribution, or a Gaussian, if there is no extraordinary large modulations. The theorem can be proved rigorously for a process consisting of many random variables, none of which is extraordinarily large (see Chapter I for the proof of the central limiting theorem). Since every fluctuation of protein structure involves many variables which are randomly changing, including solvent degrees of freedom, the distribution of fluctuation becomes Gaussian, or the normal distribution, characterized with the first moment (average structure) and the second moment (variance-covariance matrix). The "Gaussian distribution" is synonymous to the linearity of the fluctuation as is well regarded in physics. [Hirata 2015] It is also well regarded that the Gaussian distribution is one of the most prevailing laws in nature. It is not surprising that the robustness of a living system is sustained by the most prevailing law of nature.

Although we have discussed so far about the linearity of the structural fluctuation of protein, applicability of the theory is not necessarily limited to a linear process. The theory can be readily extended to a non-linear regime by means of *analytical continuation*. The mathematical theorem can be realized by dividing the course of relaxation into several time steps in such a way that each step stays in the convergence radius. In the actual calculation, we have to restart with a revised initial-condition for the structure in each time step. We suppose that the convergence radius or the time step can be reasonably large, because there is no singularity expected in the time course of structural relaxation.

The Literature

The book edited by Wax (1954) includes six review articles concerning phenomenological theories of Brownian motion. One of the paper written by Wang and Uhlenbeck (1945) provides a complete mathematical description of coupled harmonic oscillators in viscous media. The translational dynamics of an entire polymer in a confined space is described in terms of the so called "reptation model" in the book written by Doi and Edwards (1986).

CHAPTER VI

Applications of the Theories to *In-silico* Drug Discovery

Introduction: Possibility of Medical Application

Application of the theory and method developed in the previous chapters is not limited just to a basic research in the field of life science. It has a great perspective to find applications in the field of medicine or drug discovery. In this chapter, we briefly outline how the methodologies developed in the previous chapters can be applied to the problems in the pharmaceutical design. In general, drugs are chemical compounds that promote or inhibit a function of biomolecules such as protein and DNA. Typically, a drug compound promotes or inhibits a function of protein by binding the active site of the biomolecule. For example, a drug called "Tamifulu" binds at the active site of neuraminidase in influenza virus to inhibit the enzymatic reaction of the protein, which is vital for the micro-organism to survive.

There are many steps in the pharmaceutical design: (1) searching a target protein molecule on which a drug attacks, (2) determining the structure of the target protein, (3) screening the candidate compounds of the drug, (4) designing transportability of drug molecules, (5) synthesizing the drug compound, (6) clinical test of the drug, including the analysis of side-effects, and so on.

Step (1) is almost equivalent to investigating functions of protein, and a comprehensive survey, empirical as well as computational, is required, since it is the role of drug to promote or suppress a function of protein. For example, if a protein is specific to some bacteria or virus, and failure of the protein is fatal for the micro-organism, the protein is a good candidate of the drug target. On the other hand, if a protein is generic, and is functioning both in micro-organisms and in human beings, the protein may not be a good target, because the drug will malfunction not only the protein in a micro-organism but also that in human body. If a protein in a micro-organism is a far remote homologue from that of human in terms of the evolution tree, the protein may be a good target, because the drug won't harm the homologue in human.

Step (2) has been carried out currently by means of the two empirical methods, the X-ray crystallography and the NMR spectroscopy, and a huge amount of structural data in atomic resolution is stored in the data base such as Protein Data Bank (PDB). There also are the libraries storing structural information of chemical compounds that become candidates of a drug. However, the data will never be complete due to several reasons. A problem shared by all the empirical methods is the position of water molecules and small ions which are residing inside protein working as a substrate or a co-catalyst (see Chapter IV). Those species are usually diffusive in space, thereby it is quite difficult to resolve their position by any empirical method. Nevertheless, those small molecules bound inside the active site of protein play crucial roles in the process of drug binding, since those molecules should be removed, or "desolvated," from the reaction pocket upon the drug binding. The 3D-RISM theory can play a complementary role in the problem, since the position of water molecules as well as ions is determined rather easily by the method.

Step (3) is the step in which the theories developed in this book may give potentially an important contribution to discovery of a new drug. It is the process to find a chemical compound that has high affinity to the acitive site of a target protein, so that the activity or function of the protein may be inhibited. There are several requirements for a compound to be a successful drug against microorganisms. The affinity to the active site should be sufficiently high, so that the native activity of a protein in the microorganism is effectively suppressed. The drug compound should be effective with minimum dose, otherwise it may be toxic or even fatal to patient. A successful drug should be free from side effects. There are potentially millions of compounds that are candidates of a drug. It is so called "drug screening" that narrows a drug compound down from the huge number of candidate compounds that meet the requirements stated above. Actually, most of the candidate compounds can be thrown away using primitive computational-tools, so-called "docking," that are based on the concept of *key and lock*. So, the real challenge is to narrow candidate-compounds down further to few compounds which satisfy the requirements for a drug. The process in which a drug molecule is bound at the active site of a biomolecule is nothing but the *molecular recognition,* described in Chapter IV in this book. So, it is quite reasonable to expect that the 3D-RISM/RISM theory may have a great perspective in this step of drug discovery. Some of the recent work concerned with the drug screening, which were carried out by the author and his collaborators in the past few years, are introduced in the following sections.

Step (4) is concerned with transportation of drug molecules in our body. A drug molecule dosed in our body via mouth or by injection should be transported to a sick organ through the vascular system. There are several problems associated with this issue. The first is the solubility of a drug compound to aqueous solution. Many drug compounds are some organic molecules that in general do not have great solubility to aqueous solutions. So, some modification should be devised to make a drug compound to be soluble in aqueous solutions. One of such devices is the inclusion of drug compounds by some amphiphlic molecules such as cyclodextrin, protein, micelles, and so on. There is a concern related to the inclusion process. A drug compound included by amphiphilic molecules should be released from the carrier at the organ where the drug is needed, otherwise it won't work as is expected: for example, a drug for stomach disease should work under the strong acidic condition, pH $\sim$ 2. So, the affinity or free energy of inclusion should be controlled in such a way that the drug molecule has high affinity to the carrier under the normal physiological solution, while it should have lower affinity under the solution condition in the sick organ. The process is again the *molecular recognition* that requires ability to discriminate precisely the solution condition such as pH, kind and concentration of salts dissolved in water, and so on. The theory developed in the previous chapters can be applied to the problem.

Along the course of the transportation, the drug should pass through several membrane-barriers consisting of lipid molecules. If not, the drug molecule may not reach the cell where it is needed, and it will be disposed from the body without working as a drug. So, the permeability of a drug through membrane is one of a serious concern in the drug discovery. An important problem highlighted by the community is the transportation of a drug to the *brain*, because the brain is specially prevented from an attack by foreign substances due to the blood-brain barrier. The problem is intimately related to the solubility of a drug, or a complex of drug and a carrier, to the two environments, aqueous solution and membrane.

Step (5) has currently been carried out experimentally by means of organic synthesis. So called "combinatorial chemistry" has been providing a new possibility of high through-put synthesis. Nevertheless, wet chemistry is still laborious as well as expensive. It will be highly desirable if one can predict the possibility of synthesizing the candidate compound by means of computational chemistry. It of course requires the quantum chemistry, because any chemical reaction involves the change of electronic structure. However, it is not the only requirement. The quantum calculation has to be carried out in *solvent*, because most of the experiments of organic synthesis have been carried out in solvent. The RISM-SCF method described in Chapter II-5 in this book meets such a demand.

Step (6) is far beyond the scope of this book, since it involves animals and human beings. However, the *in-silico* drug design should use the information fed back from the clinical experiment.

The following sections are devoted to present selected topics concerning the application of 3D-RISM/KH theory to the problems related to Step (3), or the screening of drug candidates, which has been carried out by the author and his collaborators. The screening is classified into two different categories, depending on the structural information of the complex of protein and drug, or ligand, available in advance. If the structural information, including pose of the ligand, is known more or less for a complex, and if the purpose of the screening is to find a compound which has the best affinity to the target protein, the main task of the screening is to calculate the binding free energy, or the change of the free energy upon the complex formation, for a variety of combinations of protein or ligand. The following two Sections (VI-2, VI-3) are devoted to examples of application of the 3D-RISM/RISM method to the calculation of the binding free energy. On the other hand, if there is no structural information available for the protein-ligand complex, the task of screening is to find the structure of protein-ligand complex including pose of the ligand at the binding site. The problem is quite difficult by any means, because the structure of both protein and ligand is fluctuating around their equilibrium structure. The last two Sections, VI-4 and VI-5, in this chapter are concerned with such attempts to find the binding mode of drug compound in receptor.

VI-1 The Binding Free Energy of Drug Compounds to a Protein

The binding affinity of a drug compound to a protein can be defined in term of the free energy by,

$$K_A = \exp\left[-\beta \Delta G_{bind}\right] \tag{VI-1.1}$$

where K_A is the association constant that can be measured by experiments, and ΔG_{bind} denotes the binding free energy, or the change of the chemical potential upon binding of the two species,

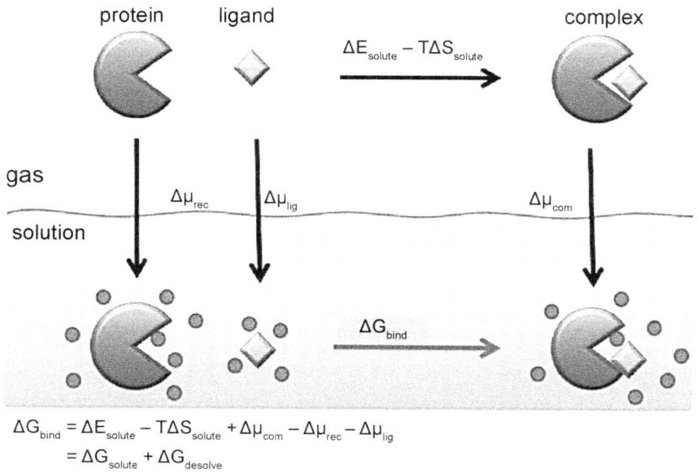

Fig. VI-1.1: Thermodynamic cycle to determine the binding free energy.

protein and ligand. The binding free energy can be calculated from the 3D-RISM/KH theory, based on the thermodynamic cycle illustrated in Fig. VI-1.1. The thermodynamic cycle gives the following equation.

$$\Delta G_{bind} = \Delta G_{solute} + \Delta G_{solv} \tag{VI-1.2}$$

In the equation, ΔG_{solute} denotes the contributions from intramolecular degrees of freedom of solute itself to the binding free energy, and ΔG_{solv} is that from the solvation free energy. The ΔG_{solute} is defined as,

$$\Delta G_{solute} = \Delta E_{solute} - T\Delta S_{solute} \tag{VI-1.3}$$

where ΔE_{solute} and ΔS_{solute} are the changes of the interaction energy among atoms in the solutes upon binding and of the conformational entropy of the solute defined, respectively, by

$$\Delta E_{solute} \equiv E_{complex} - \left(E_{recepter} + E_{ligand}\right) \tag{VI-1.4}$$

$$\Delta S_{solute} \equiv S_{complex} - \left(S_{recepter} + S_{ligand}\right) \tag{VI-1.5}$$

where the subscripts, *complex*, *receptor*, and *ligand*, specify respective species in the solution. Those quantities can be calculated by means of the molecular mechanics (MM) with a special care to the structural fluctuation of protein.

The ΔG_{solv} is the change in the solvation free energies of the solute species upon the complexation, and is defined by

$$\Delta G_{solv} = \Delta\mu_{complex} - \left(\Delta\mu_{recepter} + \Delta\mu_{lignad}\right) \tag{VI-1.6}$$

The quantities, $\Delta\mu_{complex}$, $\Delta\mu_{protein}$, and $\Delta\mu_{ligand}$ denote the solvation free energies of the respective species indicated by the subscript. Those quantities can be calculated by means of the 3D-RISM/RISM method, again, with special care about the structural fluctuation.

The structural fluctuation of protein, mentioned above, can be evaluated in principle by means of the generalized Langevin theory and the linear response theory introduced in Chapter V of this book. Unfortunately, however, the numerical recipe of the equation is largely under construction. Therefore, we employ the molecular dynamics simulation tentatively to evaluate the structural fluctuation of protein.

VI-2 Binding Efficiency of the Oseltamivir (Tamifulu) to the Wild-type and Mutant forms of the Viral Influenza B Neuraminidase

Influenza has been a threat to our health for a long time. There are three types of influenza viruses currently known, influenza A/H1N1, A/H3N2, and influenza B [Barr et al. 2014]. The two type-A-viruses have caused pandemic several times in the past, and the preventative measures for the disease, including vaccination, have been relatively well developed. On the other hand, the investigation for the influenza B virus is considerably less developed compared to the A-viruses, may be because it does not have the pandemic potential. Nevertheless, it still is a threat to human health, and the possibility of the virus becoming more dangerous may not be denied.

Although all three types of virus share many features and viral activities, influenza B virus harbors unique genetics and can only infect humans [Fields et al. 2007, Taylor and Russel 2010]. Its surface membrane has two glycoproteins, hemaglutinin and neuraminidase, which recognize the terminal sialic acid on host cell membrane components. Neuraminidase plays a role for virus replication by cleaving the sialic acid residues and then releasing the new virions from the infected cell in order to infect new cells. Currently, neuraminidase inhibitors have been employed as anti-influenza agents, either influenza A or B strains [McKimm-Breschkin 2002, Gubareva et al. 2000], in which the structure of their active-site and framework site are almost identical and conserved (Table S1). Hence, we can assume that the effective drug against influenza B virus could be applicable also against influenza A virus, although their similarity in terms of the amino acid homology and RMSD is relatively low.

To date, the FDA-approved neuraminidase inhibitors are zanamivir (Relenza® marketed by GlaxoSmithKline) and oseltamivir (Tamiflu® from Roche). The orally administered drug oseltamivir has been widely used and stockpiled for the treatment both of influenza A and B viruses more often than zanamivir [Sugaya et al. 2007, Kawai et al. 2008]. Although oseltamivir is less effective for treatment of influenza B, its resistance is commonly reported in both influenza types. The high level of oseltamivir treatment worldwide has led to the rapid spread of mutations of the viral neuraminidase gene that produces drug resistance. The neuraminidase mutations, which lead to oseltamivir resistance in both *in vitro* and *in vivo* experiments and/or isolation from patients with resistance, are listed as follows. The seasonal influenza H1N1 and avian influenza H5N1 viruses carry single substitutions in neuraminidase at H274Y or N294S, whereas the pandemic H1N1 2009 (pH1N1) neuraminidases have three single mutations, H274Y, I223R and S246N, plus the S246N and H274Y double mutation that confer resistance to oseltamivir. In H3N2, the oseltamivir-resistant variants contain the single mutations of E119V, R292K and N294S. For influenza B virus, the emergence of the reduced neuraminidase inhibitor sensitivities imparted by the E119G, R152K and D198N single mutations has been isolated from patients after drug treatment, which is not found on the neuraminidase gene of influenza A viruses [Gubareva et al. 1998, Barnett et al. 1999, Gubareva et al. 2000]. From an *in vivo* study, the R152K and D198N neuraminidase mutants have resulted a 100- and 9-fold lower susceptibility to oseltamivir inhibition (in terms of IC_{50} values), respectively, compared to the wild-type strain [Mishin et al. 2005]. Using reverse genetics, oseltamivir was found to have a 31- and 252-fold lowered efficiency against influenza B viruses with the E119G and R152K neuraminidase mutants, respectively, compared to that for the wild-type [Jackson et al. 2005].

In the present section, the physicochemical origin of the binding affinity between oseltamivir and the influenza B neuraminidase is clarified theoretically using the 3D-RISM/RISM method. The study is further extended to clarify how the inhibition efficiency is reduced by the amino-acid substitutions, R512K, E119G, and D198: R152K at the catalytic site, and E119G and D198N at the framework site (Fig.VI-2.1).

Two types of information should be provided for clarifying the mechanism of a drug to inhibit a receptor: one of those is structural information of the receptor, and the other is the binding affinity of a ligand to the receptor. The structure of a receptor in the wild type is available in many cases from the protein data bank. However, it is not necessarily the case for mutants, since it takes considerable amount of time to determine the structure of a protein from the X-ray or NMR measurement. Fortunately, the structural change due to a mutation should not be so large in nature that the receptor is still able to keep its intrinsic function. It will become possible in future to evaluate such a structural change due to amino-acid substitution by means of the 3D-RISM/RISM theory, given that a numerical algorithm for the linear response theory described in Chapter V is provided. However, in a tentative solution, we can use the molecular simulation for considering the structural fluctuation. We believe the molecular dynamics (MD) simulation is good enough to produce the protein structure perturbed by mutation, starting from the structure of wild type as a template. On the other hand, it is quite difficult for MD to evaluate the binding affinity of a ligand to the target protein. We calculate the binding affinity or the free energy based on the thermodynamic cycle illustrated in Fig. VI-1.1. The binding affinity is determined by difference in the free energy between the bound and unbound states of the molecules in which the free energy of a molecule or a molecular complex consists of the intramolecular interactions and the solvation free energy. It is the solvation free energy that puts

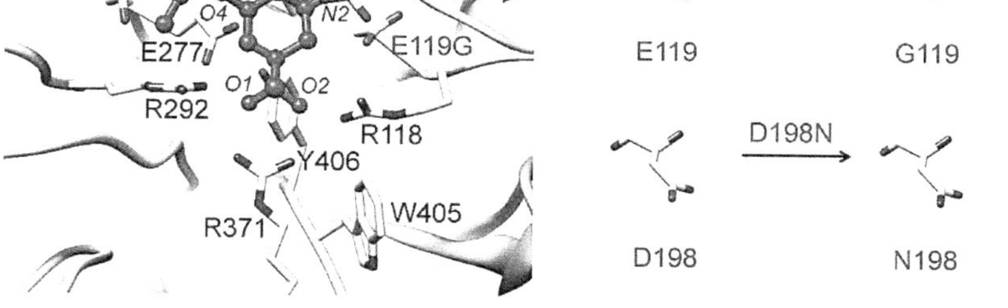

Fig. VI-2.1: Illustration of the positions of amino-acid substitution. Copyright (2015) The Protein Society.

Color version at the end of the book

MD in trouble, because the sampling of water configuration, especially at the internal cavity of protein, is extremely difficult for the method. We employ the 3D-RISM/RISM method to evaluate the solvation free energy, which is free from the sampling problem inherent to the molecular simulation.

Binding free energy of wild type:

The free energy components determined by the 3D-RISM and MD methods are listed in Table VI-2.1 (see Section VI-1 for the definition of each component of the binding free energy).

Table VI-2.1: Free energy change upon ligand binding. Copyright (2015) The Protein Society.

Free energy components (kcal/mol)	complex	receptor	ligand	
ΔE_{solute}	–30068.8	–29582.8	282.9	–203.1
$-T\Delta S_{solute}$	–3986.0	–3962.6	–45.3	21.9
ΔG_{solute}	–34054.8	–33545.4	–328.2	–181.2
ΔG_{solv}	5486.3	5329.5	–24.1	180.9

The calculation gives –0.3 kcal/mol for ΔG_{bind} ($\equiv \Delta G_{solute} + \Delta G_{solv}$), which is rather small compared to an empirical estimate represented by IC_{50}, which is –11 kcal/mol. So, in a glance, the theoretical estimate is not so great in terms of the "number." However, the results have great implication in terms of the "physics" of binding affinity. Two important points should be noted. The most important point is that the contribution from the intramolecular degrees of freedom, or ΔG_{solute}, to the binding free energy, or ΔG_{bind}, is comparable to the desolvation free energy, or ΔG_{solv}, which largely cancels out. The result asserts the following: *any method of the* in silico *drug screening that does not account for the desolvation free energy, may end up with a failure.* Since the desolvation free energy is determined by the microscopic distribution of solvent species including water around ligand, receptor, and their complex, any method that cannot account for the distribution of solvent will be fatal.

The other point to be noted is that the contribution from the structural entropy, or $-T\Delta S_{solute}$, represents the loss of the structural degrees of freedom of solute molecules due to the binding. The contribution is not so large compared to the other contributions, or ΔE_{solute} and ΔG_{solv}. However, it has a *right to make decision* in a sense, because the two big contributions ΔE_{solute} and ΔG_{solv} balance each other. The structural *entropy* is closely related to the structural *fluctuation* of the both solute species under concern, receptor and ligand. In the case of the receptor without ligand, some of the side-chains consisting the active site will be solvated by solvent molecules, and they are relatively freely fluctuating with solvent molecules. On the other hand, in the complex form, the active-site is packed by a drug molecule that largely restricts the degrees of freedom concerning the side-chains. On the other side, drug molecules in bulk solution, which are usually large organic molecules, have considerable degrees of freedom concerning the dihedral angles. When the molecule is confined in a cavity of receptor, the degrees of freedom will be largely restricted, most likely, to a single conformation.

Solvent distribution around ligand and receptor

The importance of the desolvation free energy to determine the binding affinity was emphasized above. It implies that solvent molecules inside the binding site of receptor and those around ligand are largely removed upon the binding. So, it is of great interest to look at the change of solvent distribution around ligand as well as receptor upon the binding. Depicted in Fig. VI-2.2 are the three dimensional distribution of hydrogen atoms of water around the ligand (top), and the radial distribution function $g(r)$ of hydrogen atoms of water around the selected atoms of ligand molecules (bottom). The change of the distribution can be readily seen in the radial distributions, or in the lower figures: the radial distribution functions of hydrogen around the atoms are largely reduced, representing the *desolvation.*

3D-RISM RDF of hydrogen-bonded pairs between oseltamivir heteroatoms and water molecules before and after the complexation calculated from the free oseltamivir and oseltamivir in complex with wild-type neuraminidase (Upper) 3D-distribution function of water O-atom and H-atom around oseltamivir via 3D-RISM calculation with $g(r) > 3.0$ before and after the complexation. (Lower) 3D-RISM RDF of hydrogen-bonded pairs between oseltamivir heteroatoms and water molecules before and after the complexation calculated from the free oseltamivir and oseltamivir in complex with wild-type neuraminidase. (Lower) 3D-RISM RDF of hydrogen-bonded pairs between oseltamivir heteroatoms and water molecules before and after the complexation calculated from the free oseltamivir and oseltamivir in complex with wild-type neuraminidase.

Similar phenomena can be observed in the receptor side as depicted in Fig. VI-2.3, the 3D-distribution function and RDF of water molecules around the residues in the active-site of the wild-type receptor before and after the complexation. It is distinctive that in most of the cases, water molecules hydrogen-bonded to the amino acid residues in the bulk solution either disappeared or largely reduced after the complexation, in particular the distribution around the

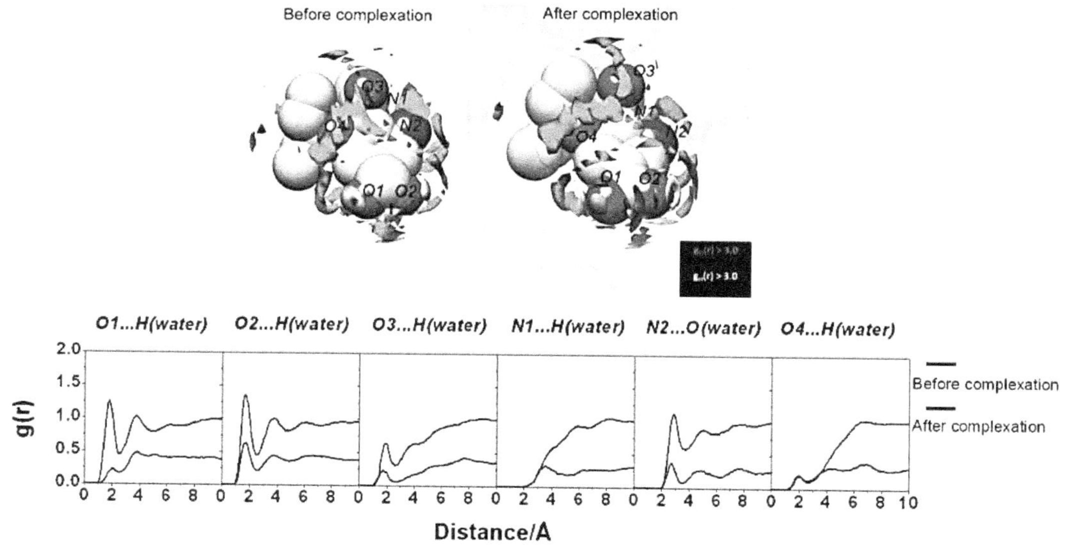

Fig. VI-2.2 3D-distribution function of water O-atom and H-atom around oseltamivir via 3D-RISM calculation with $g(r) \geq 3.0$ before and after the complexation. 3D-distribution function of water O-atom and H-atom around oseltamivir via 3D-RISM calculation with $g(r) \geq 3.0$ before and after the complexation (Upper). The RDF of hydrogen-bonded pairs between oseltamivir heteroatoms and water molecules before and after the complexation calculated from the free oseltamivir and oseltamivir in complex with wild-type neuraminidase (Lower). Copyright (2015) The Protein Society.

Color version at the end of the book

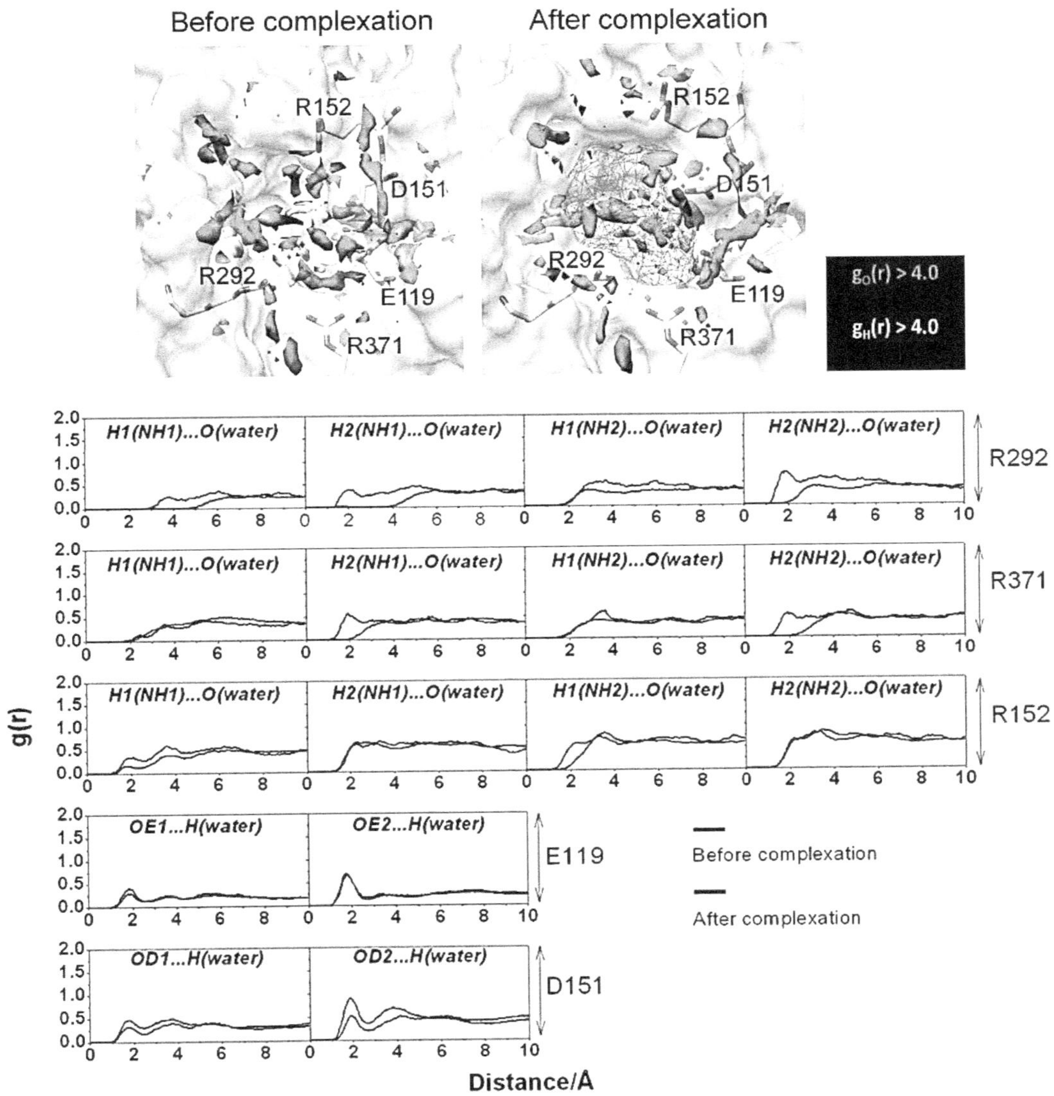

Fig. VI-2.3: 3D-distribution function of O-water (green) and H-atoms (yellow) with $g_O(r) \geq 4.0$ within 7.0 Å of oseltamivir (depicted as blue mesh sphere). (Upper) 3D-RISM RDF of hydrogen-bonded pairs between water molecules and five amino acid residues before and after complexation (Lower). Copyright (2015) The Protein Society.

Color version at the end of the book

residues of Arg292 and Arg371. So, it is again obvious that the change in the hydrogen-bond with water molecules upon the complexation is an important source of the dehydration penalty concerning the receptor.

However, it is worthwhile to note that the change in the water distribution upon the ligand binding shows more complexity depending on the position in the binding site: some are virtually intact, and some are even increased, for example, OD1...H(water) and OD2...H(water) in D151. Such increase in the water distribution may be due to a water bridge formed between the residue and the ligand. This indicates that there are cases in which the dehydration causes gain in the binding free energy instead of penalty.

Binding affinity of oseltamivir to the mutants

The energetics related to the binding affinity of oseltamivir to the different mutants of neuraminidase are listed in Table VI-3.2 along with the experimental binding free energy (ΔG^{IC50}) converted from IC_{50} data. The 3D-RISM binding free energy is the summation of the contribution from solute molecules (ΔG_{solute}) and solvation free energy (ΔG_{solv}), where ΔG_{solute} is a combination of interaction energy (ΔE_{solute}) and structural entropy of protein ($-T\Delta S_{solute}$) as is defined previously. The relative free energy $\delta\Delta G_{bind}$ of mutant strains is the difference of binding free energy in comparison with that of the wild type. The computational results are roughly in accordance with the experimental trend, but for the D198N mutant the affinity is relatively close to the wild type. The reason why the binding affinity of the D198N mutant does not change much from the wild type would be in the location of the mutation that is framework of the active site residue. In the following, let us look closely at the change of water distribution around the mutated residues in each mutant.

E119G mutant: In the mutant, one of the hydrogen-bonding sites is lost entirely due to the mutation from the negatively charged and long side chain of glutamate to the non-polar and short side chain of glycine. The loss of the hydrogen-bond site brings two effects into the binding affinity. Firstly, the hydrogen-bonding between the residue and the ligand is lost, which shifts the binding-free energy toward the positive side. Secondly, the hydrogen-bond between the residue and water molecules is lost, which makes the dehydration penalty of the receptor less. The substitution of the large residue (E) by a small residue (G) introduces another effect into the binding affinity, which is concerned with the dehydration penalty of the ligand. Shown in Fig. VI-2.4(b)) is the RDF of O-water atoms and heteroatoms of the drug in the E119G-drug complex, compared with that of the wild type (Fig. VI-2.4(a)). The figure clearly indicates that the RDF around the -NH_3^+ group in this mutant is not reduced upon binding as much as that in the wild type. It is because the substitution created a cavity for a water molecule hydrating the ligand to be accommodated in the binding site without dehydrating. The effect will lead to the reduction of the dehydration penalty of the ligand. The reduction in the binding affinity

Table VI-3.2: The average MM/3D-RISM binding free energies and their components of four influenza B neuraminidase complex are in kcal/mol. The relative binding free energy from the prediction ($\delta\Delta G_{bind}$) and experiment converted from IC_{50} ($\delta\Delta G^{IC50}$) was compared for mutated systems. Copyright (2015) The Protein Society.

Energetics	Wild-type	E119G	R152K	D198N
ΔE_{solute}	−203.1	−171.0	−199.0	−200.2
$-T\Delta S_{solute}$	21.9	22.1	21.4	22.0
ΔG_{solute}	−181.2	−148.9	−177.7	−178.2
ΔG_{solv}	180.9	151.8	178.6	178.3
ΔG_{bind}	−0.3	2.9	0.9	0.1
$\delta\Delta G_{bind}$	–	3.2	1.3	0.4
$\delta\Delta G^{IC50}$	–	2.0[a]	3.3[a]/2.7[b]	1.3[b]

[a] Jackson et al. 2005; [b] Mishin et al. 2005

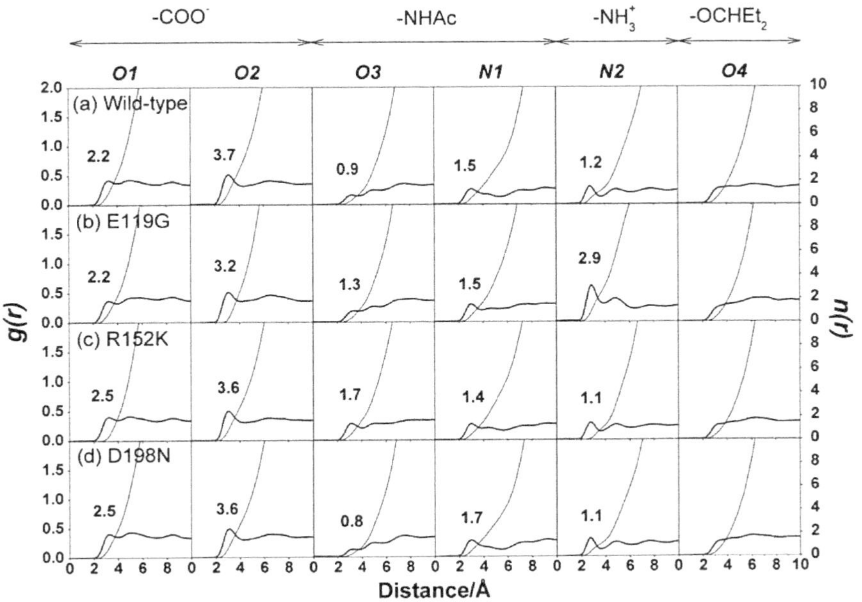

Fig. VI-2.4: 3D-RISM radial distribution function ($g(r)$) between the heteroatoms of oseltamivir (*O1-O4*, *N1* and *N2*; see Fig. VI-2.1) and O-water atoms is illustrated. The occupation numbers of water molecules integrated up to the first minimum are also given (a) wild type, (b) E119G, (c) R152K, (d) D198N. Copyright (2015) The Protein Society.

due to the mutation is determined by loss of the hydrogen bonding between the receptor and the ligand, which exceeds the reduction of the dehydration penalty both in the receptor and the ligand.

***R152K mutant*:** In the mutant, one of the basic residues, arginine, is replaced by the other basic residue lysine. The mutation causes a mild effect on the electrostatic interactions of solute free energy, because both residues bear charges with the same positive sign. The interaction depends apparently on the geometry of the residues. Arginine is slightly bigger in size than lysine. The smaller size of lysine makes the receptor-ligand interaction unfavorable due to the increased separation to the hydrogen-bond partner in the drug. On the other hand, the dehydration penalty for lysine is less than that for arginine (Table VI-3.2). The results can be understood from the RDF of water around the -NHAc group in the drug molecule (Fig. VI-2.4(c)). The hydrogen-bond peak around the O3 atom of the -NHAc group in the R152K mutant is not reduced as much as that in the wild type. This indicates that some portion of water molecules around the residue and the drug can remain upon binding without dehydration.

***D198N mutant*:** This is the mutation of a charged residue to a polar residue. The effect of the mutation on the structure should be significant, because a net charge is removed from the protein. Nonetheless, the effect on the binding affinity would be minor, because the location of the substitution is at framework of the active site. In fact, the RDFs of water around the drug and the residues do not suffer at all from the mutation (in Fig. VI-2.4(d)). This might be the reason why the affinity of the drug to the mutant does not change much compared to the wild type.

VI-3 Screening of Many Drug Candidates

In the preceding section, we looked at an example of application of the 3D-RISM/RISM method to evaluate the binding affinity of a drug compound, or oseltamivia, to different mutants of protein called neuraminidase of the influenza virus. The purpose of the work was to find the physicochemical reason as to why and how the virus acquires the resistance to the drug by mutation. In the present section, we will apply the method to screen a drug compound from many candidate compounds by the 3D-RISM/RISM method. The main task in the screening is to find a compound that scores the best affinity to a target protein. In the study, a protein called "P1-kinase" is chosen as the target protein, and "triazolopyridazine" and its derivatives are examined as inhibitors [Hasegawa et al. 2017].

The Pim-1 kinase belongs to Pim (Proviral Integration-site MulV) family along with Pim-1, 2, and it is a Serine/Threonine Kinase. The 3D structure of a Pim-1 kinase (PDBID: 3BGQ) with an inhibitor is shown in Fig. VI-3.1a. The enzyme is expressed widely in our body due to its phosphorylation activity of substrates concerning a variety of bio-functions such as cell cycle, apoptosis, and differentiation [Bachmann and Möröy 2005, Merkel et al. 2012, Zhukova et al. 2011, Liang and Li 2014]. In particular, the enzyme is expressed excessively in malignant tumor such as leukemia, lymphoma, and prostatic carcinoma [Tursynbay et al. 2015, Shah et al. 2008, Valdman et al. 2004]. On the other hand, a mouse, the Pim-1 kinase of which is knocked out, does not show any indication as a phenotype [Laird et al. 2003, Domen et al. 1993]. Therefore, it is considered that such tumor may be treated by an inhibitor of the enzyme, with minor side effects. Although there have been some reports about the compounds that tightly bind to the Pim-1 kinase, they have not been commercialized yet as an actual cancer drug [Burger et al. 2015, Wan et al. 2013].

Preparation of the initial coordinate for the MD simulation

The initial condition is prepared based on a PDB structure of a Pim-1 kinase (PDBID: 3BGQ) with a triazolo pyridazine inhibitor termed as VX2 in the PDB. The chemical structure of the

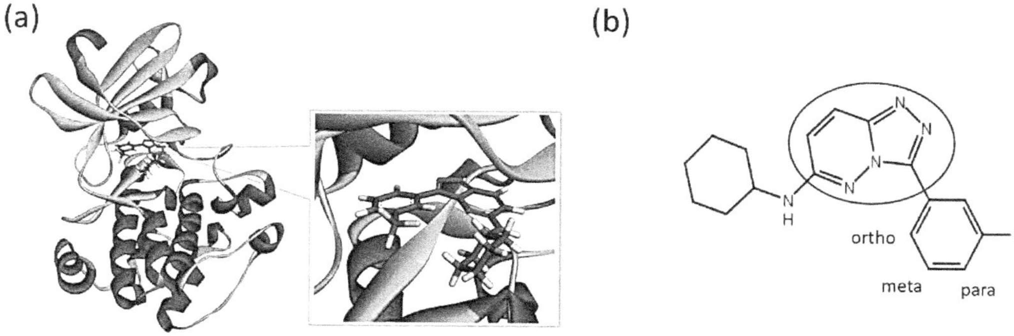

Fig. VI-3.1: (a) A 3D structure of the Pim-1 kinase and triazolo pyridazine inhibitor termed as VX2 in the PDB. The PDB code is 3BGQ. We refer to the VX2 as ligand 1, in this study, for the sake of simplicity of the terminology. (b) Chemical structure of the VX2. Circled part of the ligand is termed as triazolo[4.3-b]pyridazine scaffold. This is a common part of all the ligands which is applied in this study. Copyright (2018) American Chemical Society.

Color version at the end of the book

VX2 is shown in the Fig. VI-3.1(b). Other 15 ligands [Grey et al. 2009] are modeled based on the structure of VX2 and initial binding mode of the 15 ligands is determined as superimposing the ligands to VX2. We refer to the VX2 as Ligand 1, in this study, and other ligands are referred as Ligand 2 to 16 for the sake of simplicity of the terminology. The ligands are listed in Table VI-3.1. All the ligands commonly have a triazolo[4.3-b]pyridazine scaffold. Cyclo-alkane and phenyl ring show some variety. The cyclo-alkane is substituted by cyclo-propane, cyclo-butane, and cyclo-hexane. The meta or para position of phenyl ring are substituted by H, F, CF3, and OMe. The ff12SB and GAFF force field is applied to the protein and ligand structures and partial charges of the ligands are parameterized with AM1-BCC method [Jakalian et al. 1999, 2002]. For determining the initial coordinates of the solvent molecules around and inside the protein-ligand complex, we employ the 3D-RISM/KH theory and a software called "Solutionmap." Solutionmap is a software to place the water molecules to appropriate position and orientations, in consistence with the predicted distribution function of the water molecules [Sindhikara and Hirata 2013]. We first calculate the 3D-distribution of water molecules around the complex based on the 3D-RISM/KH theory. Then, Solutionmap is applied to place 150

Table VI-3.1: The list of ligands and its inhibition constant. K_i is the inhibition constant [Grey et al. 2009]. ΔG_{bind_exp} is the experimental value of the binding free energy estimated from $-RT\ln K_i$ where the temperature is regarded as 300K. $\Delta\Delta G_{bind\_exp}$ is the binding free energy relative to that of Ligand 1. Copyright (2018) American Chemical Society.

No.	Structure	Ki (nM)	ΔG_{bind_exp} (kcal/mol)	$\Delta\Delta G_{bind\_exp}$ (kcal/mol)	No.	Structure	Ki (nM)	ΔG_{bind_exp} (kcal/mol)	$\Delta\Delta G_{bind\_exp}$ (kcal/mol)
1	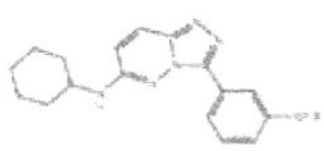	11	-10.93	0.00	9	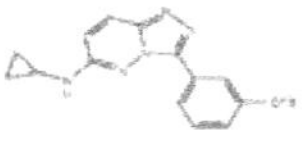	18	-10.63	0.30
2	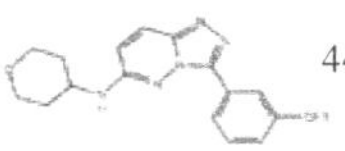	44	-10.10	0.83	10	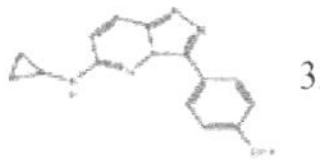	320	-8.92	2.01
3	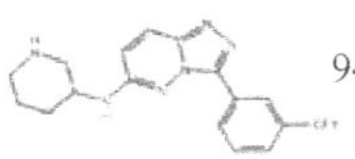	94	-9.65	1.28	11	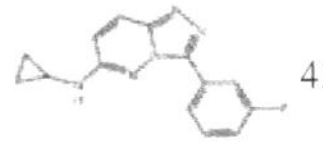	430	-8.74	2.19
4	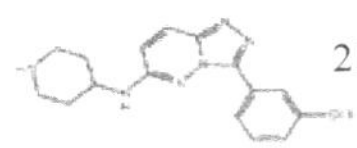	21	-10.54	0.39	12	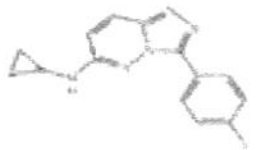	210	-9.17	1.76
5	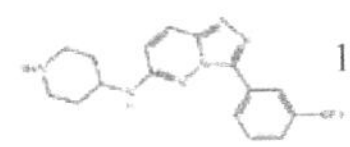	160	-9.33	1.60	13	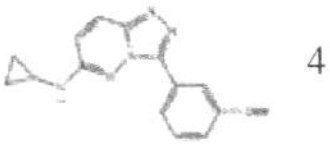	410	-8.77	2.16
6	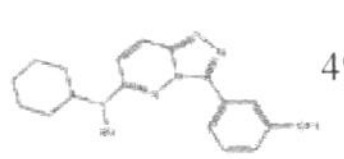	49	-10.03	0.89	14		980	-8.25	2.68
7	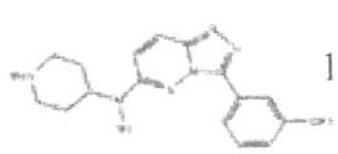	100	-9.61	1.32	15	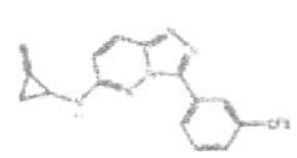	50	-10.02	0.91
8	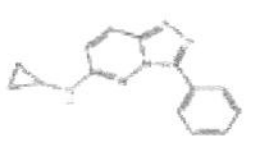	1800	-7.89	3.04	16	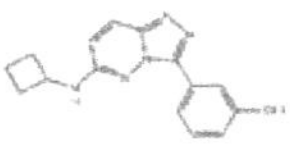	54	-9.98	0.95

water molecules at positions where the distribution has higher peaks. The complex of Pim-1 kinase and its inhibitor, *solvated* by water molecules with "Solutionmap", is immersed in a TIP3P truncated octahedral water box with a margin of 10 Å from the complex with the pre-distributed water molecules. Thirteen sodium ions are also distributed as counter ions to neutralize the negative charges of residues.

Protocol for the MD simulation

In this study, the system is minimized along three steps. The harmonic constraint is applied to coordinates of the Pim-1 kinase, changing the force constant in each step; 500, 50, and 20 kcal/molÅ^2. Each minimizing step includes 1000 steps of the steepest descent method and 2000 steps of the conjugate gradient method. The minimized structure is heated in the NVT ensemble from 0 K to 300 K taking 20 ps with applying the harmonic restraint for a coordinate of a Pim-1 kinase with a force constant of 10 kcal/molÅ^2. After that the system is relaxed in the NPT ensemble at 1 bar for a 1 ns. After the relaxation process, 200 ns of a MD simulation in the NPT ensemble at 1 bar is applied. All hydrogen atoms are restrained by a SHAKE algorithm. This protocol is applied to the sixteen protein-ligand complexes prepared in the preceding procedure.

Free energy calculation

The binding free energy consists of the potential energy change of the solute molecules, ΔE_{solute}, the contribution from the entropy change of the solute molecules to the free energy change, $-T\Delta E_{solute}$, and the desolvation free energy, ΔG_{desolv}. The desolvation free energy is estimated by the 3D-RISM/KH theory as the difference of the solvation free energies between the bound and isolated states. The 3D-RISM/KH calculation is preceded by the 1D-RISM calculation. The binding energy and desolvation free energy are calculated using 2000 snap shots extracted from a trajectory of the MD simulation for 200 ns. The standard error is calculated as a standard deviation of the average values of the $\Delta E_{solute} + \Delta G_{desolv}$ in each 10 ns of the trajectory. The entropy change of the solute molecules consists of two contributions, the external entropy and the vibrational entropy, both of which are calculated using the tools implemented in AMBER. The external entropy, or the translational and rotational entropies, are estimated by the classical statistical mechanical formulas, based on a rigid rotor assumption [McQuarrie 1976]. The vibrational part is estimated by means of the quasi-harmonic method using 100000 snapshots extracted from the trajectory of the MD simulation for 200 ns [Brooks et al. 1995]. We define a hypothetical state in which the solvent and the solutes (protein, ligand and their complex) are separated in space as the standard (or reference) state of the thermodynamic cycle. The solvation free energy of each solute species is defined by the free energy change associated with the process of transferring the species from the gas phase to the infinite dilution of a solution phase, at a fixed position and orientation. So, the standard state used in the thermodynamic cycle and that employed in the experimental data are different concerning the external entropy. The correction to the difference in the standard state is included in the external entropy calculated from the AMBER tool that takes the experimental standard state, 1 mol/l, into consideration. ΔE_{solute} and $-T\Delta S_{solute}$ are calculated using MMPBSA.py in the AMBER 14 software package. ΔG_{desolv} is calculated using the in-house software.

Comparison between the theoretical and experimental values

The binding free energies of the sixteen compounds to the target protein, calculated based on the MM/3D-RISM/KH method, are shown in Table VI-3.2 and plotted against the corresponding experimental values [Grey et al. 2009] in Fig. VI-3.2(a). A rather high correlation (R = 0.69) is observed between the theoretical and experimental results. This result demonstrates that the MM/3D-RISM/KH method is applicable to the problem of compounds screening and lead optimization, where relative affinity among the compounds has significance. However, the theoretical results show some systematic deviation from the experimental values toward the positive side. There are several causes conceivable for the systematic deviation: insufficient sampling of the conformational space of the protein, insufficient accuracy of the 3D-RISM/KH theory for estimating the solvation free energy, inadequate solution conditions, and so on. In contrast to the method that directly estimates the binding free energy, the method based on a thermodynamic cycle requires several theoretical methodologies for estimating each component of the binding free energy: inter-atomic interactions of solute, solvation free energy, conformational entropy, and external entropy. Since each method for estimating the component of the binding free energy has its own approximation, each method may systematically under- or over-estimate the thermodynamic quantity. These systematic errors seem to give rise to the positive value of the binding free energy.

Among those components of the binding free energies, the conformational entropy deserves special attention, because the scientists in the community of molecular dynamics simulation

Table VI-3.2: Binding free energy and its components obtained from MM/3D-RISM/KH method. ΔG_{bind_cal}, ΔG_{bind_exp}, ΔE_{solute}, $-T\Delta S_{solute}$, ΔG_{solv} denote the binding free energy, the potential energy change of the solute molecule, the entropy change of the solute molecule, and the desolvation free energy, respectively. The predicted binding free energy is shown with its average value and standard error. The binding free energy estimated from K_i, ΔG_{bind_exp}, is also shown [Grey et al. 2009].
Copyright (2018) American Chemical Society.

No.	ΔE_{solute} (kcal/mol)	$-T\Delta S_{solute}$ (kcal/mol)	ΔG_{solv} (kcal/mol)	ΔG_{bind_calc} (kcal/mol)		ΔG_{bind_exp} (kcal/mol)
				Ave.	Std. err.	
1	−56.18	8.06	48.02	0.26	1.33	−10.93
2	−51.81	8.05	45.39	1.62	1.94	−10.10
3	−42.22	12.85	34.76	5.38	1.27	−9.65
4	−50.34	7.95	42.56	0.17	2.96	−10.54
5	−57.74	12.83	50.09	5.17	1.69	−9.33
6	−60.98	12.57	52.86	4.45	1.41	−10.03
7	−60.04	15.29	52.37	7.62	1.37	−9.61
8	−44.7	9.96	39.27	4.53	2.13	−7.89
9	−42.49	6.54	36.64	0.69	1.99	−10.63
10	−54.48	11.11	48.08	4.71	2.46	−8.92
11	−43.69	9.79	39.51	5.61	1.27	−8.74
12	−44.39	8.87	39.81	4.29	1.07	−9.17
13	−55.57	9.6	0.61	4.63	1.67	−8.77
14	−52.04	10.11	46.95	5.02	1.16	−8.25
15	−59.06	13.3	49.45	3.7	1.35	−10.02
16	−53.51	9.88	46.16	2.53	1.86	−9.98

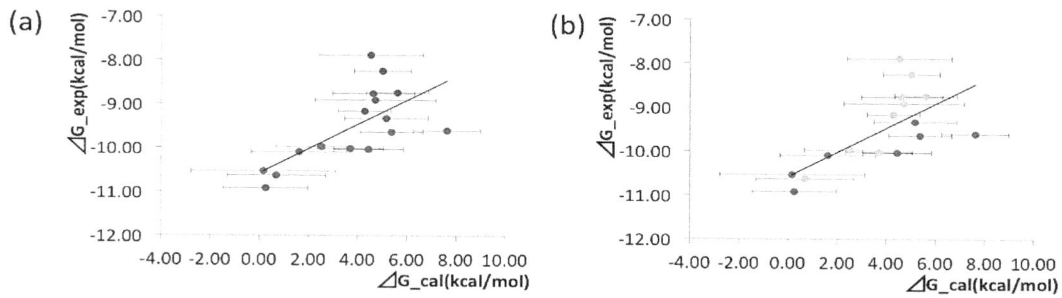

Fig. VI-3.2: Correlation between calculated and experimental values of binding free energy for all the ligands. Correlation coefficient R = 0.69. The figures are colored in different manner based on the structural feature of the ligands. (a) Ligands which include CF_3 on the meta position of the phenyl ring are colored with red. Other ligands are colored with blue. (b) Ligands which have cyclo-hexane, cyclo-butane, and cyclo-propane are colored with purple, orange, and green, respectively. Copyright (2018) American Chemical Society.

Color version at the end of the book

are experiencing a hard time to find converged results for the quantity, even after sampling a sub-micro-second length of trajectory. The situation reminds us of Levithal's estimate of the conformational degrees of freedom that amounts to ~ 10^{50} for a small protein having ~ 100 amino acids, which was discussed in the Section 5 in Chapter V. It may be impossible to find the converged results for the conformational entropy by the standard method of the simulation. An analytical approach based on the 3D-RISM/RISM method combined with the generalized Langevin theory described in Chapter V may have an advantage in that respect.

VI-4 Distribution of Ligand Molecules Bound at an Active Site of a Receptor: uu-3D-RISM Approach

In the preceding sections, Hasegawa et al. applied the 3D-RISM method to the drug screening in the case where the binding mode, or the structure of a complex between ligand and receptor, is more or less known empirically in advance [Hasegawa et al. 2017]. In that case, our main task is to evaluate the binding affinity or free energy to find a molecule among the candidate compounds, which gives the best score in terms of the affinity. However, if the binding mode between receptor and ligand is not known *at all*, such a strategy does not work. In such a case, finding the binding mode should be the first concern in the drug design. It should be reminded that the *binding* is nothing but the *molecular recognition* of a ligand by a receptor, and that the 3D-RISM theory provides good description to the molecular recognition (see Chapter IV). So, the 3D-RISM has a potential capability of describing the binding mode of a ligand to a receptor. Nevertheless, the application may not be so straightforward due to the following reason.

The numerical procedure for solving the conventional 3D-RISM equation consists of two steps. In step 1, we solve ordinary RISM (or 1D-RISM) equations for a solvent mixture including target ligands in order to obtain the density pair correlation functions (PCF) among molecules in the solution. Then, we solve the 3D-RISM equation for a solute-solvent system to find three dimensional density distribution functions (3D-DDF) of solvent species around a protein, using PCF obtained in the first step. (The terminologies "one dimensional density distribution function (1D-DDF)" and "three dimensional density distribution function (3D-DDF)" are used instead of "pair correlation function (PCF) and "spatial distribution function" (SDF) in order to make the difference of those quantities unambiguous.) A key to the success of the method was to regard a target ligand as one of "solvent" species. Unfortunately, the success is limited due to a difficulty of solving the 1D-RISM equation for a solvent mixture including *large* ligand molecules such as drug compounds. In order to resolve the problem, and to make the method applicable to large ligand molecules such as drug, two methodologies have been proposed: the uu-3D-RISM theory, and the fragment-based drug design [Kiyota et al. 2011, Imai et al. 2009]. In the present section, let us introduce the uu-3D-RISM method [Kiyota et al. 2011] (The fragment-based drug design is presented in the following section).

uu-3D-RISM approach

In this approach, we solve a solute-solute 3D-RISM equation for a protein-ligand system in which both proteins and ligands are regarded as "solutes" at infinite dilution. The 3D- and 1D-RISM equations are solved for protein-solvent and ligand-solvent systems, respectively, in order to obtain the 3D-DDF and 1D-DDF of solvent around the solutes, which are required for solving the solute-solute 3D-RISM (uu-3D-RISM) equation.

The procedure to obtain the receptor-ligand distribution function is shown in Scheme 1. First, vv-DF is evaluated by vv-1D-RISM, where solvent includes water, electrolyte, organic solvent and so on. The vv-DDF is used both in uv-3D-RISM and in uv-1D-RISM calculations. The uv-3D-RISM and uv-1D-RISM calculations are carried out to obtain receptor-solvent and ligand-solvent DDF, respectively. By inserting these two DDFs, uu-3D-RISM can be solved to get receptor-ligand DDF. From receptor-ligand DDF, a molecular recognition (MR) analysis such as a binding pocket search can be performed.

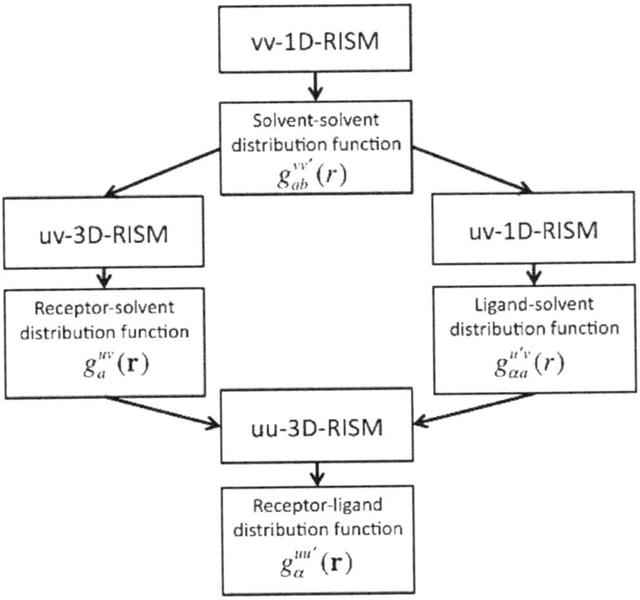

Scheme 1: Scheme of uu-RISM method.

Binding mode of aspirin at active-site in PLA2

In order to demonstrate the capability of the uu-RISM method, the binding mode of aspirin at the active site of a receptor called Phospholipase A2 (PLA2) is examined [Dennis 1994, Nicolas et al. 1997, Argiolas and Pisano 1983]. Since aspirin is a rather large ligand having 14 specific interaction-sites (Fig. VI-4.1) in the neutral state, it may not be treated readily with the ordinary uv-3D-RISM due to the difficulty stated above. So, this is a good example to demonstrate capability of the uu-3D-RISM approach. Aspirin, acetylsalicylic acid, is a weak acid in aqueous solution. We employed a neutral state which was shown in Fig. VI-4.1, because the affinity of the neutral state to binding site is much higher than a charged state.

The 3D-DDF of aspirin at the non-protonated state is shown in Fig. VI-4.2 with the threshold $g_\gamma(\mathbf{r}) > 2$. The red, gray, and blue surfaces are the distributions of carboxyl group, COOH (number of sites is 4), the aromatic ring (number of sites is 6), and the acetoxy group, $OCOCH_3$ (number of sites is 4), respectively. We can easily observe these distributions not only inside the binding site, but also around the protein.

As you see the Fig. VI-4.2, the distributions are jumbled inside the binding site. Since the final goal of our study is to establish a method to probe a large ligand molecule recognized by protein using 3D-RISM, we must determine the location of binding site which has the highest affinity of ligand in the target protein. The 3D-DDFs or the Potential of Mean Force (PMF) are a good indicator to evaluate the affinity. However, since 3D-DDF is the distribution function of an individual site consisting a ligand molecule, it is difficult to evaluate the affinity of whole ligand molecule directly.

Then, we introduce a function for Distribution Center (DC) of ligand to measure the affinity of ligand molecule. The function of DC is defined as:

$$f_{DC}(\mathbf{x}) = \begin{cases} \left(\prod_{\gamma}^{N} \frac{1}{V_{box} - V_{protein}(\mathbf{x})} \int_{V_{box}(\mathbf{x})} g_\gamma(\mathbf{r}) d\mathbf{r} \right)^{\frac{1}{N}} & \text{for } V_{box} - V_{protein}(\mathbf{x}) \geq V_{ligand} \\ 0 & \text{for } V_{box} - V_{protein}(\mathbf{x}) < V_{ligand} \end{cases} \qquad \text{(VI-4.1)}$$

where $\mathbf{x}$ denotes the center of box, N is the total number of sites of ligand molecule for normalization, V_{box} is the volume of the box, and $V_{protein}(\mathbf{x})$ is the excluding volume of the solute protein in the box. Therefore, $V_{box} - V_{protein}(\mathbf{x})$ denotes the space where ligand can be distributed. Note that the integrations on right-hand-side of Eq. (VI-4.1) are only performed inside V_{box} centered at $\mathbf{x}$. The size of box is adjusted to the length of a ligand molecule. The uu-3D-DDF is integrated in the box and the result is projected to the center of box. If the value is larger than one, the probability of finding "an aspirin molecule" at the position is higher than bulk. Although the DC only gives us the rough estimate of the location of binding site, it is helpful to guide a further analysis concerning the binding mode in more detail using information of RDFs.

Kiyota et al. performed the calculation of DC based on Eq. (VI-4.1) in order to estimate the affinity of ligand to the binding site. The result of DC is shown in Fig. VI-4.3 with the threshold $f_{DC}(\mathbf{x}) > 1.45$. The maximum value is 1.56. The DC function does not take quite a high value because it is averaged over the sites and volume. Note that the result of DC is projected onto the center of the calculated box. In Fig. VI-4-3, Kiyota et al. observe the highest peak at the center of the binding site which is determined by the X-ray crystallography. The results demonstrate that the new approach is capable of locating the binding site in the protein and the affinity of a ligand to the site properly.

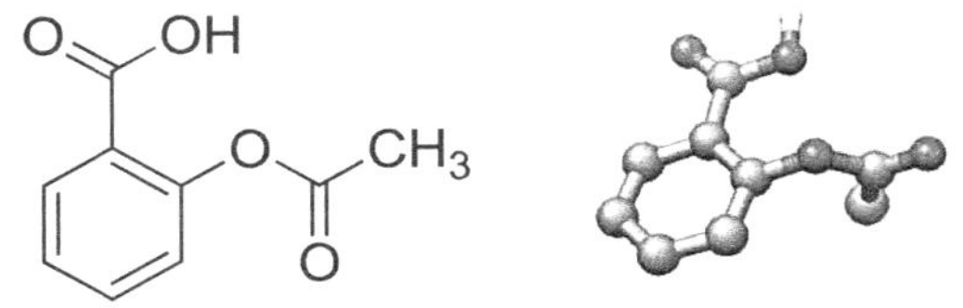

Fig. VI-4.1: The structure of aspirin depicted with two different presentations.

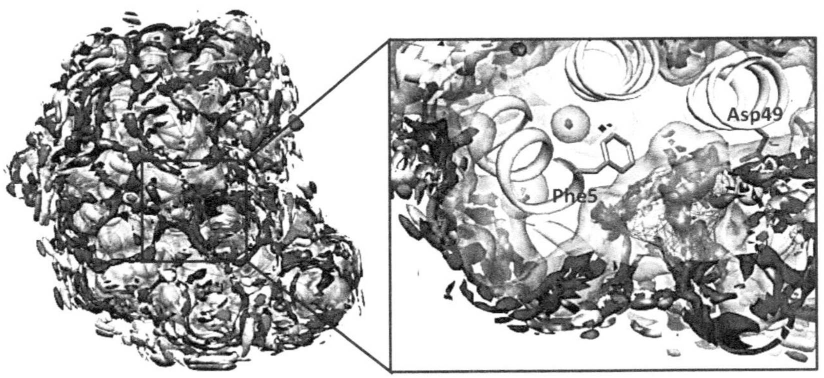

Fig. VI-4.2: The 3D-DDF of aspirin around and inside Phospholipase A2 (PA2), obtained by uu-3D-RISM with the threshold $g_\gamma(\mathbf{r}) > 2$: red, COOH; gray, aromatic ring; blue, $OCOCH_3$. The protein surfaces are represented as gray transparent surface. The location of aspirin in X-ray structure is depicted with a wire frame. Copyright (2011) American Chemical Society.

Color version at the end of the book

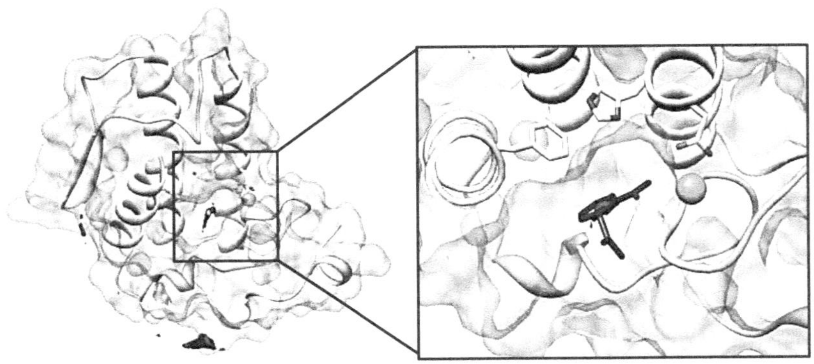

Fig. VI-4.3: The affinity of an aspirin molecule to binding site in the Phospholipase A2 estimated by the function of DC based on local potential mean force with the threshold $f_{DC}(\mathbf{x}) > 1.45$. The protein surfaces are represented as gray transparent surface. In the top view, the location of aspirin in X-ray structure is depicted with blue sticks. Copyright (2011) American Chemical Society.

Color version at the end of the book

In order to understand a mechanism of molecular recognition process, it is important to determine an explicit structure of the ligand inside the specific binding site, because we need the distinct structure to calculate some physiological properties. Actually, we can also

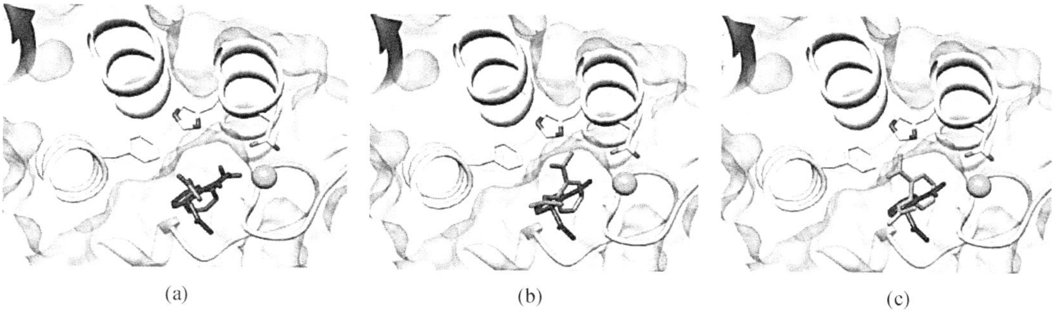

Fig. VI-4.4: Predicted structures of ligand from top three peaks of affinity based on the overlap between the ligand and its distribution function. The protein surfaces are represented as gray transparent surface. The location of aspirin in X-ray structure is depicted with blue sticks. Copyright (2011) American Chemical Society.

Color version at the end of the book

investigate the orientation of the ligand by calculating the overlap between the structure of the ligand and the 3D-DFs which are obtained by uu-3D-RISM. We define the target function for the orientation by the following equation:

$$f_{ori}(\mathbf{x}, \boldsymbol{\Omega}) = \prod_{\gamma}^{N} g_{\gamma}(\mathbf{x} + \mathbf{l}_{\gamma} \cdot \hat{\mathbf{R}}(\boldsymbol{\Omega})) \qquad \text{(IV-4.2)}$$

where $\mathbf{x}$ denotes center of box which is obtained by Eq. (VI-4.1), $\mathbf{l}_{\gamma}$ denotes the internal coordinate of site γ, and $\hat{\mathbf{R}}(\boldsymbol{\Omega})$) denotes rotational matrix with the Euler angles to search the entire orientational space. As we mentioned above, the ligand molecule just fits the size of box. It means that the center of the box coincides with the center of the molecule which is the origin of the rotational matrix at the same spatial point. The advantage of this approach is that we can measure the affinity quantitatively as the degree of overlap.

The results of this searching are summarized in Fig. VI-4.4. The figure shows the explicit ligand structures corresponding to the top three orientations by using Eq. (VI-2.2). The location and orientation of aspirin in the X-ray structure are depicted with blue sticks. It is worthwhile to note that the structure corresponding to the one with the greatest overlap (Fig.VI-4.4(a)) has the same orientation with the X-ray structure, although the rotational angle around the axis is somewhat different from each other. This orientation seems to be induced by the calcium ion located at the binding site, since the carboxylic group of aspirin faces the calcium ion. In case of the other two structures with lower score, the carboxylic groups are facing toward His48, which is positively polarized as well.

VI-5 Fragment-based Drug Design (FBDD) based on 3D-RISM

In the preceding section, we introduced the uu-3D-RISM theory to find the binding mode of a large polyatomic molecule to a receptor. The method, however, has limitation in two respects: the size and the structural fluctuation of ligand. The first limitation concerns the numerical solution of the uv-1D-RISM equation. It is well regarded from many applications of the equation to liquid mixtures that the convergence of the numerical solution becomes exceedingly slower as the molecular size becomes larger, and fails to converge. The second limitation is more essential for drug screening, because most of drug compounds have considerable internal degrees of freedom, and their structure fluctuates around the equilibrium structure according to the thermodynamic conditions. The fluctuated structures are nothing but *structural isomers* that coexist in solution with some probability corresponding to their chemical potentials. The existence of numerous isomers in solution puts a high barrier to any method of the drug screening based on the rigid molecule as a ligand, including the uu-3D-RIM, because you never know which isomer is bound preferentially to a receptor under concern. It is a quite high barrier, because the molecular recognition, or binding, is defined by the binding free energy that depends on *structures* of ligand and receptor in *atomic detail*. In this regard, the idea of "Fragment-Based Drug Design (FBDD)" will be worth paying attention.

The FBDD is recognized as one of the most promising alternatives to conventional methods of drug discovery, such as lead identification and optimization through a high-throughput screening campaign using biochemical assay techniques. In FBDD, a series of fragments of drug compounds is first screened against the target protein. Then, a few active fragments bound to the protein are linked to construct a much more active lead compound. The process is called "fragment linking". The greatest advantage of FDBB is that a significantly fewer number of the fragment compounds has to be prepared and examined than in the high throughput campaign, because the number of substructures that commonly compose drug compounds is relatively limited, and it is unnecessary to prepare all possible combination with FBDD. On the other hand, FBDD also has certain drawbacks. One of these is that the fragmentation generally reduces the binding affinity, so that more sensitive experiments or higher concentration of fragment molecules are necessary for detection. In addition, detailed information regarding molecular position and orientation in binding site is required for efficient linking, merging, or growth of the fragments. The method based on the 3D-RISM theory has a clear advantage over the other methods, empirical or computational, in this regard, since the method provides accurate information concerning the distribution of ligand in atomic detail as far as the size of the molecules are not so large. The FBDD consists of the following three steps.

1) Decomposing a drug compound into several fragments, or hypothetical molecules, taking the following requirements into consideration. (i) Each fragment should be free from dihedral rotation, so that it can be regarded as virtually a rigid molecule. (ii) Each fragment should be as small as possible so that the uv-RISM calculation can be readily converged, while it should be as large as possible in order to maximize the affinity to the receptor.
2) Finding the spatial distribution function (SDF; $g(\mathbf{r})$) of atoms of each fragment in the receptor based on the 3D-RISM method.
3) Deducing the binding mode (or pose) of the distribution of each fragment from the SDF.
4) Linking all fragments to construct the original compound.

Among (1) ~ (4) steps, step (1) should be made more or less on intuitive basis, involving trial and errors. Step (2) can be done with the 3D-RISM method in the same way as explained by many examples in Chapter IV. Steps (3) and (4) are obviously the ones that are crucial for the method to be able to predict successfully the binding affinity of a drug to a receptor. The reduction from the peaks of SDFs to ligand-binding modes is an obvious operation for small ligand molecules consisting of one or a few atoms, whereas it is a nontrivial task for polyatomic molecules because the atomic distributions are determined through an ensemble average of different combinations of molecular configurations. The task is to find the most probable binding mode (MPBM) that best matches to the SDF. An algorithm to find MPBM based on the SDF is generally called "ligand mapping." The first ligand mapping study based on 3D-RISM was proposed by Imai et al. in 2009 [Imai et al. 2009]. Enzymes called "thermolysin" and carbobenzoxy-L-Phe^{P}-L-Leu-L-Ala (ZF^{P}LA) were examined as receptor and ligand, respectively, because there was a corresponding experimental study based on the X-ray crystallography, which can be compared with the theoretical results.

Ligand mapping algorithm based on 3D-RISM

The algorithm is based on a grid search procedure in which the SDF acts as a kind of "score function" used in molecular docking simulation. It should, however, be noted that the SDF in our grid search is essentially distinguished from conventionally-used empirical score functions for the hydration effect, since the SDFs are obtained on the basis of statistical mechanics using the explicit water model and therefore naturally include the hydration effect that is equivalent to the hydration free energy. Besides, the theory can deal with ligand-water mixtures of arbitrary concentrations at a constant computational cost. The 3D-RISM-based method thus realizes the entire identification of ligand-binding sites, or "ligand mapping," on protein surfaces in a "real solution" system.

In the 3D-RISM-based ligand mapping (3D-RISM-LM), the MPBMs are determined in the flowing procedure (see Fig. VI-5-1). (1) The SDFs of ligand atoms, $g_\gamma(\mathbf{r})$, are calculated by the 3D-RISM theory. (2) Highly probable binding modes (HPBMs) of the ligand molecule are constructed from $g_\gamma(\mathbf{r})$ using the grid search algorithm. The HPBMs are defined as the modes whose "binding free energies" are less than a threshold value. The binding free energy here is approximated by the site-integrated potential of mean force (SI-PMF),

$$\bar{W}(\mathbf{R}, \tfrac{1}{2}) = -k_B T \sum_{\gamma=1}^{N_L} \ln g_\gamma\left(\mathbf{r}_\gamma(\mathbf{R}, \tfrac{1}{2})\right), \qquad \text{(VI-1-1)}$$

where $\mathbf{R}$ and $\boldsymbol{\Omega}$ denotes the position and orientation of the ligand molecule, respectively, and $\mathbf{r}_\gamma(\mathbf{R},\boldsymbol{\Omega})$ represents the absolute coordinate of ligand site γ when the ligand molecule assumes position $\mathbf{R}$ and orientation $\boldsymbol{\Omega}$. (3) HPBMs generally gather in several positions to form clusters. A MPBM is defined as the mode that possesses the lowest SI-PMF in each cluster. (See the original paper [Imai et al. 2009] for more computational details.)

In order to demonstrate the performance of 3D-RISM-LM, the result for the binding of isopropanol to thermolysin in 1 M isopropanol-aqueous solution is given in Fig. VI-5-2 [Imai et al. 2009]. The MPBMs in the active site of the protein are shown in Fig. VI-5-2(a) (MPBMs are marked with Roman numerals.). Obviously, the representation of MPBM makes the ligand-binding character much easier to recognize than SDF. For example, it is found that MPBM-I and III are stabilized by hydrogen-bond network involving zinc ion and N112, as illustrated in Fig. VI-5-2(b). It is interesting to compare the obtained MPBMs with the X-ray crystallographic molecules [English et al. 1999], which are indicated by Arabic numerals in Fig. VI-5-2(a). The MPBMs are basically in good agreement with the crystallographic positions, except for MPBM-IV, which has only a small overlap with the corresponding crystallographic molecule (#5). The discrepancy is, however, reduced when the ligand concentration is decreased to 1 mM.

It is commonly assumed in conventional docking simulations that ligand-binding modes do not change with variation in the ligand concentration. This assumption is premised on the insensitivity of the binding free-energy surface to the solvent condition. However, Imai et al. found that the ligand-binding modes were actually modified, significantly in some cases, with the ligand concentration. Figures VI-5-2(c) and (d) show MPBMs-I and III in pure isopropanol

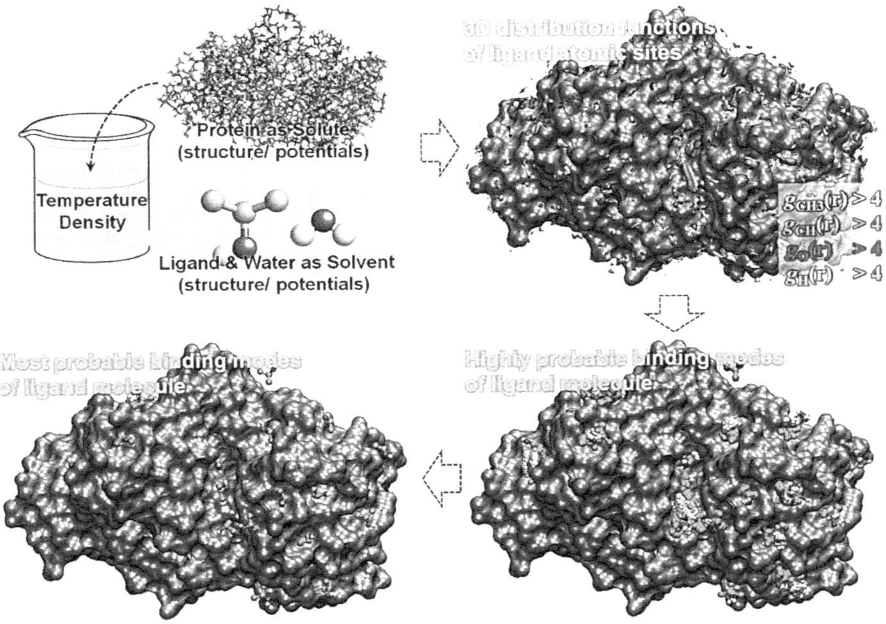

Fig. VI-5-1: Schematic illustration of the 3D-RISM-based ligand mapping method. Copyright (2009) American Chemical Society.

Color version at the end of the book

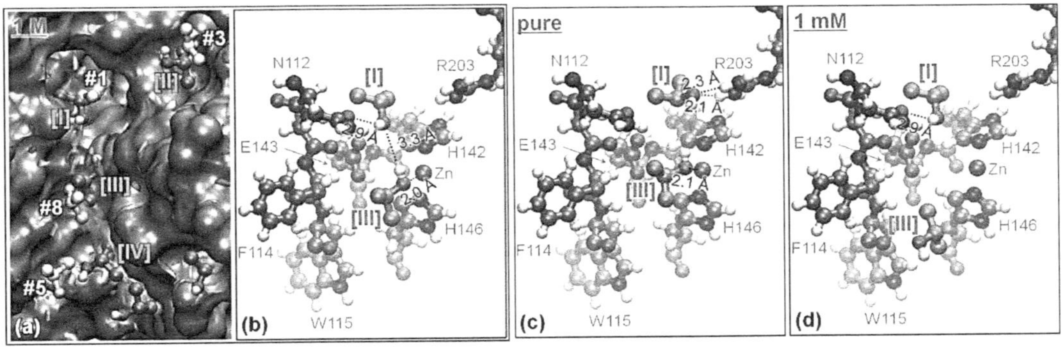

Fig. VI-5-2: Most probable binding modes (MPBMs) of isopropanol in the active site of thermolysin in 1 M isopropanol-aqueous solution. (a) Isosurface representation of $g(\mathbf{r}) > 4$ (yellow, CH_3; orange, CH; red, O; light blue, H.), MPBMs (green, carbon; red, oxygen; white, hydrogen), and X-ray crystallographic isopropanol molecules (cyan, carbon; red, oxygen). (b) Hydrogen-bond network between MPBM-I and III and its surrounding residues. (c) MPBM-I and III in pure isopropanol. (d) MPBM-I and III in 1 mM isopropanol-aqueous solution. Copyright (2009) American Chemical Society.

Color version at the end of the book

(as an extremely dense condition) and in 1 mM isopropanol-aqueous solution, respectively. The orientation of MPBM-I is significantly changed by the reduction of the ligand concentration. The isopropanol molecule preferably forms two hydrogen bonds to R203 in pure isopropanol (Fig. VI-5-2(c)), whereas it turns to N112 to make a hydrogen bond with N112 when the solution is diluted by water (in 1 M (Fig. VI-5-2(b)) and in 1 mM (Fig. VI-5-2(d)) solutions). This orientational transition is explained in terms of preferential hydration. Since a water molecule also can bind to both N112 and R203, a balance in the binding free energy between isopropanol and water determines which binding mode of isopropanol is more stable in aqueous solution. In this case, the isopropanol molecule prefers to form hydrogen bonds with R203 in pure isopropanol, while the addition of water makes it preferable to form a hydrogen bond with N112 rather than with R203 by competing with water for R203 as a hydrogen binding partner. A similar orientational transition is observed for MPBM-III. Although the stable interaction of MPBM-III with the zinc ion holds even in the 1 M aqueous solution, the isopropanol molecule concedes the interaction to a water molecule and moves to another preferable position when the concentration is reduced to 1 mM.

One of the most effective applications of 3D-RISM-LM can be found in fragment-based drug design (FBDD). In FBDD, a series of fragments of drug compounds is first screened against a target protein, and then a few active fragments are linked or merged to construct a much more active lead compound [Rees et al. 2004, Hajduk and Greer 2007, Congreve et al. 2008]. One of the greatest advantages of FBDD is that only a considerably fewer number of fragment compounds have to be prepared in the screening processes, because the substructures that commonly compose drug compounds are limited. On the other hand, FBDD has also certain problems. One is that a highly sensitive "detector" is needed because the fragmentation generally reduces the binding affinity. Additionally, detailed information regarding the fragment-binding modes is required for efficient fragment-linking, merging, or growth. 3D-RISM-LM is a promising method to solve the underlying problems in FBDD, since it can detect the binding of small molecules at the atomic resolution, as shown above.

As an indication of the effectiveness of 3D-RISM-LM in FBDD, Fig. VI-5.3 shows a comparison of selected MPBMs of small organic molecules, isopropanol, phenol, acetone,

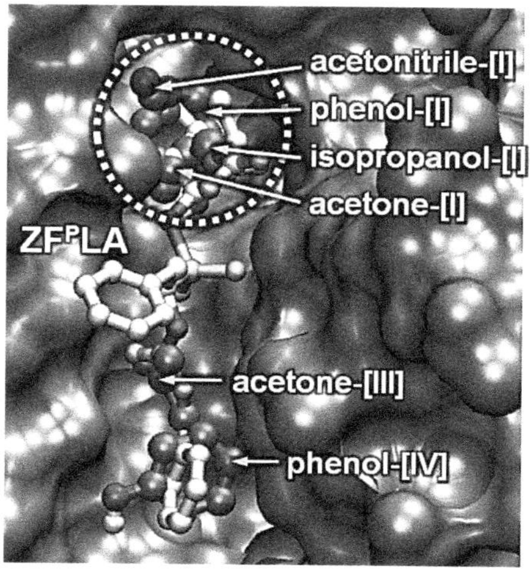

Fig. VI-5-3: Comparison of selected most probable binding modes (MPBMs) with the crystallographic structure of an inhibitor, carbobenzoxy-*Phe*^{P}-Leu-Ala (ZF^{P}LA). MPBMs are represented by ball-and-stick model (green, carbon; red, oxygen; blue, nitrogen). ZF^{P}LA is represented by the small ball-and-stick model (cyan, carbon; mauve, oxygen; blue, nitrogen; tan, phosphorus). The yellow dotted circle indicates the binding site for the common anchor motif of known inhibitors. Copyright (2009) American Chemical Society.

Color version at the end of the book

and acetonitrile, with the crystallographic structure of a known inhibitor [Holden et al. 1987], carbobenzoxy-*Phe*^{P}-Leu-Ala (ZF^{P}LA), in the active site of thermolysin. First, it is significant that every MPBM-I is found in the binding position of the head of the inhibitor (indicated by the yellow dotted circle). (Note that the site numbering hereafter is according to the original paper [Imai et al. 2009].) In fact, this part is the common anchor motif of known inhibitors; in other words, this binding site is the so-called "hot spot". This result implies that 3D-RISM-LM with a series of fragments can be used for the hot-spot search. There are additional two MPBMs that correspond to a functional moiety of the inhibitor. The MPBM-III of acetone is found to overlap with the carbonyl moiety of ZF^{P}LA. The acetone molecule forms a hydrogen bond with a peptide-hydrogen. On the other hand, the carbonyl group of ZF^{P}LA has the same hydrogen-binding pattern to stabilize the ZF^{P}LA-thermolysin complex. One of the benzene rings of ZF^{P}LA is also found at the same position as the MPBM-IV of phenol. The small difference in orientation is probably due to the hydration effect of the phenol hydroxyl group. (Note that the four ligands were not chosen as fragments of the inhibitor, but originally because their crystallographic binding positions are available for comparison.) Considering the fact that the two binding molecules of acetone and phenol could not be detected by X-ray crystallographic analysis [English et al. 2001], the result implies that 3D-RISM-LM can be used as a sensitive "detector" in FBDD.

References

Abragam A. 1961. The Principle of the Nuclear Magnetism. Clarendon Press. Oxford.

Akasaka K., Li H., Yamada H., Li R., Thoresen T. and Woodward C.K. 1999. Pressure response of protein backbone structure. Pressure induced amide 15N chemical shifts in BPTI. Protein Sci. 8: 1946–1953.

Akasaka K. 2003. Highly fluctuating protein structure revealed by variable-pressure nuclear magnetic resonance. Biochemistry. 42: 10877–10885.

Akasaka K. 2006. Probing conformational fluctuation of proteins by pressure perturbation. Chem. Rev. 106: 1814– 1835.

Akasaka K. and Matsuki H. 2015. High Pressure Bioscience—Basic Concepts, Applications and Frontiers. 2015. Springer. Dordrecht.

Akola J. and Jones R. 2003. Atp hydrolysis in water—a density functional study. J. Phys. Chem. B. 107(42): 11774–11783.

Anand R., Hoskins A.A., Stubbe J. and Ealick S.E. 2004. Domain organization of salmonella typhimurium formylglycinamide ribonucleotide amidotransferase revealed by X-ray crystallography. Biochemistry. 43: 10328–10342.

Ando H., Kuno M., Shimizu H.I.M. and Oiki S. 2005. Coupled K(+)-water flux through the HERG potassium channel measured by an osmotic pulse method. J. Gen. Physiol. 126: 529–538.

Anfinsen C.B. 1973. Principle that govern the folding of protein chain. Science. 181: 223–230.

Ansari A. 1999. Langevin mode analysis of myoglobin. J. Chem. Phys. 110: 1774.

Arai M. and Kuwajima K. 2000. Role of the molten gloubule state in protein folding. Advances in Protein Chemistry. 53: 209–282.

Argiolas A. and Pisano J.J.J. 1983. Facilitation of phospholipase-A2 activity by mastoparans, a new class of mast-cell degranulating peptides from WASP VENOM. Biol. Chem. 258: 3697–3702.

Asati V., Bharti S.K. and Budhwani A.K. 2017. 3D-QSAR and virtual screening studies of thiazolidine-2,4-Dione analogs: Validation of experimental inhibitory potencies towards PIM-1 kinase. J. Mol. Struct. 1133: 278–293.

Bachmann M. and Möröy T. 2005. The serine/threonine kinase Pim-1. Int. J. Biochem. Cell Biol. 37: 726–730.

Balucani U. and Zoppi M. 1994. Dynamics of the Liquid State. Clarendon Press. Oxford.

Barnett J.M., Cadman A., Burrell F.M., Madar S.H., Lewis A.P., Tisdale M. and Bethell R. 1999. *In vitro* selection and characterisation of influenza B/Beijing/1/87 isolates with altered susceptibility to zanamivir. Virology. 265: 286–295.

Barr I.G., McCauley J., Cox N., Daniels R., Engelhardt O.G., Fukuda K., Grohmann G., Hay A., Kelso A., Klimov A., Odagiri T., Smith D., Russell C., Tashiro M., Webby R., Wood J., Ye Z. and Zhang W. 2014. Epidemiological, antigenic and genetic characteristics of seasonal influenza A(H1N1), A(H3N2) and B influenza viruses: Basis for the WHO recommendation on the composition of influenza vaccines for use in the 2009–2010 northern hemisphere season. Vaccine. 28: 1156–1167.

Beglov D. and Roux B. 1997. An integral equation to describe the solvation of polar molecules in liquid water. J. Phys. Chem. B. 101: 7821–7826.

Bernal J.D. and Fowler R.H. 1933. A theory of water and ionic solution, with particular reference to hydrogen and hydroxyl ions. J. Chem. Phys. 1: 515–548.

Boon J.P. and Yip S. 1980. Molecular Hydrodynamics. McGraw-Hill. New York.

Borgnia M., Nielsen S., Engel A. and Agre P. 1999. Cellular and molecular biology of the aquaporin water channels Annu. Rev. Biochem. 68: 425–458.

Born Von M. 1920. Volumen und Hydratationswarme der Ionen. Z. Phys. 1: 45–48.

Boyd D.B. and Lipscomb W.N. 1969. Electronic structures for energy-rich phosphates. J. Theor. Biol. 25(3): 403–420.

Brooks B.R., Janežič D. and Karplus M. 1995. Harmonic analysis of large systems. I. Methodology. J. Comput. Chem. 16: 1522–1542.

Bryngelson J.D., Onuchic J.N., Socci N.D. and Wolynes P.G. 1995. Proteins, funnels, pathways, and the energy landscape of protein folding: a synthesis. Proteins. 21: 167–195.

Burger M.T., Nishiguchi G., Han W., Lan J., Simmons R., Atallah G., Ding Y., Tamez V., Zhang Y., Mathur M., Muller K., Bellamacina C., Lindvall M.K., Zang R., Huh K., Feucht P., Zavorotinskaya T., Dai Y., Basham S., Chan J., Ginn E., Aycinena A., Holash J., Castillo J., Langowski J.L., Wang Y., Chen M.Y., Lambert A., Fritsch C., Kauffmann A., Pfister E., Vanasse K.G. and Garcia P.D. 2015. Identification of N-(4-((1 R,3 S,5 S)-3-Amino-5-Methylcyclohexyl)Pyridin-3-Yl)-6-(2,6-Difluorophenyl)-5-Fluoropicolinamide (PIM447), a potent and selective proviral insertion site of moloney murine leukemia (PIM) 1, 2, and 3 kinase inhibitor in clinical trials for hematological malignancies. J. Med. Chem. 2(58): 8373–8386.

Burghardt T.P., Josephson M.P. and Ajtai K. 2011. Single myosin cross-bridge orientation in cardiac papillary muscle detects lever-arm shear strain in transduction. Biochemistry. 50: 7809–7821.

Cantor C.R. and Scimmel P.R. 1980. Biophysical Chemistry III, W.H. Freeman and Company, San Fransisco.

Chandler D. and Andersen H.C. 1972. Optimized cluster expansions for classical fluids. II. Theory of molecular liquids. J. Chem. Phys. 57: 1930–1931.

Chandler D. 1977. The dielectric constant and related equilibrium properties of molecular fluids: Interaction site cluster theory analysis. J. Chem. Phys. 67: 1113–1124.

Chandler D. 1987. Introductionto Modern Statistical Mechanics. Oxford, New York.

Chandrasekhar S. 1943. Stochastic problems in physics and astronomy. Rev. Mod. Phys. 15: 1–89.

Chen G., Fenimore P.W., Frauenfelder H. and Mezai F. 2008. Philosophical magazine. 88: 3877–3883.

Chen H., Wu Y. and Voth G.A. 2007. Proton transport behavior through the influenza A M2 channel: Insights from molecular simulation. Biophys. J. 93: 3470–3479.

Chen H., Ilan B., Wu Y., Zhu F., Schulten K. and Voth G.A. 2007. Charge delocalization in proton channels, I: The aquaporin channels and proton blockage. Biophys. J. 92: 46–60.

Chiles R. and Rossky P.J. 1984. Evaluation of reaction free-energy surfaces in aqueous-solution–an integral-equation approach. J. Am. Chem. Soc. 106: 6867–6868.

Chipot C. and Pohorille A. (eds.). 2007. Free energy calculations. Theory and Applications in Chemistry and Biology. Springer. Berlin.

Chizhmakov I.V., Geraghty F.M., Ogden D.C., Hayhurst A., Antoniou M. and Hay A.J. 1996. Selective proton permeability and pH regulation of the influenza virus M2 channel expressed in mouse erythroleukaemia cells. J. Physiol. 494: 329–336.

Chong S-H. and Hirata F. 1998a. Interaction-site-model description of collective excitations in classical molecular fluids. Phys. Rev. E. 57: 1691–1701.

Chong S-H. and Hirata F. 1998b. Mode-coupling theory for molecular liquids based on the interaction-site model. Phys. Rev. E. 58: 6188–6189.

Chong S-H. and Hirata F. 1998c. Dynamics of solvated ion in polar liquids: An interaction-site- model description. J. Chem. Phys. 108(17): 7339–7349.

Chong S-H. and Hirata F. 1999a. Interaction-site-model descritption of collective excitations in liquid water I: Theoretical study. J. Chem. Phys. 111: 3083–3094.

Chong S-H. and Hirata F. 1999b. Interaction-site-model descritption of collective excitations in liquid water II: Comparison with simulation results. J. Chem. Phys. 111: 3095–3104.

Chong S-H. and Hirata F. 1999c. Dynamics of Ions in liquid water: An interaction-site-model description. J. Chem. Phys. 111: 3654–3667.

Clementi C., Nymeyer H. and Onuchic J.N. 2000. Topological and energetic factors: What determines the structural details of the transition state ensemble and "en-route" intermediates for protein folding? An investigation for small globular proteins. J. Mol. Biol. 298: 937–953.

Clementi C., García A.E. and Onuchic J.N. 2003. Interplay among tertiary contacts, secondary structure formation and protein foldng mechanism: All-atom reperesentation studyof protein link. J. Mol. Biol. 326: 933–954.

Colvin M., Evleth E. and Akacem Y. 1995. Quantum-chemical studies of pyrophosphate hydrolysis. J. Am. Chem. Soc. 117(15): 4357–4362.

Congreve M., Chessari G., Tisi D. and Woodhead A.J. 2008. Recent developments in fragment-based drug discovery. J. Med. Chem. 51: 3661–80.

Cowan J.A. 2002. Structure and catalytic chemistry of magnetisium-dependent enzymes. Biometals. 15: 225–235.

Crick F.H.C. 1970. Central dogma of molecular biology. Nature. 227: 561–563.
Cumming P. and Stell G. 1982. Interaction site models for molecular fluids. Mol. Phys. 46: 383–426.
de Groot B.L. 2001. Grubmüller. Science. 294: 2353.
Dennis E.A. 1994. Diversity of group types, regulation, and function of phospholipase A2. 269: 13057–13060.
Doi M. and Edwards S.F. 1986. The Theory of Polymer Dynamics. Oxford Unversity Press.
Domen J., van der Lugt N.M.T., Laird P.W., Saris C.J.M., Clarke A.R., Hooper M.L. and Berns A. 1993. Impaired interleukin-3 response in Pim-1-Deficient bone marrow-derived mast cells. Blood. 82: 1445–1452.
Doster W., Cusack S. and Petry W. 1989. Dynamical transition of myoglobin revealed by ineleastic neutron scattering. Nature. 337: 754–756.
Doyle D.A. 1998. The structure of the potassium channel: Molecular basis of K+ conduction and selectivity. Science. 280: 69–77.
Eglestaff P.A. 1967. An Introduction to the Theory of the Liquid State. Academic Press. New York.
Eisenberg D. and Kauzman W. 1969. The Structure and Properties of Water. Clarendon, Oxford.
Englander S.W. and Mayne L. 2017. The case for defined protein folding pathways. PNAS. 114: 8253–8258.
English A.C., Done S.H., Caves L.S., Groom C.R. and Hubbard R.E. 1999. Locating interaction sites on proteins: the crystal structure of thermolysin soaked in 2% to 100% isopropanol. Proteins Struct. Funct. Genet. 37: 628–640.
English A.C., Groom C.R. and Hubbard R.E. 2001. Experimental and computational mapping of the binding surface of a crystalline protein. Protein Eng. 14: 47–59.
Ermak D.L. and McCammon J.A. 1978. Brownian dynamics with hydrodynamic interaction. J. Chem. Phys. 69: 1352.
Ewing G.J. and Maestas S. 1970. The thermodynamics of bsorption of xenon by myoglobin. J. Phys. Chem. 74: 2341–2344.
Feig M. (ed.). 2009. Modeling Solvent Environment. Wiley-VCH, 2009.
Fields B.N., Knipe D.M. and Howley P.M. 2007. Fields Virology. Wolters Kluwer Health/Lippincott Williams & Wilkins. Philadelphia.
Franks F. and Ives D.J.G. 1966. The structural properties of alcohol-water mixtures. Quat. Rev. Chem. Soc. 20: 1–44.
Franks F. 1972. Water: Compehensive Treatise. Plenum, New York.
Friedman H.L. 1964. Physica. 30: 509.
Friedman H.L. 1985. A Course in Statistical Mechanics. Prentice-Hall. New Jersey.
Friedman H.L., Reineri F.O., Hirata F. and Perng B-C. 1995. Surrogat hamiltonian description of solvation dynamics. Site number density and polarizable charge density formulation. J. Stat. Phys. 78: 239–266.
Feynman R.P., Leighton R.B. and Sands M. 1963. The Feynman Lectures on Physics. Addison-Wesley.
Feynman R.P. and Hibbs A.A. 1965. Quantum Mechanics and Path Integrals. McGraw-Hill, New York.
George P., Witonsky R.J., Trachtma M., Wu C., Dorwart W., Richman L., Richman W., Shurayh F. and Lentz B. 1970. Squiggle-h2o - an enquiry into importance of solvation effects in phosphate ester and anhydride reactions. Biochim. Biophys. Acta. 223(1): 1–15.
Grey R., Pierce A.C., Bemis G.W., Jacobs M.D., Moody C.S., Jajoo R., Mohal N. and Green J. 2009. Structure-based design of 3-Aryl-6-Amino-Triazolo[4,3-B]Pyridazine inhibitors of Pim-1 kinase. Bioorg. Med. Chem. Lett. 19: 3019–3022.
Grigorenko B.L., Rogov A.V. and Nemukhin A.V. 2006. Mechanism of triphosphate hydrolysis in aqueous solution: Qm/mm simulations in water clusters. J. Phys. Chem. B. 110(9): 4407–4412.
Groll D.H., Jeltsch A., Selent U. and Pingoud A. 1997. Does the restriction endonuclease EcoRV employ a two-metal-ion mechanism for DNA cleavage? Biochemistry. 36: 11389–11401.
Gubareva L.V., Matrosovich M.N., Brenner M.K., Bethell R.C. and Webster R.G. 1998. Evidence for Zanamivir resistance in an immunocompromised child infected with influenza B virus. J. Infect. Dis. 178: 1257–1262.
Gubareva L.V., Kaiser L. and Hayden F.G. 2000. Influenza virus neuraminidase inhibitors. Lancet. 355: 827–835.
Gubareva L.V. 2004. Molecular mechanisms of influenza virus resistance to neuraminidase inhibitors. Virus Res. 103: 199–203.
Gunnarson E., Axehult G., Baturina G., Zelenin S., Zelenina M. and Aperia A. 2005. Lead induces increased water permeability in astrocites expressing aquaporin 4. Neuroscience. 136: 105–114.
Habbard and Onsager. 1977. Dielectric-dispersion and dielectric friction in electrolyte-solutions. 1. J. Chem. Phys. 67: 4850–4857.
Hajduk P.J. and Greer J.A. 2007. Decade of fragment-based drug design: Strategic advances and lessons learned. Nat. Rev. Drug Discov. 6: 211–9.
Hammond C., Kartenbeck J. and Helenius A. 1992. Effects of dithiothreitol on beta-cop distribution and golgi to er membrane traffic. Mol. Biol. Cell. 3: A35–A35.

Hansen, J.P. and McDonald I.R. 1986. Theory of Simple Liquids. Academic Press, London.
Harvey S. and McCammon J.A. 1987. Dynamics of Proteins and Nucleic Acids. Cambridge University Press. Cambridge.
Harano Y., Sato H. and Hirata F. 2000a. Solvent effects on a diels-alder reaction in supercritical water: RISM-SCF study. J. Am. Chem. Soc. 122(10): 2289–229.
Harano Y., Sato H. and Hirata F. 2000b. Theoretical study on diels-alder reaction in ambient and supercritical water: Viewing solvent effect through the frontier orbitals. Chem. Phys. 258: 151–161.
Harashima, A. 1954. Theory of Liquids. Iwanami. Tokyo (written in Japanese).
Hasegawa T., Sugita M., Kikuchi, T. and Hirata F. 2017. A systematic analysis of the binding affinity between the pim-1 kinase and its inhibitors based on the MM/#D-RSIM/KH method. J. Chem. Inf. Model. 57: 2789–2798.
Helenius, A. 1992. Unpacking the incoming influenza virus. Cell. 69: 577–578.
Heltzberg O. and James M.N. 1985. Structure of the calcium regulatory muscle protein troponin-C at 2.8A resolution. Nature. 313: 653–659.
Henzler-Widman and Kern. 2007. Dynamic personality of protein. Nature. 450: 964–972.
Hill T.L. and Morales M.F. 1951. On high energy phosphate bonds of biochemical interest. J. Am. Chem. Soc. 73(4): 1656–1660.
Hille B. 2001. Ion Channels of Excitable Membranes. 3rd edn. Sinauer.
Hirata F. and Rossky P.J. 1981. An extended RISM equation for molecular polar fluids. Chem. Phys. Lett. 329–334.
Hirata F., Pettitt M.B. and Rossky P.J. 1982. Application of an extended RISM equation to dipolar and quadrupolar fluids. J. Chem. Phys. 77: 509–520.
Hirata F., Rossky P.J. and Pettitt M.B. 1983. The interionic potential of mean force in a molecular polar solvent from an extended RISM equation. J. Chem. Phys. 78: 4133–4144.
Hirata F. 1992. Interaction-site representation of the smoluchowski-vlasov equation: The space-time correaltion functions in a molecular liquid. J. Chem. Phys. 96: 4619–4624.
Hirata F., Munakata T., Raineri F. and Freidman H.L. 1995. An interaction-site representation of the dynamic structure factor of liquid and solvation dynamics. J. Mol. Liquids. 65/66: 15–22.
Hirata F. 2003. Molecular Theory of Solvation. Kluwer-Springer. Dordrecht.
Hirata F. 2015. In High Pressure Bioscience–Basic Concepts, Applications and Frontiers. Chapter 7. Springer. 2015.
Hirata F. and Akasaka K. 2015. Structural fluctuation of proteins induced by thermodynamic perturbation. J. Chem. Phys. 142: 044110–044118.
Hirata H. and Kim B. 2016. Multi-scale dynamics simulation of protein based on the generalized langevin equation combined with 3D-RISM theory. J. Mol. Liq. 217: 23–28.
Hirata F., Sugita M., Yoshida M. and Akasaka K. 2018. Structural fluctuation of protein and anfinsen's thermodynamic hypothesis. J. Chem. Phys. 148: 020901–020909.
Hirata F. 2018. On the interpretation of the temperature dependence of the mean sqaure of displacement (MSD) of protein, obtained from the incoherent neutron scattering. J. Mol. Liq. 270: 218–226.
Hofmann K.P. and Zundel G. 1974. Large hydration structure changes on hydrolyzing atp. Experientia. 30(2): 139–140.
Holden H.M., Tronrud D.E., Monzingo A.F., Weaver L.H. and Matthews B.W. 1987. Slow- and fast-binding inhibitors of thermolysin display different modes of binding: crystallographic analysis of extended phosphonamidate transition-state analogues. Biochemistry. 26: 8542–53.
Hong J., Yoshida N., Chong S.-H., Lee C., Ham S. and Hirata F. 2012. Elucidating the molecular origin of hydrolysis energy of pyrophosphate in water. J. Chem. Theory Comput. 8: 2239–2246.
Horton N.C., Newberry K.J. and Perona J.J. 1998. Metal ion-mediated substrate-assisted catalysis in type II restriction endonucleases. Proc. Natl. Acad. Sci. USA. 13489–13494.
Horton N.C. and Perona J.J. 2001. Making the most of metal ions. Nature Structural Biology. 290–293.
Horton N. and Perona J.J. 2004. DNA cleavage by EcoRV endonuclease: Two metal ions in three metal ion binding sites. Biochemistry. 6841–6857.
Hoye J.S. and Stell G. 1974. Statistical mechanics of polar systems: Dielectric constant for dipolar fluids. J. Chem. Phys. 61: 562–572.
Hu J., Fu R., Nishimura K., Zhang L., Zhou H., Busath D.D., Vijayvergiya V. and Cross T.A. 2006. Histidines, heart of the hydrogen ion channel from influenza A virus: Toward an understanding of conductance and proton selectivity. Proc. Natl. Acad. Sci. U.S.A. 103: 6856–6870.
Huang X.H. and Boxer S.G. 1994. Discovery of new ligand binding pathways in myoglobin by random mutagenesis. Struct. Biol. 1: 226–229.

Huang X., Holden H.M. and Raushel F.M. 2001. Channeling of substrates and intermediates in enzyme-catalyzed reactions. Annu. Rev. Biochem. 70: 149–180.

Iida K. and Sato H. 2012a. An extended formula of site-site Smoluchowski-Vlasov equation for electrolyte solution and infinitely dilute solution. J. Chem. Phys. 137: 034506.

Iida K. and Sato H. 2012b. A theory for time-dependent solvation structure near solid-liquid interface. J. Chem. Phys. 136: 244502.

Ikeguchi M., Ueno J., Sato M. and Kidera A. 2005. Protein structural change upon ligand binding. Phys. Rev. Lett. 94: 078102.

Ikura M., Clore G.M., Gronenborn A.M., Zhu G., Klee C.B. and Bax A. 1992. Solution structure of a calmodulin-target peptide complex by multidimensional NMR. 256: 632–638.

Ikuta Y. and Hirata F. 2019. First principle determination of dielectric constant of polar liquids by the extended RISM theory with the bridge diagrams concerning dipole-dipole interaction. https://doi.org/10.1016/j.molliq.2019.111567.

Ilan B., Tajkhorshid E., Schulten K. and Voth G. 2004. The mechanism of proton exclusion in aquaporin channels. Protein-structure Function and Bioinformatics. 55: 223–228.

Imai T., Kinoshita M. and Hirata F. 2000. Theoretical study for partial molar volume of amino acids in aqueous solution: Implacation of ideal fluctuation volume. J. Chem. Phys. 112: 9469–9478.

Imai T., Kovalenko A. and Hirata F. 2004 Solvation thermodynamics of protein studied by the 3D-RISM theory. Chem. Phys. Lett. 395: 1–6.

Imai T., Kovalenko A. and Hirata F. 2005a. Partial molar volume of proteins studied by the three-dimensional reference interaction site model theory. J. Phys. Chem. B. 109: 6658–6665.

Imai T., Hiraoka R., Kovalenko A. and Hirata F. 2005b. Water molecules in a protein cavity detected by a statistical-mechanical theory. J. Am. Chem. Soc. 127: 15334–15335.

Imai T., Hiraoka R., Kovalenko A. and Hirata F. 2007a. Locating missing water molecules in protein cavities by the three-dimensional reference interaction site model theory of molecular solvation. Proteins-Structure Function and Bioinformatics. 66: 804–813.

Imai T., Hiraoka R., Seto T., Kovalenko A. and Hirata F. 2007b. Three-dimensional distribution function theory for the prediction of protein-ligand binding sites and affinities: application to the binding of noble gases to hen egg-white lysozyme in aqueous solution. J. Phys. Chem. B. 111: 11585–11591.

Imai T., Ohyama S., Kovalenko A. and Hirata F. 2007. Theoretical study of the partial molar volume change associated with pressure-induced structural transition of ubiquitin. Protein Science. 16: 1927–1933.

Imai T., Oda K., Kovalenko A., Hirata F. and Kidera A. 2009. Ligand mapping on protein surfaces by the 3D-RISM theory: Toward computational fragment-based drug design. J. Am. Chem. Soc. 131: 12430–12440.

Intharathep P., Laohpongspaisan C., Rungrotmongkol T., Loisruangsin A., Malaisree M., Decha P., Aruksakunwong O., Chuenpennit K., Kaiyawet N., Sompornpisut P., Pianwanit S. and Hannongbua S. 2008. How amantadine and rimantadine inhibit proton transport in the M2 protein channel. J. Mol. Graph. Model. 27: 342–348.

Ishida T., Hirata F., Sato H. and Kato S. 1998. Molecular theory of solvent effect on keto-enol tautomers of formamide in aprotic solvents: RISM-SCF approach. J. Phys. Chem. B. 102: 2045–2050.

Ishida T., Hirata F. and Kato S. 1999a. Thermodynamics analysis of the solvent effect on tautomerization of acetylacetone: An Ab initio approach. J. Phys. Chem. 110: 3938–3945.

Ishida T., Hirata F. and Kato S. 1999b. Solvation dynamics of benzonitrile excited state in polar solvents: A time-dependent reference interaction site model self-consistent field approach. J. Chem. Phys. 110: 11423–11432.

Jackson D., Barclay W. and Zürcher T. 2005. Characterization of recombinant influenza B viruses with key neuraminidase inhibitor resistance mutations. J. Antimicrob. Chemother. 55: 162–169.

Jakalian A., Bush B.L., Jack D.B. and Bayly C.I. 1999. Fast, efficient generation of high-quality atomic charges. AM1-BCC Model: I. Method. J. Comput. Chem. 21: 132–146.

Jakalian A., Jack D.B. and Bayly C.I. 2002. Fast, efficient generation of high-quality atomic charges. AM1-BCC Model: II. Parameterization and validation. J. Comput. Chem. 23: 1623–1641.

Jiang J., Daniels B.V. and Fu D. 2006. Crystal structure of AqpZ tetramer reveals two distinct Arg-189 conformations associated with water permeation through the narrowest constriction of the water-conducting channel. J. Biol. Chem. 281: 454–460.

Johansson I., Karlsson M., Johanson U., Larsson C. and Kjellbom P. 2000. The role of aquaporins in cellular and whole plant water balance. Biochimica et Biophysica Acta. 1465: 324–342.

Kamerlin S.C.L. and Warshel A. 2009. On the energetics of atp hydrolysis in solution. J. Phys. Chem. B. 113(47): 15692–15698.

Kass I. and Arkin T. 2005. How pH opens a H+ channel: The gating mechanism of influenza A M2, Structure. 13: 1789–1798.

Kataoka M., Nishi I., Fujisawa T., Ueki T., Tokunaga F. and Goto Y. 1995. Structural characterization of the molten globule and native states of apomyoglobin by solution X-ray scattring. J. Mol. Biol. 249: 215–228.

Kataoka M. and Nakagawa H. 2011. Effect of Hydration on Protein Dynamics. in Water, The Forgotten Biological Molecule. PanStanford Publishing. Singapore.

Kawai N., Ikematsu H., Iwaki N., Maeda T., Kanazawa H., Kawashima T., Tanaka O., Yamauchi S., Kawamura K., Nagai T., Horii S., Hirotsu N. and Kashiwagi S. 2008. A comparison of the effectiveness of zanamivir and oseltamivir for the treatment of influenza A and B. J. Infect. 56: 51–57.

Kawata M., Ten-no S., Kato S. and Hirata F. 1996. Solvent effect on acidity: A hybrid approach based on the RISM and hartree-fock equation. J. Phys. Chem. 100: 1111–1117.

Kawata M., Tenno S., Kato S. and Hirata F. 1995. Irregular order in basicities of methylamines in aqueous solutions: A RISM-SCF study. J. Am. Chem. Soc. 117: 1638–1640.

Kim B. and Hirata F. 2013. Structural fluctuation of protein in water around its native state: A new statistical mechanics formulation. J. Chem. Phys. 138: 054108–054111.

Kim J.-G., Muniyappan S., Oang K.-Y., Kim T.-W., Yang C., Kim K.-H., Kim J. and Ihee H. 2016. Cooperative protein structural dynamics of homodimeric hemoglobin linked to water cluster at subunit interface revealed by time-resolved X-ray solution scattering. Structural Dynamics. 3: 023610.

King L.S., Kozono D. and Agre P. 2004. From structure to disease: The evolving tale of aquaporin biology. Nat. Rev. Mol. Cell Biol. 5: 687–698.

Kirkwood J.G. and Buff F.P. 1951. The statistical mechanical theory of solutions. J. Chem. Phys. 19: 774–777.

Kitao A., Hirata F. and Go N. 1991. The effect of solvent on the conformation and the collective motions of protein: normal mode analysis and molecular dynamics simulations of melittin in water and in vacuum. Chem. Phys. 158: 447–472.

Kiyota Y., Hiraoka R., Yoshida N., Maruyama Y., Imai T. and Hirata F. 2009. Theoretical study of CO escaping pathway in myoglobin with the 3D-RISM theory. J. Am. Chem. Soc. 131: 3852–3853.

Kiyota Y., Yoshida N. and Hirata F. 2011. A new approach for investigating the molecular recognition of protein: Toward structural-based drug design based on the 3D-RISM theory. J. Chem. Theor. Comp. 7: 3803–3815.

Klaehn M., Rosta E. and Warshel A. 2006. On the mechanism of hydrolysis of phosphate monoesters dianions in solutions and proteins. J. Am. Chem. Soc. 128(47): 15310–15323.

Koga N. and Takada S. 2001. Roles of native topology and chain-length scaling in protein folding: A simulation study with a go-like model. J. Mol. Biol. 313: 171–180.

Koide, S. 1969. Quantum Mechanics (Japanese), Shokabo, Tokyo.

Koneshan S., Rasaiah J.C. and Lynden-Bell R.M. 1998. Solvent structure, dynamics, and ion mobility in aqueous solutions at 25°C. J. Phys. Chem. B. 102: 4193–4204.

Kostrewa D. Winkler. 1995. Mg2+ Binding to the active site of EcoRV endonuclease endonuclease: A crystallographic study of complexes with substrate and product DNA at 2Å resolution. Biochemistry. 683–696.

Kottlam J. and Case D.A. 1990. Langevin mode of macromolecules: Applications to crambin and DNA hexamers. Biopolymers. 29: 1409–1421.

Koua F.H.M. and Kandori H. 2016. Light-induced structural change during early photo-intermediates of the eubacterial Cl-pump fulvimarina rhodopsin observed by FTIR difference spectroscopy. RSC Advances. 6: 383–392.

Kovalenko A. and Hirata F. 1998. Three-dimensional density profiles of water in contact with a solute of arbitraly shape: A RISM approach. Chem. Phys. Letter. 290: 237–244.

Kovalenko A. and Hirata F. 1999. Self-consistent description of a metal-water interface by the kohn-sham density functional theory and the three-dimensional reference interaction site model. J. Chem. Phys. 110: 10095–10112.

Kubo R. 1957. Statistical-mechanical theory of irreversible processes. I. J. Phys. Soc. Jpn. 12: 570–585.

Kubo R., Toda M. and Hashitsume N. (eds.). 1992. Statistical Physics II, Nonequilibrium Statistical Mechanics, Springer, Berlin.

Kubota H., Tanaka Y. and Makita T. 1987. Volumetric behavior of pure alcohols and their mixtures under high-pressure. Int. J. Thermophys. 8: 47–70.

Kuroki R. and Yutani K. 1998. Structural and thermodynamic responses of mutations at a Ca2+ binding site engineered into human lysozyme. J. Biol. Chem. 273: 34310–34315.

Laird P.W., van der Lugt N.M.T., Clarke A., Domen J., Linders K., McWhir J., Berns A. and Hooper M. 2003. *In vivo* analysis of Pim-1 deficiency. Neucleic Acids Res. 21: 4750–4755.

Lamm G. and Szabo A. 1986. Langevin mode of macrmolecules. J. Chem. Phys. 85: 7334.

Lamb R.A., Holsinger L.J. and Pinto L.H. 1994. In receptor-mediated virus entry into cells. pp. 303–321. *In*: Wimmer E. (ed.). Cold Spring Harbor Laboratory Press, Cold Spring Harbor, NY.

Landau L.D. and Lifshitz E.M. 1957. Mechanics, Addison-Wesley, Reading, Mass.

Landau L.D. and Lifshitz E.M. 1958. Statistical Physics, Addison-Wesley, Reading, Mass.

Leach A.R. 2001. Molecular Modelling. Principles and Applications. Second Edition. Prentice-Hall. Harlow. England.

Lear J.D. 2003. Proton conduction through the M2 protein of the influenza A virus; a quantitative, mechanistic analysis of experimental data. FEBS Letters. 552: 17–22.

Lee J.Y., Yoshida N. and Hirata F. 2006. Conformational equilibrium of 1,2-dichloroethane in water: Comparison of PCM and RISM-SCF methods. J. Phys. Chem. B. 110: 16018–16025.

Lehn J.-M. 1990. From molecular recognition towards molecular informaiton processing and self-organization. Angewante Chemie. 29: 1304–1319

Lehn J.-M. 1995. Supramolecular Chemistry. Willy-VCH, Weinheim.

Levinthal C. 1968. Are there pathways for protein folding? J. Chim. Phys. 65: 44–45.

Liang C. and Li Y.-Y. 2014. Use of regulators and inhibitors of Pim-1, a serine/threonine kinase, for tumour therapy (Review). Mol. Med. Rep. 9: 1–10.

Lipmann F. 1941. Metabolic generation and utilization of phosphate bond energy. Adv. Enzymol. Rel. S Bi. 1: 99–162.

Makowski L., Rodi, D. and Mandava S. 2008. Molecular crowding inhibits intramolecular breathing motions in proteins. J. Mol. Bio. 375: 529–546.

Massiere F. and Badet-Denisot M.A. 1998. The mechanism of glutamine-dependent amidotransferases. Cell. Mol. Life Sci. 54: 205–222.

McHaourab H.S., Oh K.J., Fang C.J. and Hubbel W.L. 1997. Conformation of T4 lysozyme in solution. Hinge-bending motion and the substrate-induced conformational transition stucied by site-directed spin labling. Biochemistry. 36: 307–316.

McKimm-Breschkin J.L. 2002. Neuraminidase inhibitors for the treatment and prevention of influenza. Expert. Opin. Pharmacother. 3: 103–112.

McQuarrie. 1976. D.A. Statistical Mechanics. Harper and Row. New York.

Merkel A.L., Meggers E. and Ocker M. 2012. PIM1 kinase as a target for cancer therapy. Expert. Opin. Invest. Drugs. 21: 425–436.

Meyerhof O. and Lohmann K. 1932. Energetic exchange connections amongst the volume of phosphoric acetic acid in muscle extracts. Biochem. Z. 253: 431–461.

Michaelis L. and Menten M. 1913. Die kinetik der invertinwirkung. Biochem. Z. 49: 333–369.

Milani M., Pesce A., Bolognesi M., Bocedi A. and Ascenzi P. 2003. Substrate channeling: Molecular bases. Biochemistry and Molecular Biology Education. 31: 228–233.

Miles E.W., Rhee S. and Davies D.R. 1999. The molecular basis of substrate channeling. J. Biol. Chem. 274: 12193–12196.

Mishin V.P., Hayden F.G. and Gubareva L.V. 2005. Susceptibilities of antiviral-resistant influenza viruses to novel neuraminidase inhibitors. Antimicrob. Agents. Chemother. 49: 4515–4520.

Mizutani Y. and Kitagawa T. 1997. Direct observation of cooling of heme upon photodissociation of carbonmonoxy myoglobin. Science. 278: 443–446.

Morais-Cabral J.H., Zhou Y.F. and MacKinnon R. 2001. Energetic optimization of ion conduction rate by the K+ selectivity filter. Nature. 414: 37–42.

Morita T. and Hiroike K. 1960. A new approach to the theory of classical fluids. I. Prog. Theor. Phys. 23: 1003–1024.

Morita T. and Hiroike K. 1961. A new approach to the theory pf classical fluids. III. General Treatment of Classical Systems. Prog. Theor. Phys. 25: 537–578.

Mouilleron S. and Golinelli-Pimpaneau B. 2007. Conformational changes in ammonia-channeling glutamine amidotransferases. Curr. Opin. Struct. Biol. 17: 653–664.

Mould J.A., Li H., Dunlak C.S., Lear J.D., Pekosz A., Lamb R.A. and Pinto L.H. 2000a. Mechanism for proton conduction of the M2 ion channel of influenza A virus. J. Biol. Chem. 275: 8592–8599.

Mould J.A., Drury J.E., Frings S.M., Kaupp U.B., Pekosz A., Lamb R.A. and Pinto L.H. 2000b. Permeation and activation of the M2 ion channel of influenza A virus. J. Biol. Chem. 275: 31038–31050.

Naka K., Sato H., Morita A., Hirata F. and Kato S. 1999. RISM-SCF study for the free energy profile of menshutkin type reaction NH3 + CH3Cl →NH3CH3+ + Cl- in Aqueous Solution. Theor. Chem. Acc. 102: 165–169.

Nakagawa H., Kamikubo H., Tsukushi I., Kanaya T. and Kataoka M. 2004. Protein dynamical heterogeneity derived from neutron incoherent elastic scattering. J. Phys. Soc. Jpn. 73: 491–495.

Nakagawa H., Joti Y., Kitao A. and Kataoka M. 2008. Hydration affects both harmonic and anharmonic nature of protein dynamics. Biophys. J. 95: 2916–2923.

Nakagawa H. and Kataoka M. 2010. Percolation of hydration water as a control of protein dynamics. J. Phys. Soc. Jpn. 79: 083801.

Nemeth-Cahalan K.L., Kalman K. and Hall J.E. 2004. Molecular basis of pH and Ca2+ regulation of Aquaporin water ermeability. J. Gen. Physiol. 123: 573–580.

Nemethy G. and Scheraga H.A. 1962. Structure of water and hydrophobic bonding in protein. I. A model for the thermodynamic properties of liquid water. J. Chem. Phys. 36: 3382–3340.

Nicolas J.P., Lin Y., Lambeau G., Ghomashchi F., Lazdunski M. and Gelb M.H. 1997. Localization of structural elements of bee venom phospholipase A2 involved in n-type recepter binding and neurotoxicity. J. Biol. Chem. 272: 7173–7181.

Nimigean C.M. and Allen T.W. 2011. Origins of ion selectivity in potassium channels from the perspective of channel block. J. Gen. Physiol. 137: 405–413.

Nishihara Y., Sakakura M., Kimura Y. and Terazima M. 2004. The escape process of carbon monoxide from myoglobin to solution at physiological temperature. J. Am. Chem. Soc. 126: 11877–11888.

Nishiyama K., Hirata F. and Okada T. 2000a. Importance of acoustic solvent mode and solute-solvent radial distribution functions in solvation dynamics: Studied by RISM theory. J. Chin. Chem. Soc. 47: 837–842.

Nishiyama K., Hirata F. and Okada T. 2000b. Relaxation of average energy and rearrangement of solvent shells in various polar solvents in connection with solvation dynamics: Studied by RISM theory. Chem. Phys. Lett. 330: 125–131.

Nishiyama K., Hirata F. and Okada T. 2001. Nonlinear response of solvent molecules induced by instantaneous change of solute electronic structure: Studied by RISM theory. J. Mol. Struct. 565: 31–34.

Nishiyama K., Hirata F. and Okada T. 2003. Solute-structure dependence of solvation dynamics studied by reference interaction-site model theory. J. Chem. Phys. 118: 2279–2285.

NIshiyama K.,Yamaguchi T. and Hirata F. 2009. Solvation dynamics in polar solvents studied by means of RISM/ Mode-Coupling theory. J. Phys. Chem. B. 113: 2800–2804.

Nishiyama K., Watanabe Y., Yoshida N. and Hirata F. 2012. Solvent effects on electronic structures of coumarin 153: Parallel studies by means of spectroscopy and RISM-SCF calculations. J. Phys. Soc. Jpn. 81: SA016.

Noguti T. and Go N. 1982. Collective variable description of small-amplitude conformational fluctuations in a globular protein. Nature. 296: 776–778.

Noji H., Yasuda R., Yoshida M. and Kinoshita K. Jr. 1997. Direct observation of F1-ATPase. Nature. 386: 299–302.

Nomura H. and Kawaizumi F. Ed. 1994. Special Issue: Structure, Fluctuation, and Relaxation in Solutions—Proceedings of the Yamada Conference XXXXII. Elsevier Science Publications and Yamada Science Foundations. Amsterdam.

Okada A., Miura T. and Takeuchi H. 2001. Protonation of histidine and histidine-tryptophan interaction in the activation of the M2 ion channel from influenza A virus. Biochemistry. 40: 6053–6060.

Omelyan I., Kovalenko A. and Hirata F. 2003. Compressibility of tert-Butyl alchohol-water mixtures: The RISM theory. J. Theor. Comp. Chem. 2: 192–203.

Onishi I., Sunaba S., Yoshida N. and Hirata F. 2018. Role of Mg^{2+} ions in DNA hydrolysis by EcoRV, studied by the 3D-reference interaction site model and molecular dynamics. J. Phys. Chem. B. 122: 9061–9075.

Panchenko A.R., Schulten Z.L., Cole R. and Wolynes P.G. 1997. The Foldon Universe: A survey of structural similarity and self-reorgnaization of independently folding units. J. Mol. Biol. 272: 95–105.

Pange T., Schiltz M., Permot L., Colloc'h N., Longhi S., Bourguet W. and Fourme R. 1998. Exploring hydrophobic site in proteins with xenon or krypton. Proteins: Struct. Funct. Genet. 30: 61–73.

Perkyns J.S. and Pettitt B.M. 1992a. A dielectrically consistent interaction site theory for solvent electrolyte mixtures. Chem. Phys. Lett. 190: 626–630.

Perkyns J.S. and Pettitt B.M. 1992b. A site site theory for finite concentration saline soutions. J. Chem. Phys. 97: 7656–7666.

Perutz M.F. and Matthews F.S. 1966. An X-ray study of azide methaemoglobin. J. Mol. Biol. 21: 199–202.

Pettitt M.B. and Rossky P.J. 1982. Integral-equation predictions of liquid-state structure for water like intermolecular potentials. J. Chem. Phys. 77: 1451–1457.

Phongphanphanee S., Yoshida N. and Hirata F. 2007. The statistical-mechanics study for the distribution of water molecules in aquaporin. Chem. Phys. Lett. 449: 196–201.

Phongphanphanee S., Yoshida N. and Hirata F. 2008. On the proton exclusion of aquaporins: A statistical mechanics study. J. Am. Chem. Soc. 130: 1540–1541.

Phongphanphanee S., Yoshida N. and Hirata F. 2010a. Molecular selectivity in aquaporin channels studied by the 3D-RISM theroy. J. Phys. Chem. B. 114: 7967–7973.

Phongphanphanee S., Rungrotmongkol T., Yoshida N., Hannonbua S. and Hirata F. 2010b. Proton transport through the influenza A M2 channel: Three-dimensional reference interaction site model study. J. Am. Chem. Soc. 132: 9782–9788.

Phongphanphanee S., Yoshida N., Oiki S. and Hirata F. 2014a. The 'Ambivalent' Snug-fit sites in the KcsA potassium channel probed by 3D-RISM theory. Pure Appl. Chem. 86: 97–104.

Phongphanphanee S., Yoshida N., Oiki S. and Hirata F. 2014b. Distinct configurations of cations and water in the selective filter of the KcsA potasium channel probed by the 3D-RISM theory. J. Mol. Liquid. 200: 52–58.

Pinto L.H., Holsinger L.J. and Lamb, R.A. 1992. Influenza virus M2 protein has ion channel activity. Cell. 69: 517–528.

Pople J.A. 1951. A theory of the structure of water. Proc. R. Soc. London A. 205: 163.

Preston G.M., Carroll T.P., Guggino W.B. and Agre, P. 1992. Appearance of water channels in Xenopus oocytes expressing red cell CHIP28 protein. Science. 256: 385–387.

Raushel F.M., Thoden J.B. and Holden H.M. 1999. The amidotransferase family of enzymes: Molecular machines for the production and delivery of ammonia. Biochemistry. 38: 7891–7899.

Rees D.C., Congreve M., Murray C.W. and Carr R. 2004. Fragment-based lead discovery. Nat. Rev. Drug Discov. 3: 660–672.

Raineri F.O., Resat H., Perng B.C., Hirata F. and Friedman H.L. 1994. A molecular theory of solvation dynamics. J. Chem. Phys. 100: 1477–1491.

Ricci M.A., Rocca D., Rucco G. and Vallauri R. 1989. Theoretical and computer-simulation study of the density-fluctuation in liquid water. Phys. Rev. A. 40: 7226–7238.

Rossky P.J. and Dale W.D.T. 1980. Generalized recursive solutions to ornstein-zernike integral-equation. J. Chem. Phys. 73: 2457–2464.

Sakakura M., Yamaguchi S., Hirota, N. and Terazima M. 2001. Dynamics of structure and energy of horse carboxymyoglobin after photodissociation of the carbon monoxide. J. Am. Chem. Soc. 123: 4286–4294.

Samoilov O.Y. 1965. Structure of Aqueous Electrolyte Solutions and the Hydration of Ions. Consultants Bureau, New York.

Sato H., Hirata F. and Kato S. 1996. Analytical energy gradient for the reference interaction site model multiconfigurational self-consistent-field method: Application to 1,2-Difluoroethylene in aqueous solution. J. Chem. Phys. 105: 1546–1551.

Sato H. and Hirata F. 1998. Theoretical study for autoionization of liquid water: Temperature dependence of the ionic product (pK(w)). J. Phys. Chem. A. 102: 2603–2608.

Sato H., Hirata F. and Myers A.B. 1998. Theoretical study of the solvent effect on triiodide ion in solutions. J. Phys. Chem. 102: 2065–2070.

Sato H. and Hirata F. 1999a. The Syn-/Anti-conformational equilibrium of acetic acid in water studied by the RISM-SCF/MCSCF method. THEOCHEM. 461: 113–120.

Sato H. and Hirata F. 1999b. Revisiting the acid-base equilibrium in aqueous solutions of hydrogen halides: Study by the Ab Inition electronic structure theory combined with the statistical mechanics of molecular liquids. J. Am. Chem. Soc. 121: 3460–3467.

Sato H. and Hirata F. 1999c. Ab Initio study on molecular and thermodynamic properties of water: A theoretical prediction of pKw over a wide range of temperature and density. J. Phys. Chem. B. 103: 6596–6604.

Sato H., Kobori Y., Tero-Kubota S. and Hirata F. 2003. Theoretical study on electronic and solvent reorganization associated with a charging process of organic compounds: I. Molecular and atomic level description of solvent reorganization. J. Chem. Phys. 119: 2753–2760.

Sato H., Kobori Y., Tero-Kubota S. and Hirata F. 2004. Theoretical study on electronic and solvent reorganization associated with a charging process of organic compounds: 2. A new decomposition procedure into electrostatic and nonelectrostatic responses. J. Phys. Chem. B. 108: 11709–11715.

Schotte F., Soman J., Olson J.S., Wulff M. and Anfinrud A. 2004. Picosecond time-resolved X-ray crystallography. J. Struc. Biol. 147: 235–246.

Scott E.E., Gibson Q.H. and Olson J.S. 2001. Mapping the pathways for O2 entry into and exit from Myoglobin. J. Bio. Chem. 276: 5177–5188.

Shah N., Pang B., Yeoh K.-G., Thorn S., Chen C.S., Lilly M.B. and Salto-Tellez M. 2008. Potential roles for the PIM1 kinase in human cancer—A molecular and therapeutic appraisal. Eur. J. Cancer. 44: 2144–2151.

Shen T., Hofmann C.P., Oliveberg M. and Wolynes P.G. 2005. Scanning malleable transition state ensembles: Comparing theory and experiment for folding protein U1A. Biochemistry. 44: 6433–6439.

Shih C.C. and Georghiou S. 2000. Harmonic analysis of DNA dynamics in a viscous medium. J. Biomol. Struct. Dyn. 17: 921–32.

Sindhikara D.J. and Hirata F. 2013. Analysis of biomolecular solvation sites by 3D-RISM theory. J. Phys. Chem. B. 117: 6718–6723.

Singer S.J. and Chandler D. 1985. Free-energy functions in the extended RISM approximation. Mol. Phys. 55: 621–625.

Smith W.W., Schreck C.F., Hashem N., Soltani S., Nath A., Rhoades E. and O'Hern C.S. 2012. Phys. Rev. E. 86: 041910–041919.

Spivey H.O. and Ovádi J. 1999. Substrate channeling. Methods. 19: 306–321.

Sugase K., Dyson H.J. and Wright P.E. 2007. Mechanism of coupled folding and binding of an intrinsically disordered protein. Nature. 447: 1021–1025.

Sugaya N., Mitamura K., Yamazaki M., Tamura D., Ichikawa M., Kimura K., Kawakami C., Kiso M., Ito M., Hatakeyama S. and Kawaoka Y. 2007. Lower clinical effectiveness of oseltamivir against influenza B contrasted with influenza A infection in children. Clin. Infect. Dis. 44: 197–202.

Sullivan D.E. and Gray C.G. 1981. Evaluation of angular-correlation parameters and the dielectric-constant in the RISM approximation. Mol. Phys. 42: 443–454.

Tabellout M., Lanceleur P. and Emery Jr. Hayward D. and Pethrick R.A. 1990. Dielectric, ultrasonic and C-13 Nuclear-magnetic-resonance relaxation measurements of Tert-Butyl alcohol water mixtures. J. Chem. Soc. Faraday Trans. 86: 1493–1501.

Takahashi H., Matubayasi N., Nakahara M. and Nitta T. 2004. A quantum chemical approach to the free energy calculations in condensed systems: the qm/mm method combined with the theory of energy representation. J. Chem. Phys. 121(9): 3989–3999.

Takahashi H., Kawashima Y., Nitta T. and Matubayasi N. 2005. A novel quantum mechanical/molecular mechanical approach to the free energy calculation for isomerization of glycine in aqueous solution. J. Chem. Phys. 123: 124504.

Takeuchi H., Okada A. and Miura T. 2003. Roles of the histidine and tryptophan side chains in the M2 proton channel from influenza A virus. FEBS Lett. 552. 35–38.

Tang Y., Zaitseva F., Lamb R.A. and Pinto L.H. 2002. The gate of the influenza virus M2 proton channel is formed by a single tryptophan residue. J. Biol. Chem. 277: 39880–39886.

Tanwar A., Sindhikara D., Hirata F. and Anand R. 2015. Determination of the formylglycinamide ribonucleotide amidotransferase ammonia pathway by combining 3D-RISM theory with experiment. ACS Chem. Biol. 10: 698–704.

Taylor G. and Russell R. 2010. Influenza Virus Neuraminidase Inhibitors. pp. 103–110. *In*: Bradshaw R.A. and Dennis E.A. (eds.). Handbook of Cell Signaling (Second Edition). Academic Press. San Diego.

Tenno S., Hirata F. and Kato S. 1993. A hybrid approach for the solvent effect on the electronic structure of a solute based on the RISM and hartree-fock equations. Chem. Phys. Lett. 214: 391–396.

Tenno S., Hirata F. and Kato S. 1994. Reference interaction site model self-consistent field study for solvation effect on carbonyl compounds in aqueous solution. J. Chem. Phys. 100: 7443–7453.

Terazima M. 2000. Translational diffusion of organic radicals in soluton. Acc. Chem. Res. 33: 687–694.

Terazima M. 2015. Molecular Science of Fluctuations Toward Biological Functions. Springer.

Terazima M. 2016. Molecular Science of Fluctuations Toward Biological Functions. Springer. Japan.

Thiessen W.E. and Narten A.H. 1982. Neutron diffraction study of light and heavy water mixtures at 25 C. J. Chem. Phys. 77: 2655–2662.

Thompson A.N., Kim I., Panosian T.D., Iverson T.M., Allen T.W. and Nimigean C.M. 2009. Mechanism of potassium-channel selectivity revealed by Na(+) and Li(+) binding sites within the KcsA pore. Nat. Struct. Mol. Biol. 16: 1317–U1143. Doi: 10.1038/Nsmb.1703.

Tilton R.F., Kuntz I.D. and Petsk G.A. 1984. Cavities in proteins: Structure of a metmyoglobin-xenon complex solved to 1.9 A. Biochemistry. 23: 2849–2857.

Törnroth-Horsefield S., Wang Y., Hedfalk K., Johanson U., Karlsson M., Tajkhorshid E., Neutze R. and Kjellbom P. 2006. Structural mechanism of plant aquaporin gating. Nature. 439: 688–694.

Tursynbay Y., Zhang J., Li Z., Tokay T., Zhumadilov Z., Wu D. and Xie Y. 2015. Pim-1 kinase as cancer drug target: An Update (Review). Biomed. Rep. 4: 1–7.

Valdman A., Fang X., Pang S.-T., Ekman P. and Egevad L. 2004. Pim-1 expression in prostatic intraepithelial neoplasia and human prostate cancer. The Prostate. 60: 367–371.

Venkataraman P., Lamb R.A. and Pinto L.H. 2005. Chemical rescue of histidine selectivity filter mutants of the M2 ion channel of influenza A virus. J. Biol. Chem. 280: 21463–21472.

Vijayvergiya V., Wilson R., Chorak A., Gao P.F., Cross T.A. and Busath D.D. 2004. Proton conductance of influenza virus M2 protein in planar lipid bilayers. Biophys. J. 87: 1697–1704.

Vinothkumar K.R., Montgomery M.G., Liu S. and Walker J.E. 2016. Structure of the mitochondrial atp synthase from pichia angusta determined by electron cryo-microscopy. Proceedings of the National Academy of Sciences of the United States of America. 113(45): 12709–12714.

Vipond I.B., Baldwin G.S. and Halford S.E. 1995. Divalent metal ions at the active sites of the EcoRV and EcoRI restriction endonucleases. Biochemistry. 697–704.

Wan X., Zhang W., Li L., Xie Y., Li W. and Huang N. 2013. A new target for an old drug: Identifying mitoxantrone as a nanomolar inhibitor of PIM1 kinase via kinome-wide selectivity modeling. J. Med. Chem. 56: 2619–2629.

Wang C., Huang W.T. and Liao J.L. 2015. Qm/mm investigation of atp hydrolysis in aqueous solution. J. Phys. Chem. B. 119(9): 3720–3726.

Wang M.C. and Uhlenbeck G.C. 1945. On the theory of Brownian motion II. Rev. Mod. Phys. 17: 323 – 342.

Wax N. (ed.). 1954. Selected Papers on Noise and Stochastic Processes, ed. Dover. New York, 1954.

Wittenberg J.B. and Wittenberg B.A. 2003. Myoglobin function reassessed. J. Exp. Biol. 206: 2011–2020.

Wolynes P.G. 1978. Molecular theory of solvated ion dynamics. J. Chem. Phys. 68: 473–483.

Wilson K.P., Malcolm B.A. and Matthews B.M. 1992. Structural and thermodynamic analysis of compensating mutations within the core of chicken egg white lysozyme. J. Biol. Chem. 267: 10842.

Yamamoto T. 2010. Preferred dissociative mechanism of phosphate monoester hydrolysis in low dielectric environments. Chem. Phys. Lett. 500(4-6): 263–266.

Yanagida T., Kitamura K., Tanaka H., Iwane A.H. and Esaki S. 2000. Single molecule analysis of the actomyosin motor. Current Opinion in Cell Biology. 12: 20–25.

Yoshida K., Kovalenko A., Yamaguchi T. and Hirata F. 2002. Structure of Tert-Butyl alcohol-water mixtures studied by the RISM theory. J. Phys. Chem. B. 106: 5042–5049.

Yoshida N., Phongphanphanee S., Maruyam Y., Imai T. and Hirata F. 2006. Selective ion-binding by protein probed with the 3D-RISM theory. J. Am. Chem. Soc. 128: 12042–12043.

Yoshida N., Phongphanphanee S. and Hirata F. 2007. Selective ion-binding by protein probed with the statistical mechanical integral equation theory. J. Phys. Chem. B. 111: 4588–4595.

Yoshida N., Ishida T. and Hirata F. 2008. Theoretical study of temperature and solvent dependence of the free energy surface of the intramolecular electron transfer based on the RISM-SCF theory: Application to 1,3-dinitrobenzene radical anion in acetonitrile and methanol. J. Phys. Chem. 112: 433–440.

Yoshida N., Imai T., Phongphanphanee S., Kovalenko A. and Hirata F. 2009. Molecular recognition in biomolecules studied by statistical-mechanical integral-equation theory of liquids. J. Phys. Chem. B. 113: 873–886.

Yoshida N. and Hirata F. 2018. The Role of Water in ATP Hydrolysis Energy Transduction by Protein Machinery. Ed. Suzuki, Springer 2018.

Zacharias M. 2000. Comparison of molecular dynamics and harmonic mode calculations on RNA. Biopolymers. 54: 547–560.

Zeidel M.L., Ambudkar S.V., Smith B.L. and Agre P. 1992. Reconstitution of functional water channels in liposomes containing purified red cell CHIP28 protein. Biochemistry. 31: 7436–7440.

Zelenina M., Bondar A.A., Zelenin S. and Aperia A. 2003. Nickel and extracellular acidification inhibit the water permeability of human Aquaporin-3 in lung epithelial cells. J. Biol. Chem. 278: 30037–30043.

Zhou Y.F., Morais-Cabral J.H., Kaufman A. and MacKinnon R. 2001. Chemistry of ion coordination and hydration revealed by a K+ channel-Fab complex at 2.0 angstrom resolution. Nature. 414: 43–48.

Zhukova Y.N., Alekseeva M.G., Zakharevich N.V., Shtil A.A. and Danilenko V.N. 2011. Pim family of protein kinases: Structure, functions, and roles in hematopoietic malignancies. Mol. Biol. 45: 695–703.

Zwanzig R. 1970. Dielectric friction on a moving ion. 2. Revised Theory. J. Chem. Phys. 52: 3625–3628.

Index

B

C

N

O

P

Q

R

Z

Epilogue

"How did life begin?" "Where did life come from?" Those are questions that have been asked repeatedly since human's thought started, no matter whether it is scientific or non-scientific. From a viewpoint of religions represented by Christianity, it is something except *matter* that created life. A sentence, "Human beings are created in the image and likeliness of God", in Old Testament is a typical example of such beliefs. On the other hand, from a viewpoint of the modern science, it is nothing but *matter* that has created human in a series of evolutions. There are two essential stages in the evolution: the stage to create life from material, and the evolution in life from the most primitive one, like microorganism, to human beings. It is the latter stage that has been explored so far by the entire spectrum of natural science including biology, chemistry, physics, and so on. The central dogma governs the stage of evolution, and water plays crucial roles in the process as was explored in this book. On the other hand, the stage to create life from material is yet to be clarified, and it is entirely an open question. What is the essential gap that separates life from materials? What questions should be clarified in order to fill the gap? The answer to the question is *information*. DNA, RNA, and protein are not mere molecules as materials, but convey *sequence information* that is inherited from generation to generation. Protein that has a specific structure and a function will never be made without such information. A *protein* molecule created without a particular sequence-information is just a polymer as a material. On the other, in order for a molecule with information to be created, a molecule having specific structure and function, or an enzyme, is required to catalyze the reaction. So, the problem is a sort of "chicken-and-egg question." It is a task of science to resolve such a question without help from *God*. The question can be restated scientifically as "how have materials acquired the sequence information?" Of course, there is no definite answer to the question to date, which has been proved with experimental evidence. In the following, I present my idea to resolve the question. The idea will stay just as a fantasy till it is proved by some experiment.

Let us imagine an ancient time of the earth before life was born. It may be reasonable to assume that there was an environment in ocean to produce small organic molecules including amino acids and nucleotides from inorganic compounds such as water, carbon dioxide, and ammonia. A candidate of such an environment is *submarine volcano* that might have played a role of *reactor* for the chemical reactions to produce biomolecules. The environment of submarine volcano is considered to be ideal for chemical reactions. Water around the volcano would be in a very wide temperature range from nearly zero centigrade to supercritical condition, depending on the distance from crater. The water also contains many kinds of minerals from volcano including metal ions, which may serve as catalysts of chemical reactions. It is not very hard to imagine that all sort of chemical reactions are taking place in such a condition. So, let us assume that there was such an environment on the ancient earth, in which biomolecules are naturally produced from inorganic compounds. Among the biomolecules produced, we may find molecules by chance, that convey sequence information necessary to be reproduced. Let us estimate the probability that a molecule with a particular sequence is produced just by chance.

Let us assume a protein consists of 100 amino acids. Then, a possible amino-acid sequence of the protein is $\sim 20^{100}$, assuming that each amino acid in the sequence can be selected from the twenty essential-amino-acids. Therefore, the probability of finding a particular sequence is $\sim 1/20^{100}$ or $\sim 20^{-100}$. If we have $\sim$ 1.0 mol of the molecule, the probability of finding the particular sequence will become $\sim 20^{-100} \times 10^{23} = \sim 10^{-77}$, which is essentially zero, or no chance. However, the estimate is made based on a living system on the modern earth. In the environment of ancient earth when the most primitive life was born, the kind and number of amino acids in a protein having a biological activity would have been much less. Suppose that a protein consists of 20 amino acids, each of which can be selected from four kinds of amino acid. Then, the probability of finding a protein with a particular sequence is estimated as $\sim 4^{-20} \times 10^{23} \sim 10^{3}$. Then, the next question is if there is such a class of molecules that express biological functions with relatively small variations like four. The answer is "yes." A typical example is ribonucleic acid (RNA). RNA is a relatively small polymer consisting of just four different nucleotides carrying genetic information, but it sometimes act as an *enzyme* to catalyze chemical reactions. All those are telling that the possibility to find a molecule with a biological activity might *not* be zero on the earth before life was born.

Recently, a space project to explore small planet was launched in Japan. A mission of the project is to find the *origin of life* on the earth. The space ship called "Hayabusa" came back successfully after a long trip to a small planet named "Ryugu." Now, the scientists are in the middle of analyzing the stone to see if it contains any trace of *life*. Unfortunately, the mission will not approach an inch to the goal of mission. Why? It is because the set up of mission itself is distorted. They may find a *trace* of life in the planet. But, it will not tell anything about the *origin* of life. In order to find the origin of life, one should explore how materials acquire the genetic information. It may be much more probable to find the origin of life at the *submarine volcano* on the earth than a small planet in space.

Color Plate Section

Chapter IV

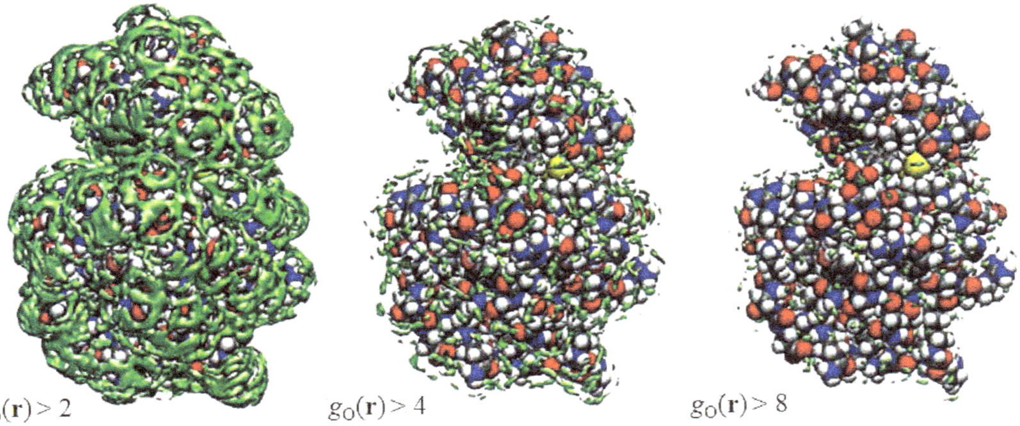

Fig. IV-3.1: Isosurface representation of spatial distribution of $g(\mathbf{r})$ of water around a lysozyme molecule.

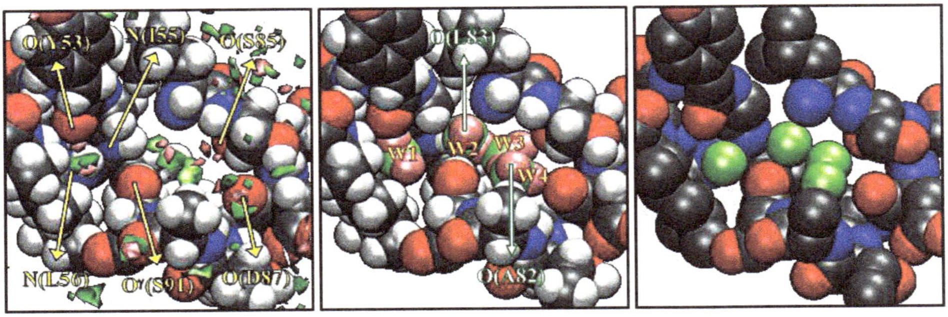

Fig. IV-3.2: The spatial distribution ($g(\mathbf{r})$) of water inside the active site of lysozyme. Green surface, oxygen; purple surface, hydrogen.

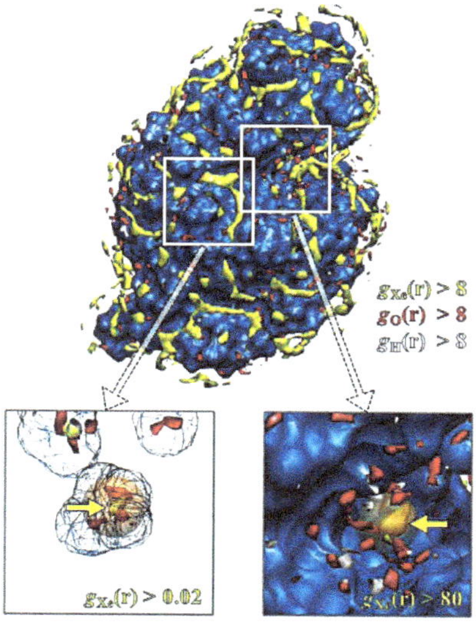

Fig. IV-3.3 The spatial distribution ($g(\mathbf{r})$) of Xe and water molecules in Lysozyme: red surface, water oxygen; white surface, water hydrogen; yellow surface, xenon. The X-ray results are painted by orange surface.

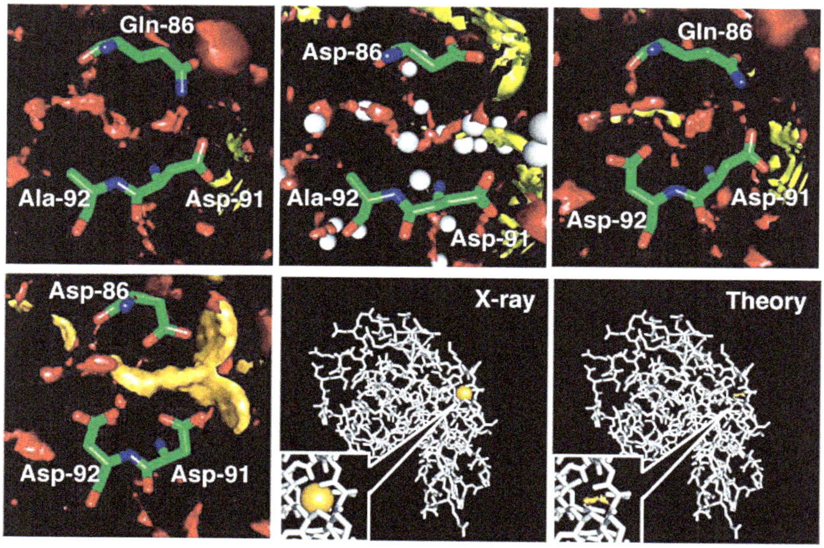

Fig. IV-3.6: The spatial distribution functions ($g(\mathbf{r})$) of ions in lysozyme.

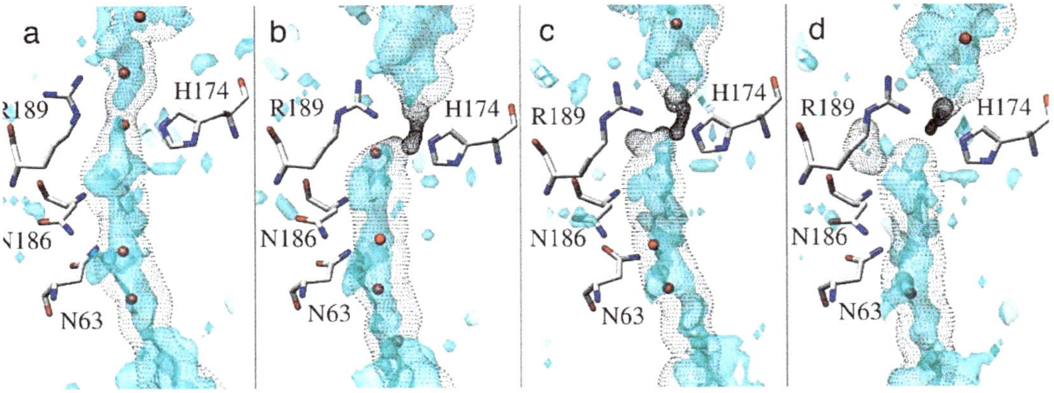

Fig. IV-4.1: Distribution of water inside AQPZ.

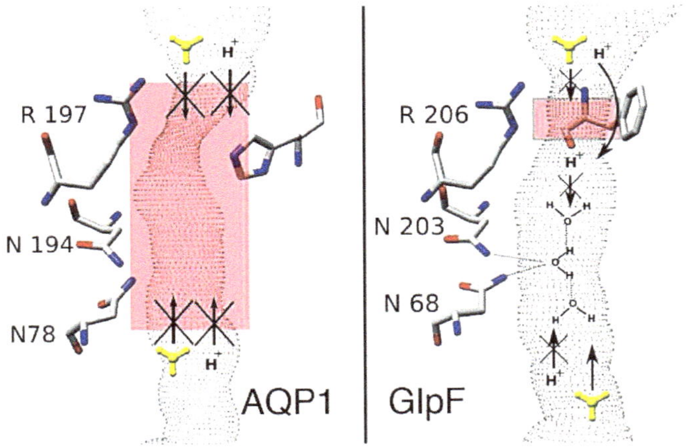

Fig. IV-4.4: Schematic illustration of the mechanisms of proton exclusion in AQP1 and GlpF. The transparent pale red surface represents the gap region in the distribution of hydronium ions. The molecules colored depict the hydronium ions.

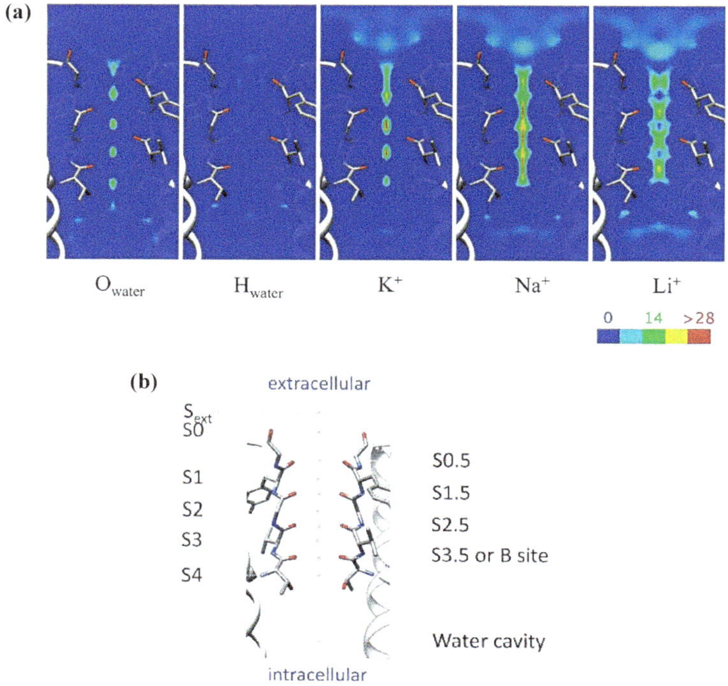

Fig. IV-4.5: (a) Contour of the spatial distribution function ($g(\mathbf{r})$) of O and H of water, K^+, Na^+, and Li^+ in the selective filter (SF) region. Red color denotes the region of high probability. (b) Illustrative view of the selective filter (SF) region. S0, S1, etc., specify the position of ion-binding sites.

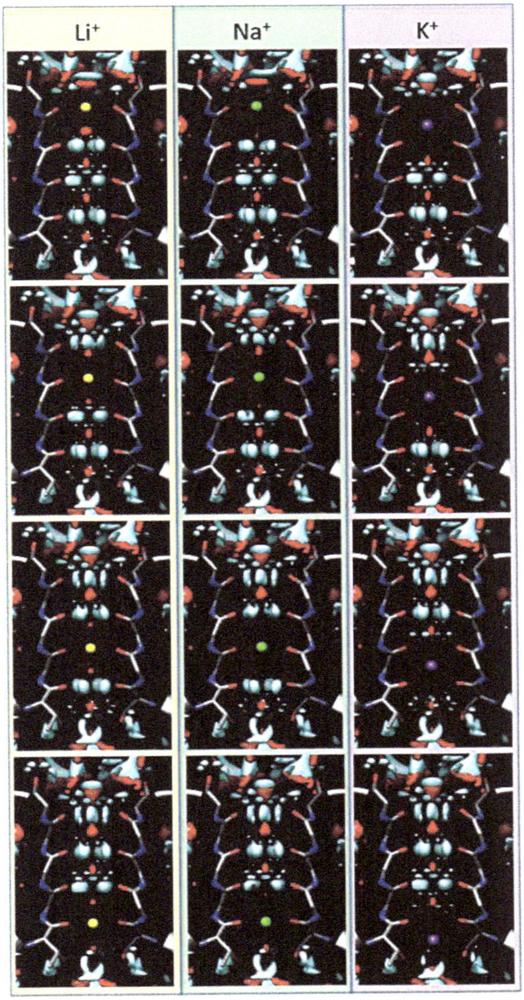

Fig. IV-4.6: Spatial distribution function ($g(\mathbf{r})$) of O and H of water around an ion fixed at the binding sites: $g_O > 3$ (red), $g_H > 2$ (light blue). Li^+, Na^+, and K^+ are colored, respectively, by yellow, green, and purple. The ions are fixed at S0.5, 1.5, 2.5 and 3.5 for Li^+ and Na^+, at S1, S2, S3, and S4 for K^+.

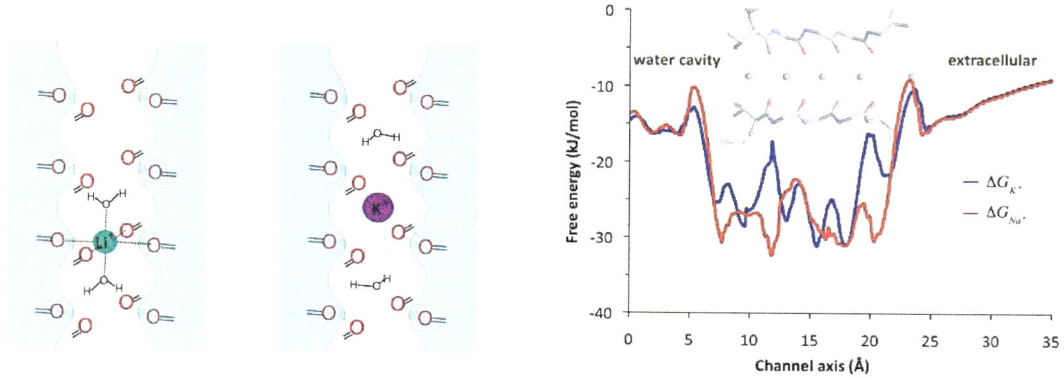

(a) (b)

Fig. IV-4.7 (a) Illustrative picture of the binding of ion in SF. Red colored O denotes the backbone carbonyl. (b) Free energy profile of K^+ (blue) and Na^+ (red) ions along the selective filter.

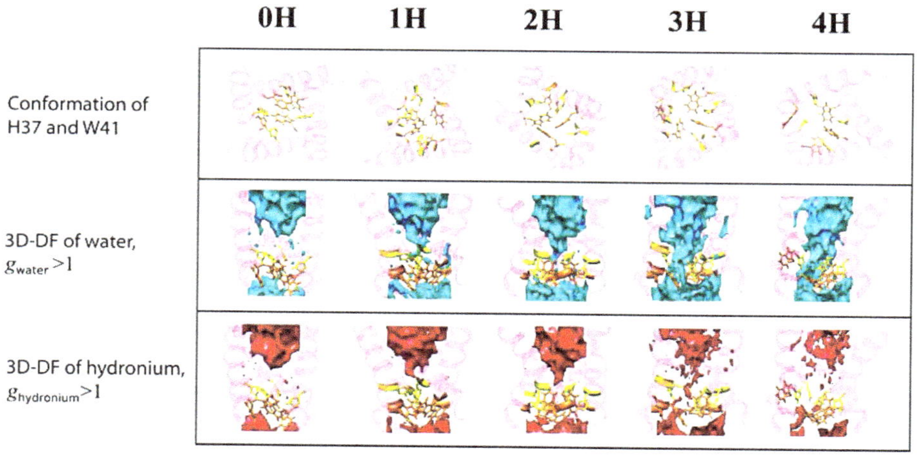

Fig. IV-4.8: Structure of Trp41gating (orange) of different protonated histidine tetrad, and SDF ($g(\mathbf{r})$) of water (cyan) in the channel and hydronium ion (red), with $g > 1$.

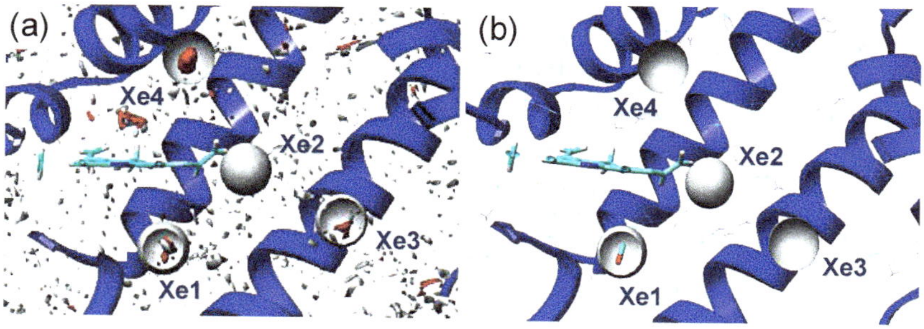

Fig. IV-5.3: (a) CO molecules in Xe1, Xe3, and Xe4 sites which are in the same plane. (b) Intermediate model for which a CO molecule exists in Xe1 ate reproduced by CO distribution.

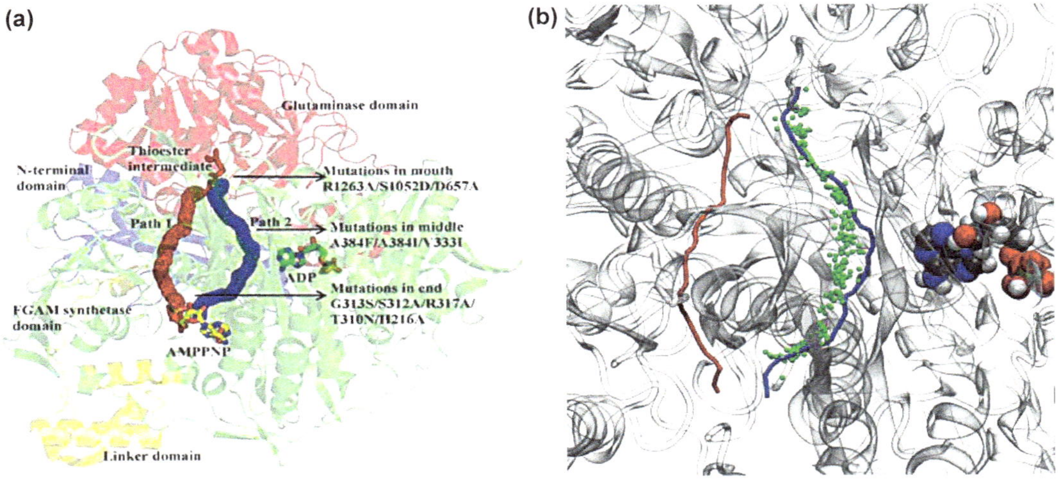

Fig. IV-6.1: Ammonia channel in StPurL: (a) Two predicted pathways for ammonia channel in StPurL shown in red and blue color mesh for path 1 and path 2, respectively. StPurL structure is depicted in cartoon with N-terminal domain in blue, linker domain in yellow, glutaminase domain in red, and FGAM synthetase domain in green color. Ligands are shown in stick representation. (b) Results of *RismPath* calculations on PurL WT trajectory. Crystallography purported Path 1 (red) and Path 2 (blue) are shown as tubes inside PurL (Cartoon, translucent).

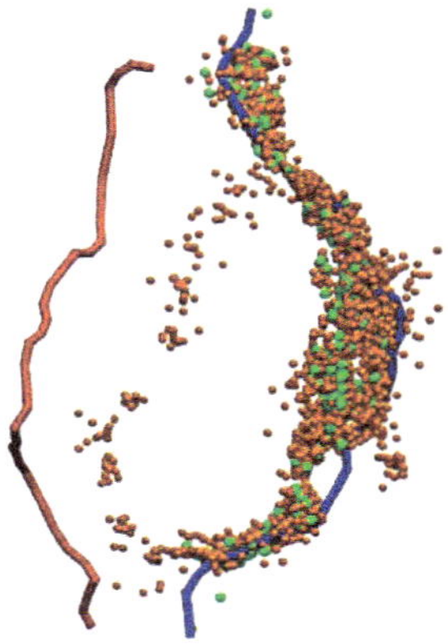

Fig. IV-6.2: NH_3 pathways in the PurL mutants, identified by 3D-RISM/KH: green, wild type; brown, mutants.

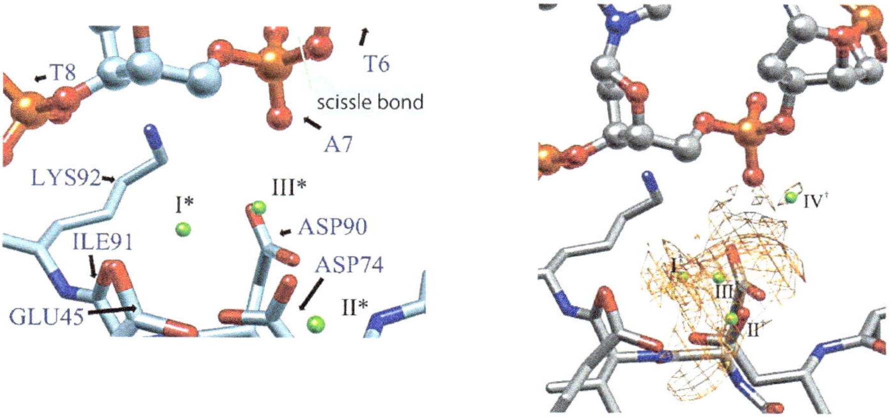

(a) **(b)**

Fig. IV-7.1: The position of Mg^{2+} at the active site of *Eco*RV: (a) X-ray crystallography, (b) 3D-RISM. The web-like surface represents the spatial distribution function $g(\mathbf{r})$ of Mg^{2+}. The surface of $g(\mathbf{r}) \geq 42$ is depicted.

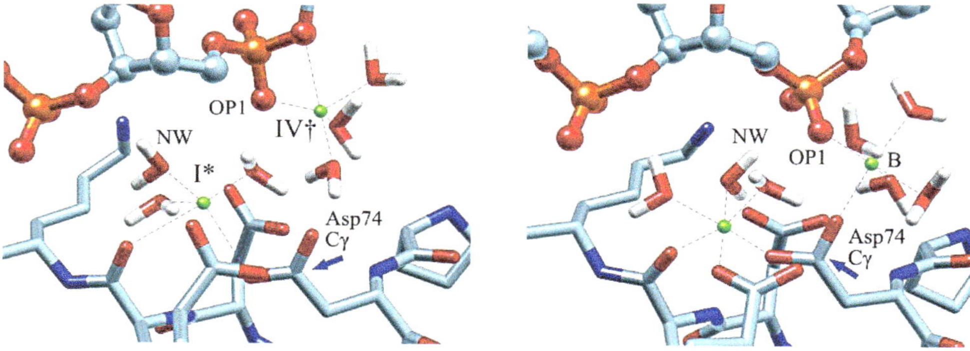

(a) **(b)**

Fig. IV-7.2: The structure of active site of *Eco*RV: (a) the initial state of the MD simulation, (b) the final state of the MD simulation.

Chapter VI

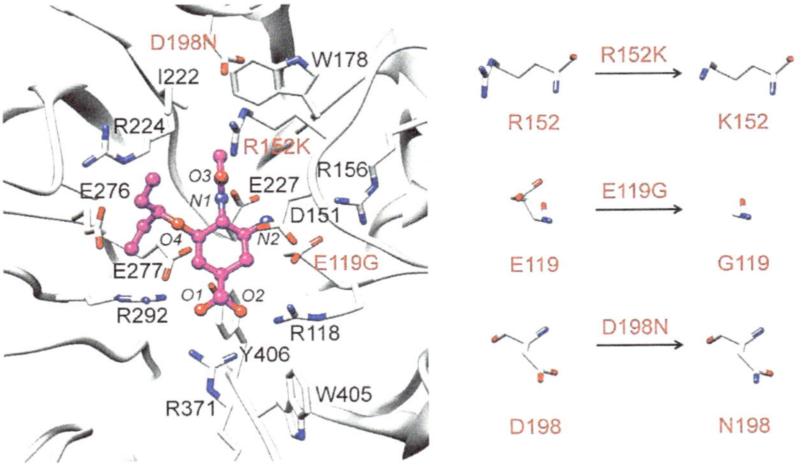

Fig. VI-2.1: Illustration of the positions of amino-acid substitution.

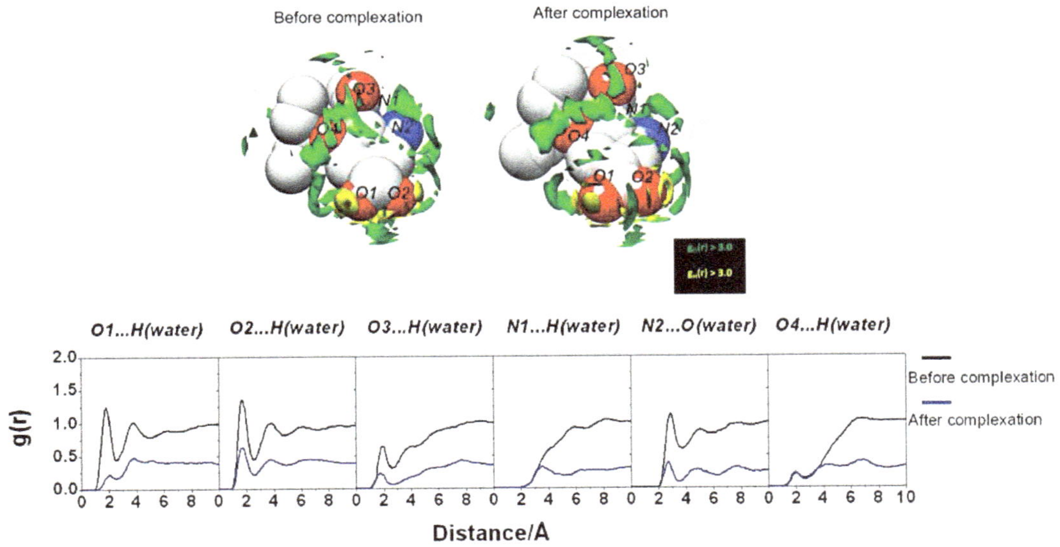

Fig. VI-2.2 3D-distribution function of water O-atom and H-atom around oseltamivir via 3D-RISM calculation with $g(r) \geq 3.0$ before and after the complexation. 3D-distribution function of water O-atom and H-atom around oseltamivir via 3D-RISM calculation with $g(r) \geq 3.0$ before and after the complexation (Upper). The RDF of hydrogen-bonded pairs between oseltamivir heteroatoms and water molecules before and after the complexation calculated from the free oseltamivir and oseltamivir in complex with wild-type neuraminidase (Lower).

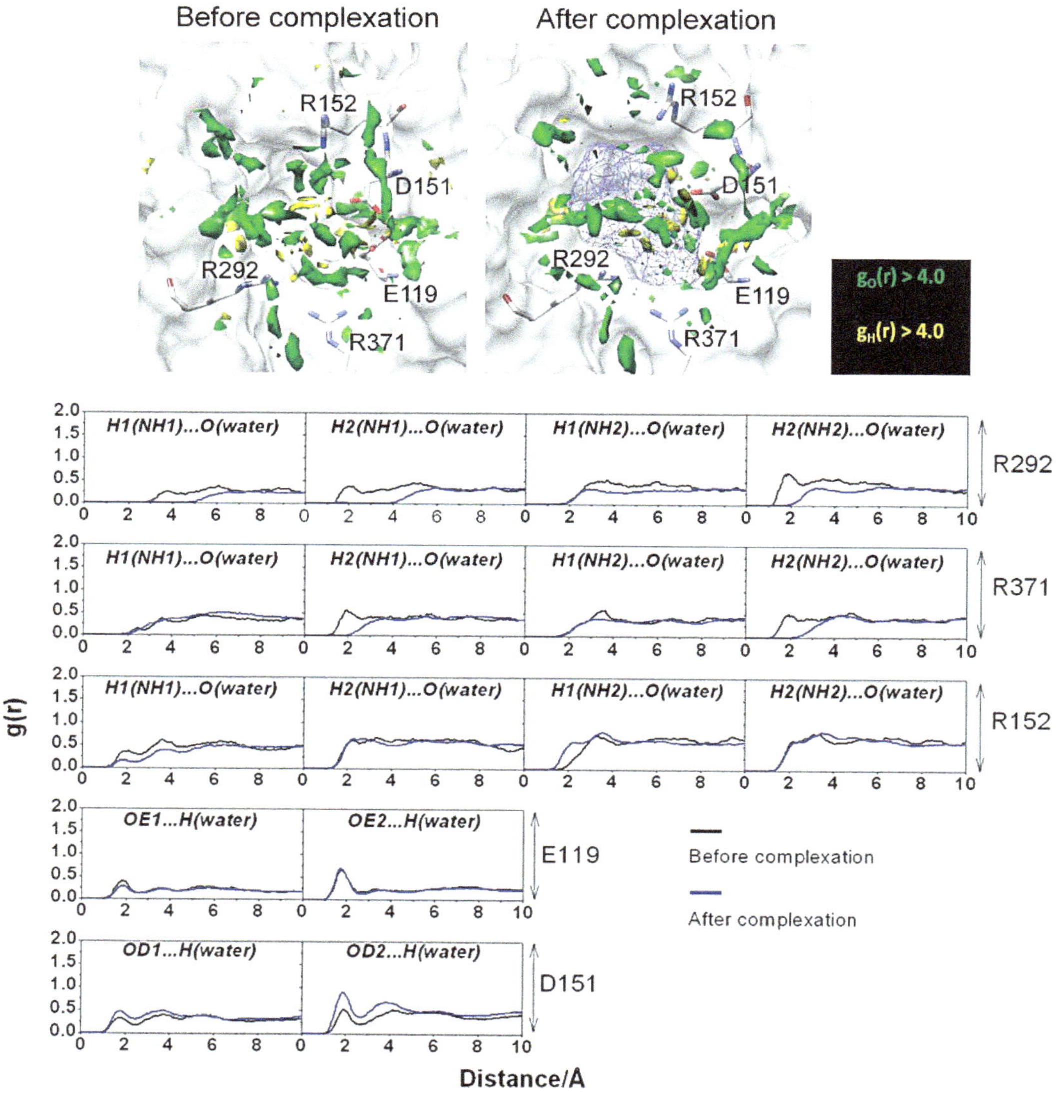

Fig. VI-2.3: 3D-distribution function of O-water (green) and H-atoms (yellow) with $g_O(r) \geq 4.0$ within 7.0 Å of oseltamivir (depicted as blue mesh sphere). (Upper) 3D-RISM RDF of hydrogen-bonded pairs between water molecules and five amino acid residues before and after complexation. (Lower)

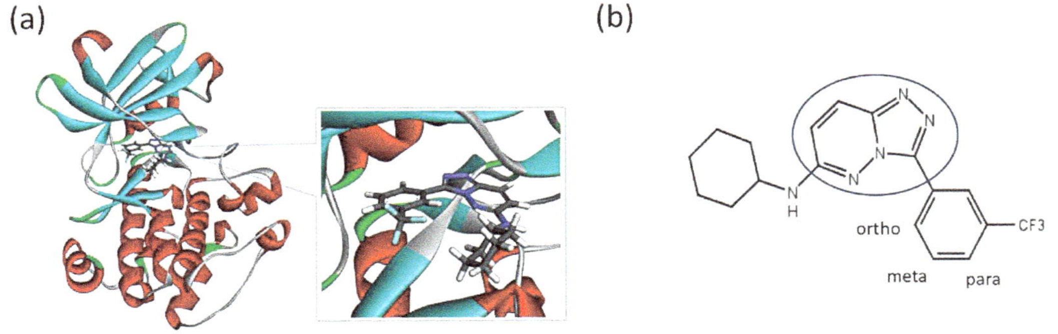

Fig. VI-3.1: (a) A 3D structure of the Pim-1 kinase and triazolo pyridazine inhibitor termed as VX2 in the PDB. The PDB code is 3BGQ. We refer to the VX2 as ligand 1, in this study, for the sake of simplicity of the terminology. (b) Chemical structure of the VX2. Circled part of the ligand is termed as triazolo[4.3-b]pyridazine scaffold. This is a common part of all the ligands which is applied in this study.

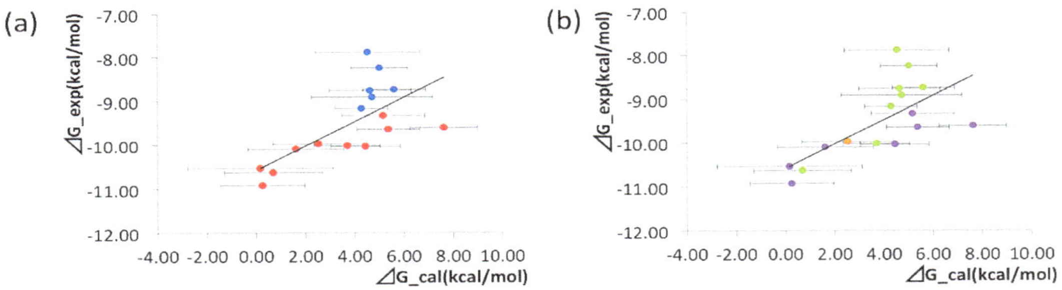

Fig. VI-3.2: Correlation between calculated and experimental values of binding free energy for all the ligands. Correlation coefficient R = 0.69. The figures are colored in different manner based on the structural feature of the ligands. (a) Ligands which include CF_3 on the meta position of the phenyl ring are colored with red. Other ligands are colored with blue. (b) Ligands which have cyclo-hexane, cyclo-butane, and cyclo-propane are colored with purple, orange, and green, respectively.

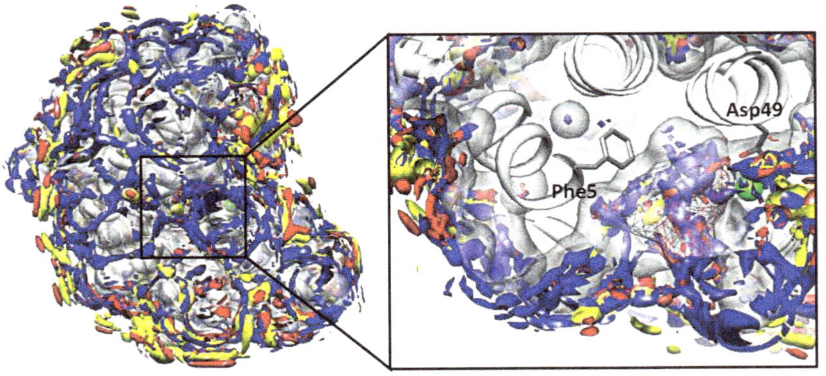

Fig. VI-4.2: The 3D-DDF of aspirin around and inside Phospholipase A2 (PA2), obtained by uu-3D-RISM with the threshold $g_\gamma(\mathbf{r}) > 2$: red, COOH; gray, aromatic ring; blue, $OCOCH_3$. The protein surfaces are represented as gray transparent surface. The location of aspirin in X-ray structure is depicted with a wire frame.

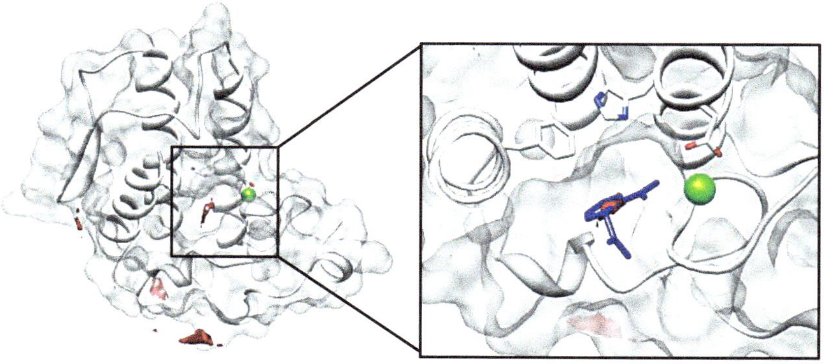

Fig. VI-4.3: The affinity of an aspirin molecule to binding site in the Phospholipase A2 estimated by the function of DC based on local potential mean force with the threshold $f_{DC}(\mathbf{x}) > 1.45$. The protein surfaces are represented as gray transparent surface. In the top view, the location of aspirin in X-ray structure is depicted with blue sticks.

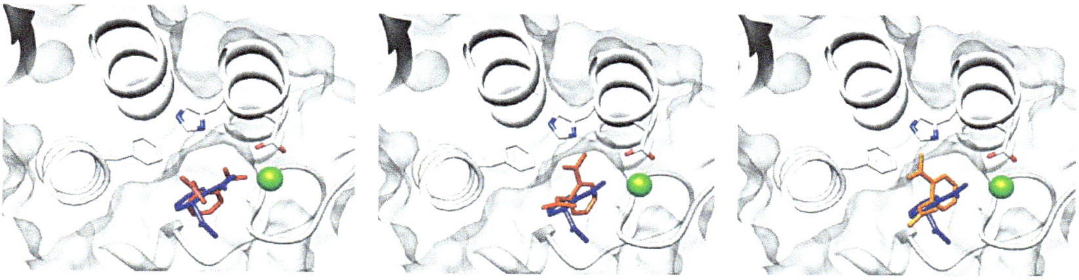

(a) (b) (c)

Fig. VI-4.4: Predicted structures of ligand from top three peaks of affinity based on the overlap between the ligand and its distribution function. The protein surfaces are represented as gray transparent surface. The location of aspirin in X-ray structure is depicted with blue sticks.

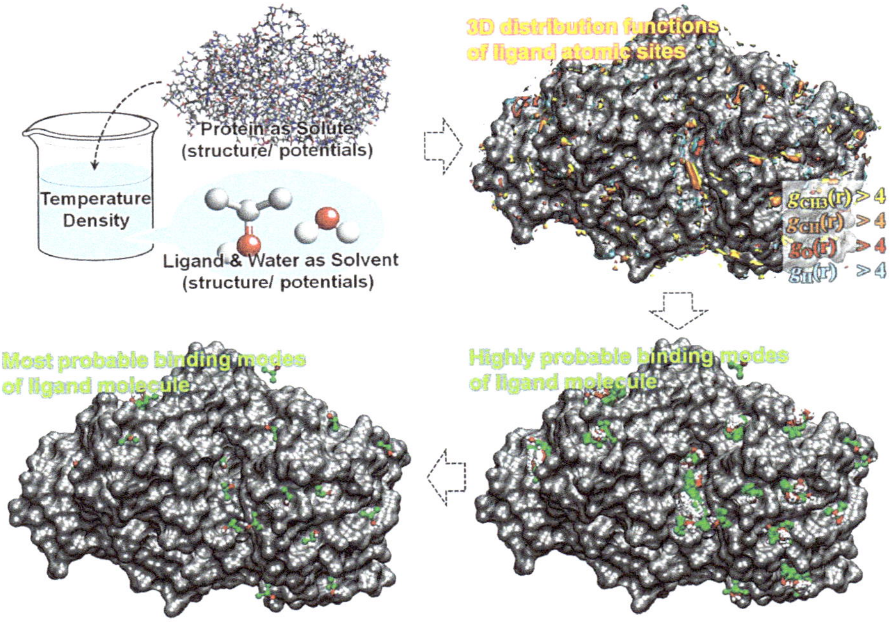

Fig. VI-5-1: Schematic illustration of the 3D-RISM-based ligand mapping method.

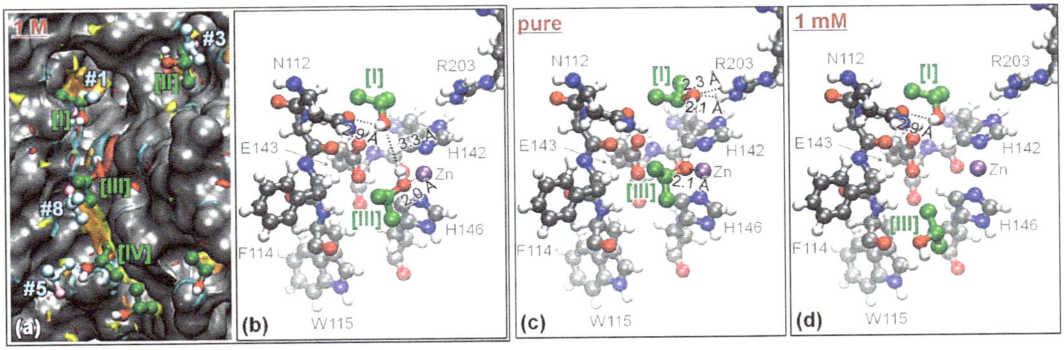

Fig. VI-5-2: Most probable binding modes (MPBMs) of isopropanol in the active site of thermolysin in 1 M isopropanol-aqueous solution. (a) Isosurface representation of $g(\mathbf{r}) > 4$ (yellow, CH_3; orange, CH; red, O; light blue, H.), MPBMs (green, carbon; red, oxygen; white, hydrogen), and X-ray crystallographic isopropanol molecules (cyan, carbon; red, oxygen). (b) Hydrogen-bond network between MPBM-I and III and its surrounding residues. (c) MPBM-I and III in pure isopropanol. (d) MPBM-I and III in 1 mM isopropanol-aqueous solution.

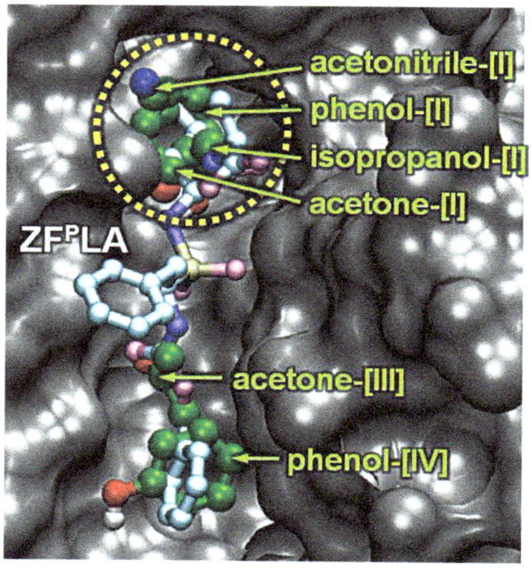

Fig. VI-5-3: Comparison of selected most probable binding modes (MPBMs) with the crystallographic structure of an inhibitor, carbobenzoxy-*Phe*^{P}-Leu-Ala (ZF^{P}LA). MPBMs are represented by ball-and-stick model (green, carbon; red, oxygen; blue, nitrogen). ZF^{P}LA is represented by the small ball-and-stick model (cyan, carbon; mauve, oxygen; blue, nitrogen; tan, phosphorus). The yellow dotted circle indicates the binding site for the common anchor motif of known inhibitors.